ITALIAN PHYSICAL SOCIETY

PROCEEDINGS
OF THE
INTERNATIONAL SCHOOL OF PHYSICS
"ENRICO FERMI"

COURSE CXLV
edited by R. A. BROGLIA and E. I. SHAKHNOVICH
Directors of the Course
and by
G. TIANA
VARENNA ON LAKE COMO
VILLA MONASTERO
11-21 July 2000

Protein Folding, Evolution and Design

2001

AMSTERDAM, OXFORD, TOKYO, WASHINGTON DC

Copyright © 2001 by Società Italiana di Fisica

All rights reserved. No part of this publication may be reproduced, stored in a retrieval system, or transmitted, in any form or any means, electronic, mechanical, photocopying, recording or otherwise, without the prior permission of the copyright owner.

ISBN 1 58603 166 X (IOS Press)
ISBN 4 274 90457 1 C3042 (Ohmsha)
Library of Congress Catalog Card Number: 2001090874

              Production Manager   Copy Editor
                A. OLEANDRI   P. MARANGON

Publisher
IOS PRESS
Nieuwe Hemweg 6B
1013 BG Amsterdam
The Netherlands
fax: +31 20 688 33 55
e-mail: order@iopress.nl

Distributor in the UK and Ireland
IOS Press/Lavis Marketing
73 Lime Walk
Headington
Oxford OX3 7AD
England
fax: +44 1865 75 0079

Distributor in the USA and Canada
IOS Press, Inc.
5795-G Burke Center Parkway
Burke, VA 22015
USA
fax: +1 703 323 3668
e-mail: iosbooks@iospress.com

Distributor in Germany, Austria and Switzerland
IOS Press/LSL.de
Gerichtsweg 28
D-04103 Leipzig
Germany
fax: +49 341 995 4255

Distributor in Japan
Ohmsha, Ltd.
3-1 Kanda Nishiki-cho
Chiyoda-ku, Tokio 101
Japan
fax: +81 3 3233 2426

Proprietà Letteraria Riservata
Printed in Italy

Supported by the European Commission, Research DG, Human Potential Programme, High-Level Scientific Conferences HPCF-CT-1999-00094

Supported by NATO under Contract LST.ASI. 976237

Supported by UNESCO VENICE OFFICE (UVO-ROSTE)

INDICE

R. A. Broglia and E. I. Shakhnovich – Preface .... pag. XIII

R. A. Broglia and E. I. Shakhnovich – *In Memoriam* .... » XV

Gruppo fotografico dei partecipanti al Corso .... » XVI

PROTEIN FOLDING, EXPERIMENTS

L. J. Lapidus, J. Hofrichter and W. A. Eaton – Dynamics of end-to-end contact formation in polypeptides .... » 3

1. Basic idea of the experiment .... » 4
2. Quenching probability and diffusion-limited rate .... » 6
3. Length dependence of end-to-end contact rate .... » 8
4. Mean-squared end-to-end distance and end-to-end diffusion constant ...... » 10
5. Concluding remarks .... » 10

M. Oliveberg – Characterisation of the transition state for protein folding: Mutational analysis taken one step further .... » 13

1. What is the folding transition state? .... » 13
2. The gross behaviour of the transition state and the parameter $b^{\ddagger}$ .... » 15
3. Phi-value analysis .... » 17
4. Transition state shifts and the broad barrier model .... » 19
5. From snap-shot to movie: scanning the activation barrier by a moving transition state .... » 20
6. Determining the folding pathway directly from the native structures .... » 22

Yu. F. Krupyanskii and V. I. Goldanskii – RSMR comparison of dynamic properties of various proteins .... » 25

1. Introduction .... » 25
2. Experimental part .... » 27
3. Summary and discussion .... » 32

PROTEIN FOLDING THEORY/SIMULATIONS

L. MIRNY and E. I. SHAKHNOVICH – Protein folding: Matching theory and experiment ..... pag. 37

    Abbreviations ..... » 37
1. Introduction ..... » 37
2. Basic concepts ..... » 38
    2˙1. Cooperativity: two-state thermodynamics and kinetics of protein folding ..... » 38
    2˙2. Transition state ensemble and folding nucleus ..... » 41
    2˙3. Folding funnels ..... » 42
    2˙4. Diffusion-collision mechanism ..... » 44
    2˙5. Determining TSE in experiment: protein engineering analysis ..... » 45
3. Simple lattice and topology-based models ..... » 47
    3˙1. Determining TSE in computer simulations of simple models ..... » 48
        3˙1.1. Nucleation in off-lattice models ..... » 50
    3˙2. Topology may define folding nucleus. A key finding from simulations and experiment and its evolutionary implications ..... » 50
    3˙3. Simple models: topology-dependent free-energy functionals ..... » 52
4. Evolutionary studies ..... » 55
    4˙1. Is folding nucleus conserved? ..... » 55
        4˙1.1. A successful blind prediction of folding nucleus from conservation ..... » 55
        4˙1.2. Universally conserved residues—Conservatism of conservatism ..... » 56
5. Molecular dynamics: folding pathways inferred from unfolding simulations ..... » 59
6. All-atom Monte Carlo simulations: possible folding mechanism at an atomic level of detail ..... » 61
7. Concluding remarks ..... » 62

R. A. BROGLIA and G. TIANA – Mechanism of folding and aggregation of proteins ..... » 69

1. Introduction ..... » 69
2. The model ..... » 71
    2˙1. Role of the different amino acids in the folding process ..... » 73
    2˙2. How many mutations can a designed protein tolerate? ..... » 76
3. How does a notional protein fold? ..... » 77
    3˙1. Consequences of the presence of local elementary structures in the aggregation of proteins ..... » 83
    3˙2. Statistical analysis of contact formation ..... » 86
4. Predicting the 3D structure of a notional protein from its amino acid sequence ..... » 90
    4˙1. Caveats and limitations ..... » 93
    4˙2. Hindsight ..... » 95
5. Conclusions and perspectives ..... » 95
    5˙1. Lattice model studies ..... » 95
    5˙2. Generalization of the $1D \to 3D$ algorithm to real proteins ..... » 95
    Appendix A ..... » 97
    Appendix B ..... » 98

R. A. Broglia and G. Tiana – Complementary pictures for proteins: Designability and foldability, are they equivalent? .... pag. 103

   1. Introduction .... » 103
   2. The HP model .... » 104
   3. The twenty-letter model .... » 108
   4. Designability of different conformations .... » 112
   5. Conclusions .... » 115
   Appendix A .... » 116

H. Orland – Dynamics of polymers: How fast can a protein fold? .... » 119

   1. Introduction .... » 119
   2. Rouse model .... » 120
   3. Including interactions .... » 123
      3˙1. Short-time behavior .... » 125
      3˙2. Long-time behavior .... » 125
   4. Conclusion .... » 126

J. R. Banavar, A. Maritan, C. Micheletti and F. Seno – Geometrical aspects of protein folding .... » 127

   1. Scope of the lectures .... » 127
   2. Introduction .... » 128
   3. Optimal shape of a compact polymeric chain .... » 130
   4. Fast-folding polymers and role of secondary motifs .... » 135
   5. Density of overlapping conformations for protein structures and role of native-state topology in the folding process .... » 140
      5˙1. Application to HIV-1 protease: drug resistance and folding pathways .... » 147
   6. Application of geometrical models to investigate the folding of membrane proteins .... » 150

J. Shimada, E. L. Kussel and E. I. Shakhnovich – The folding thermodynamics and kinetics of crambin using an all-atom Monte Carlo simulation . » 161

   1. Introduction .... » 161
   2. Simulation method .... » 163
   3. Results .... » 164
   4. Discussion .... » 169

E. Kussel, J. Shimada and E. I. Shakhnovich – Protein sidechain packing in ubiquitin .... » 173

   1. Introduction .... » 173
   2. Methods .... » 174
   3. Results .... » 177
   4. Discussion .... » 179

PROTEIN EVOLUTION

RICHARD A. GOLDSTEIN – Evolutionary perspectives on protein structure, stability, and functionality .... pag. 185

    1. Introduction .... » 185
    2. Principles of molecular evolution .... » 186
        2´1. Classical theory of gene dynamics .... » 186
        2´2. Finite populations and the neutral theory .... » 188
        2´3. Eigen's theory of quasi-species .... » 192
    3. Applications to proteins .... » 194
        3´1. Maximum foldability and the distribution of protein structures .... » 195
        3´2. Molecular evolution and the thermodynamic hypothesis .... » 199
        3´3. Why are proteins marginally stable? .... » 201
        3´4. Including explicit functionality .... » 203
    4. Conclusion .... » 207

L. MIRNY and M. S. GELFAND – What evolution can tell us about protein-DNA interactions .... » 211

    1. Introduction .... » 211
    2. DNA's point of view .... » 212
        2´1. Results .... » 212
    3. Proteins' point of view .... » 218
        3´1. Results .... » 218
            3´1.1. Conservation of DNA-binding residues .... » 218
            3´1.2. Specificity determinants of LacI/GalR family .... » 219

N. V. DOKHOLYAN and E. I. SHAKHNOVICH – Two models of amino acid conservation in proteins .... » 227

    1. Introduction .... » 227
    2. Methods .... » 230
        2´1. Protein model .... » 230
        2´2. The 6-letter potential .... » 230
        2´3. The measure of the information context of the sequences .... » 231
        2´4. The entropy of the protein fold families .... » 231
    3. Two models .... » 232
        3´1. Designing by Z-score minimization .... » 232
        3´2. Mean-field model .... » 235
    4. Results .... » 237
        4´1. Z-score model .... » 238
        4´2. Mean-field model .... » 240
    5. Conclusion .... » 242
    Appendix A: The maximal value of the sequence entropy .... » 243

PROTEIN DESIGN AND INTERACTION

A. V. Finkelstein – Physical selection of protein structures .... pag. 249

D. Eisenberg, E. Marcotte, M. Pellegrini, I. Xenarios and T. Yeates – New computational methods for detecting functional interactions among proteins .... » 267

E. N. Bogacheva and V. I. Goldanskii – Principles of design of the spatial structure of globular proteins .... » 269

CONNECTIONS TO OTHER FIELDS (POLYMERS, SPIN GLASSES, ETC.)

G. Parisi – The physics of the glassy systems .... » 281
    1. Introduction .... » 281
    2. Glassiness and metastability .... » 285
    3. General considerations on the coexistence of many phases .... » 286
    4. The overlap and its probability distribution .... » 288
        4˙1. Spin glasses .... » 288
        4˙2. Glasses .... » 289
    5. Off-equilibrium dynamics .... » 290
    6. The inherent structure picture for glasses .... » 293
    7. The form of the free-energy landscape .... » 294
    8. Some analytic and numerical results .... » 295
    9. Conclusions .... » 296

A. Grosberg – Protein folding in polymer physics context .... » 299
    1. Introduction .... » 299
    2. Entropy and geometry of compact conformations: Flory theorem, or what is secondary structure? .... » 300
        2˙1. Flory theorem: does compactness lead to helices? .... » 300
        2˙2. Ways to resolve the "Flory theorem against secondary structure" paradox .... » 303
    3. Entropy and geometry of conformation space .... » 306
        3˙1. Random walks and curvature of the conformation space .... » 306
        3˙2. Ergodicity and percolation in conformation space .... » 308
    4. Entropy and geometry in sequence space .... » 310

A. R. Khokhlov, A. Grosberg, P. G. Khalatur, V. A. Ivanov, E. N. Govorun, A. V. Chertovich and A. A. Lazutin – Conformation-dependent sequence design of protein-like AB-copolymers .... » 313
    1. Introduction. Bio-evolution mimetics .... » 313
    2. Protein-like AB-copolymers: computer generation and properties of the coil-globule transition .... » 315
    3. Long-range correlations in protein-like AB-copolymers .... » 316
    4. AB-copolymers which mimic membrane proteins .... » 323
    5. ABC-copolymers modeling proteins with active enzymatic center .... » 325
    6. Adsorption-tuned AB-copolymers .... » 327
    7. Conclusions .... » 328

Elenco dei partecipanti .... » 331

Preface

Proteins are the major functional molecules of life. They have evolved through selection pressure to perform specific functions. The functional properties of proteins depend upon their three-dimensional structures. One of the great unsolved problems of science is the prediction of the three-dimensional structure of a protein from its amino acid sequence.

The CXLV Course of the International School of Physics "Enrico Fermi" brought together scientists from a variety of areas of protein science and of physics. It provided a unique oportunity for discussions across the borders, at a time in which physicists addicted to simplifications are beginning to learn how much small details can matter in describing the properties of proteins, and molecular biologists are starting to appreciate the usefulness of having numbers attached to arrows of their cartoons and equations relating the numbers.

During the School the subject of protein folding and of protein design and evolution were discussed, both experimentally (W. Eaton, D. Eisenberg, A. Fersht, L. Serrano, V. Goldanskii and M. Oliveberg) and theoretically (R. A. Broglia, A. Finkelstein, A. Maritan, L. Mirny, H. Orland E. Shakhnovich, G. Tiana), as well as their connection to the general subject of polymers and of spin glasses (A. Grosberg, A. Khokhlov, G. Parisi). The audience made out of an extremely gifted group of young students and practitioners, contributed in a very active fashion to the School not only through questions and discussions, but also through the presentation of short, highly articulated seminars, touching on subjects lying at the forefront of protein research.

The lectures started with a review of the field and a discussion of the latest experimental progress in the subject of protein engineering and of protein design by Alan Fersht and Luis Serrano, respectively. Their lectures were central to the School, but are the only ones missing from the Proceedings. If one should look for a reason for this fact, one is reminded that Alan, at the time of the conference, had just published the third edition of his authorative monography *Structure and Mechanism in Protein Science* (W. H. Freeman Co., New York), while Luis was helping at creating a start-up biomedical venture. It is not easy to write a synthesis of a masterpiece, nor to have much time left over for relaxed writing when one goes private. While these facts did not make any easier the life of the editors of the present Proceedings, they reflect the vitality and breadth of interests the subject of protein folding and design has awaken in the scientific comunity.

On behalf of all the participants to the course we would like to express our gratitude to the Italian Physical Society for providing hospitality in the exquisite setting of Villa Monastero. We want also to thank several funding agencies (NATO, European Union, Unesco) for providing the funds needed to make the whole project a reality.

In the organization of the course it has been a great pleasure to work with Prof. F. Bassani, President of the Italian Physical Society. We would also like to express our thanks to Dr. G. Tiana, who acted as Scientific Secretary of the course, and to pay tribute to the kindness and efficiency of the secretarial staff headed by C. Vasini and B. Alzani.

R. A. BROGLIA and E. I. SHAKHNOVICH

In Memoriam

While the present Proceedings were in preparation we heard sad news that Vitalii Goldanskii, a world-renowned physical chemist and a very important contributor to the 2000 Varenna Summer School had passed away. Vitalii Goldanskii was an internationally recognized expert and leader in several fields of sciences, most notably physical chemistry, biochemical physics, nuclear and radiation chemistry and nuclear physics. He is well known for his applications of nuclear methods to chemistry and biophysics as well as to studies of low-temperature solid-state chemical reactions. He was extremely broad in his intellectual interests, which also included the origin of life and of homochirality as well as of international relations, security and arms control. Besides his seminal contributions to science, he will be remembered by many of us as a dear colleague and friend, who had great passion for Russian culture —literature and poetry, who knew many old Russian songs and enjoyed singing them, whose intellectual curiosity and integrity are unparalleled. It is hard to believe that Vitalii, who just last summer in Varenna was a leader in many heated discussions about science, life, politics and history, who sang and danced with students, is no longer with us. That is how we will remember him, a person of great intellectual power full of passion for life and science.

R. A. BROGLIA and E. I. SHAKHNOVICH

Società Italiana di Fisica
SCUOLA INTERNAZIONALE DI FISICA «E. FERMI»
CXLV CORSO - VARENNA SUL LAGO DI COMO
VILLA MONASTERO 11-21 Luglio 2000

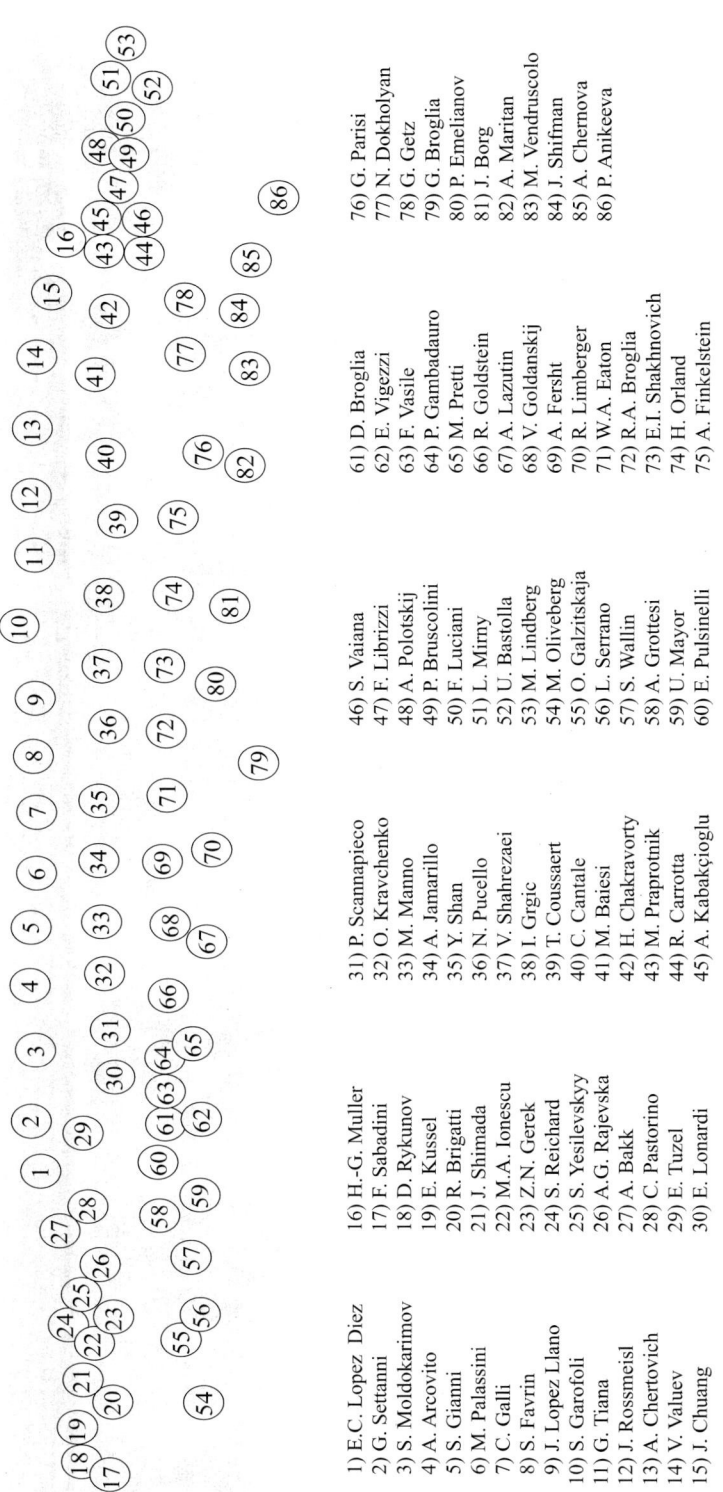

PROTEIN FOLDING, EXPERIMENTS

Dynamics of end-to-end contact formation in polypeptides

L. J. LAPIDUS, J. HOFRICHTER and W. A. EATON

Laboratory of Chemical Physics, NIDDK, National Institutes of Health
Bethesda, MD 20892-0520, USA

Understanding how proteins fold has become an attractive subject for both theoretical and experimental physical scientists. The challenge is to determine how the unique conformation of the functional biological molecule (the "native" state) is formed starting from the extremely broad distribution of structures of the unfolded polypeptide (the "denatured" state) [1-8]. Solving this problem will require an understanding of the kinetics and mechanisms for the formation of the basic structural elements of proteins. Roughly speaking, the elements are helices, sheets, strands, turns, and loops. These structures form on the nanosecond-microsecond time scale, so their investigation awaited the development of rapid initiation methods using pulsed lasers [9-13]. Studies up to now have focused on the formation of secondary structures such as the alpha-helix [14-18] and the beta hairpin [19,20], the simplest antiparallel beta sheet. The experimental results on these systems are not only key for elucidating mechanisms, but are important benchmarks for testing the accuracy of computer simulations which have the ultimate goal of folding a protein by all-atom molecular dynamics [21].

Much less is known about the kinetics of forming a contact between residues distant in sequence to make a loop or to connect secondary structural elements ("tertiary" contacts). The first glimpse into the kinetics of contact formation came with a study of unfolded cytochrome c. Photochemical triggering and transient absorption measurements showed that approximately $40\,\mu$s is required for methionines to contact the heme, about 50 residues distant along the sequence [9,22,23]. The method relies on photochemical and optical properties unique to the heme chromophore, and is therefore limited to the very small subset of proteins that contain hemes. Bieri et al. [24] recently introduced

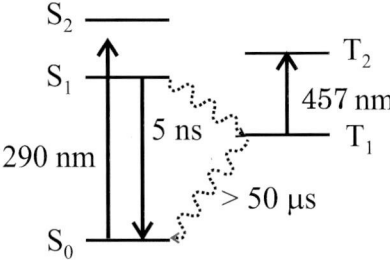

Fig. 1. – Energy levels of tryptophan. Absorption of a 290 nm photon excites tryptophan to its lowest excited singlet state, S_1, which lives for ~ 5 ns. Return to the ground electronic state (S_0) occurs by release of energy to the surroundings as heat or by emission of an ultraviolet photon. Intersystem crossing to the lowest triplet state can also occur, and the population of this triplet can be monitored by $T_1 \to T_2$ absorption with an argon ion laser operating at 457 nm. The lifetime of this triplet state in the absence of quenchers is $> 50\,\mu$s.

a method based on transfer of triplet excitation energy between two organic ring systems (from thioxanthone to naphthalene) attached to a polypeptide. However, this technique requires chemical modification of the peptide as well as the use of ethanol-water mixtures.

A more desirable method is one which utilizes only properties of naturally occurring amino acids. This lecture describes a technique to measure the rate of forming a contact between tryptophan and cysteine. The technique holds considerable promise as a generic method for measuring the kinetics of contact formation in unfolded proteins and to monitor contact formation in proteins as they fold [25]. Here we show how it has been used to determine important dynamical quantities for a flexible peptide. These include the end-to-end contact rate to form a loop, the dependence of this rate on chain length, and, using the simplest theoretical description of polymer dynamics, the mean-squared end-to-end distance and the end-to-end diffusion coefficient.

1. – Basic idea of the experiment

The method takes advantage of the fact that excitation of tryptophan with an intense laser pulse populates a long-lived, excited (triplet) electronic state which returns to the ground state with a high probability upon contact with cysteine. The rate of decay of this triplet state is therefore close to the rate of forming a contact between the tryptophan and cysteine. A more detailed description of the events subsequent to optical excitation is shown in fig. 1. Following absorption of an ultraviolet photon there is very rapid (subpicosecond) relaxation to the lowest vibrational levels of the first excited singlet state, labeled S_1. Several processes may now occur. The molecule can return to the ground electronic state (S_0) by emitting a photon (fluoresce), by giving up its energy as heat to the surroundings (radiationless decay), or by some kind of chemical interaction with solvent molecules (quenching). By changing the net spin, the molecule can also populate the lowest triplet state, labeled T_1 (intersystem crossing). The lifetime of the

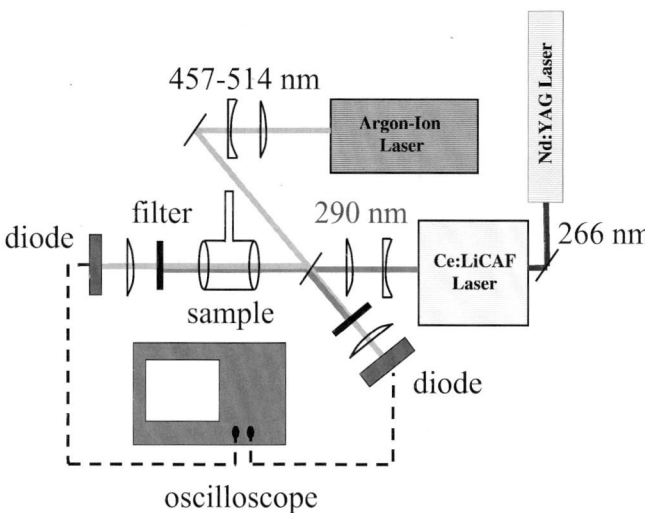

Fig. 2. – Schematic of instrument. Optical excitation of tryptophan is achieved with a ∼ 8 ns, ∼ 1 mJ, 290 nm pulse from a Nd:YAG-pumped Ce:LiCAF laser. The absorption probe is one of the 8 lines, usually 457 nm, from an argon ion laser, which is focused inside the region of the sample excited with the 290 nm pulse. The excitation beam is blocked with a filter, permitting only the probe light to reach the diode detector. Drift in the laser intensity is compensated by comparing with the intensity of a reference beam. The intensity from the sample and reference beams were recorded with a digital oscilloscope.

singlet state is only a few nanoseconds, while the lifetime of the triplet state in the complete absence of quenchers is hundreds of microseconds. Like the singlet, return to the ground state can occur by emitting a photon (phosphoresce), by a radiationless decay, or by quenching. The energy difference between T_1 and the next lowest triplet state (T_2) places the $T_1 \to T_2$ absorption spectrum in the blue-green region of the spectrum, making it convenient to monitor the population of T_1 by optical absorption measurements with an argon ion laser. Figure 2 shows the layout of the apparatus used in the experiment.

The important property of cysteine is that of all the amino acids it is by far the most effective quencher of the tryptophan triplet [26]. These unique properties—the long-lived triplet state of tryptophan and its effective quenching by cysteine—permit the rate of forming a specific contact between cysteine and tryptophan to be measured in a heteropolypeptide that may contain any of the other 18 amino acids. Although the mechanism of quenching by cysteine is not yet established, it is believed to occur by short-range transfer of an electron from the tryptophan triplet to the cysteine sulfur to form cation and anion radicals, followed by recombination to recover the ground-state molecules.

For a peptide with tryptophan attached at one end and cysteine at the other, quenching of the tryptophan triplet can be described as a two-step process, as shown in fig. 3a. The first is a purely diffusive process to form a so-called "encounter complex" or "con-

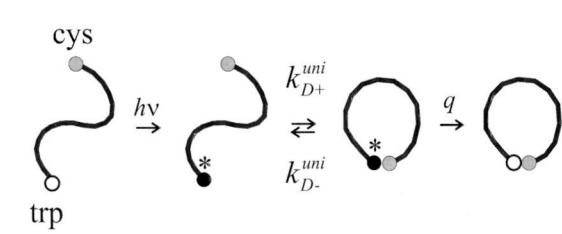

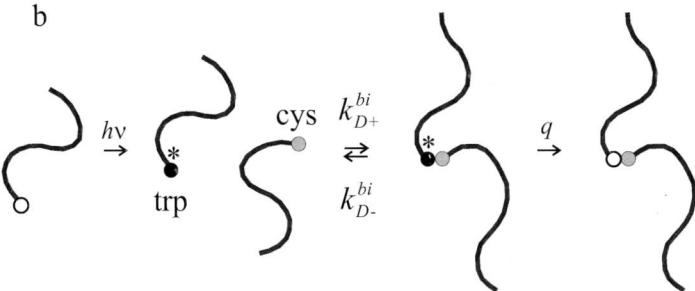

Fig. 3. – Two-step mechanism for quenching of tryptophan triplet by cysteine. (a) Unimolecular case; (b) bimolecular case (see legend to fig. 5). See text for description.

tact pair" between cysteine and tryptophan that has been optically excited to its triplet state. The second is depopulation of the triplet state within the contact pair. The overall observed rate of quenching (k^{uni}) is given by

$$(1) \qquad k^{\text{uni}} = k^{\text{uni}}_{D+} \left(\frac{q}{k^{\text{uni}}_{D-} + q} \right) \equiv k^{\text{uni}}_{D+} \phi^{\text{uni}},$$

where k^{uni}_{D+} is the rate of forming the contact pair, k^{uni}_{D-} is the rate of dissociation of the contact pair, and q is the rate of quenching. In this model both the rate of forming and dissociating the contact pair are diffusion limited. ϕ^{uni} is the probability of quenching once the contact pair is formed. If the overall rate of quenching k^{uni} is diffusion limited (*i.e.*, $q \gg k^{\text{uni}}_{D-}$) then the measured quenching rate is also the rate of contact formation, k^{uni}_{D+}. If it is not diffusion-limited, then the probability of quenching ϕ^{uni} must be separately determined in order to obtain the contact formation rate, k^{uni}_{D+}.

2. – Quenching probability and diffusion-limited rate

The quenching probability ϕ^{uni} can be determined by studying both the unimolecular and bimolecular quenching processes using quenchers with different values of q. In the bimolecular process a peptide containing tryptophan but no quencher is mixed with an

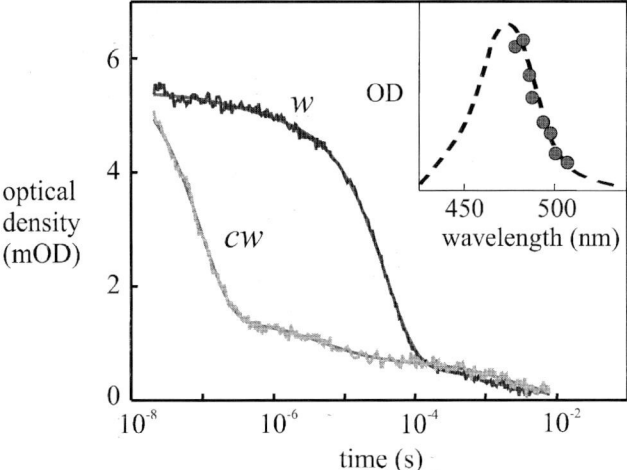

Fig. 4. – Decay of tryptophan triplet → triplet absorption following nanosecond optical excitation. Curve w is for the peptide: gln-(ala-gly-gln)$_3$-trp, while curve cw is for the peptide: cys-(ala-gly-gln)$_3$-trp, showing the dramatic reduction in lifetime due to quenching by cysteine. The inset shows the amplitude for the exponential part of curve (a) at 8 wavelengths compared with the absorption spectrum measured by Bent and Hayon [28].

excess of a peptide containing quencher but no tryptophan (fig. 3b). For quenching to occur the ends of two different peptides must first diffuse together (fig. 3b) (eq. (1) applies with "uni" replaced by "bi"). Making the reasonable assumption that the rate of quenching q at contact is the same in the unimolecular and bimolecular case, the measured unimolecular reciprocal rate $1/k^{\mathrm{uni}}$ is given by

$$\frac{1}{k^{\mathrm{uni}}} = \frac{1}{k^{\mathrm{uni}}_{D+}} + \frac{K^{\mathrm{bi}}}{K^{\mathrm{uni}}}\left(\frac{1}{k^{\mathrm{bi}}} - \frac{1}{k^{\mathrm{bi}}_{D+}}\right), \qquad (2)$$

where $K = k_{D+}/k_{D-}$ is the equilibrium constant for forming the contact pair, k^{bi} is the measured bimolecular rate, and k^{bi}_{D+} is the diffusion-limited bimolecular rate. For the simplest bimolecular diffusion-limited process both the rates and equilibrium constant can be easily calculated theoretically. In this case the quenching rate q is assumed to be effectively instantaneous when the tryptophan and quencher diffuse to within a center-to-center distance a, and is zero at longer distances (a more complete treatment would include the distance dependence of q [27], as well as excluded-volume effects). Smoluchowski showed that the rate is simply

$$k^{\mathrm{bi}}_{D+} = 4\pi D^{\mathrm{bi}} a, \qquad (3)$$

where D^{bi} is the sum of the diffusion coefficients of the two peptides. Since the equilibrium constant is simply the volume of the contact pair, i.e. $4/3\pi a^3$, the dissocia-

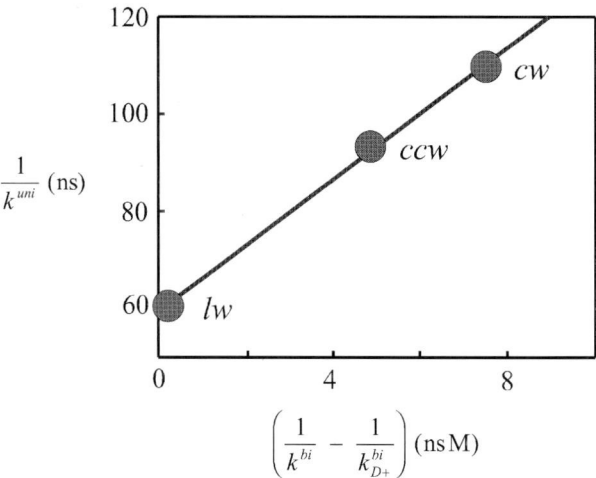

Fig. 5. – Unimolecular/bimolecular rate plot according to eq. (2) for three different quenchers. The three peptides for the unimolecular kinetics were cw = cys-(ala-gly-gln)$_3$-trp, ccw = cystine-(ala-gly-gln)$_3$-trp, and lw = lipoate-(ala-gly-gln)$_3$-trp. The corresponding peptides for the bimolecular kinetics were obtained by mixing the peptide gln-(ala-gly-gln)$_3$-trp with peptides cys-(ala-gly-gln)$_3$, cystine-(ala-gly-gln)$_3$, or lipoate-(ala-gly-gln)$_3$, respectively (see fig. 3b). From eq. (3) with $D^{bi} = 6.6 \times 10^{-3}$ cm^2/s and $a = 0.4$ nm, $k_{D+}^{bi} = 2.0 \times 10^9$ l/mol s^{-1}. The slope is 7 ± 1 l/mol and the intercept is 60 ± 1 ns.

tion rate is $3D^{bi}/a^2$. The condition for bimolecular diffusion-limited quenching is then: $q \gg 3D^{bi}/a^2$.

Equation (2) shows that the intercept of a plot of $1/k^{uni}$ vs. the difference between the observed and Smoluchowski reciprocal rates yields the rate of interest, k_{D+}^{uni}. Figure 4 shows an example of the kinetic data and fig. 5 shows the unimolecular/bimolecular rate plot for three different quenchers—cysteine, cystine (the disulfide of cysteine), and lipoate, a cyclic disulfide previously known to be a near-diffusion–limited quencher of the tryptophan triplet [28]. The diffusion coefficients for the peptides (3.3×10^{-5} cm^2/s, $D^{bi} = 6.6 \times 10^{-5}$ cm^2/s) were obtained from empirical compilations and the contact radius a was assumed to be 0.4 nm for all three quenchers. The determination of $k_{D+}^{uni} = 1.7 \times 10^7$ s^{-1} shows that the quenching probability ϕ^{uni} for cysteine in the unimolecular case is ~ 0.5. Measured unimolecular quenching rates by cysteine are therefore only about a factor of two smaller than the rate of diffusion-limited contact formation.

3. – Length dependence of end-to-end contact rate

An important dynamical property of a polymer chain is how the end-to-end contact rate depends on the chain length. We therefore studied the length dependence in a series of peptides having the sequence cys-(ala-gly-gln)$_j$-trp, with j varying from 1 to 6 (the peptide discussed up to now has $j = 3$). Assuming the same quenching probability ϕ^{uni}

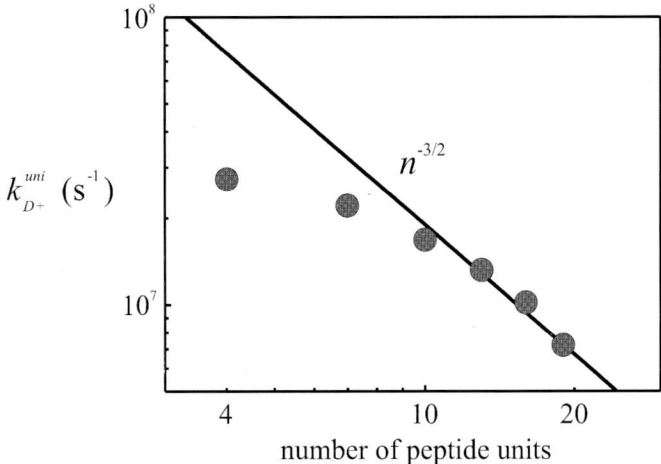

Fig. 6. – Length dependence of the end-to-end contact rate, k_{D+}^{uni}. The measured rates for the series of peptides—cys-(ala-gly-gln)$_j$-trp ($j = 1$ to 6)—were divided by the quenching probability $\phi^{\text{uni}} = 0.5$. The line is the function $k^{\text{uni}} \propto n^{-3/2}$, scaled to the observed rate for the longest peptide, $n = 19$.

for cysteine in each molecule, these experiments yield the length dependence of the end-to-end contact rate. The results are shown in fig. 6, where the contact rate is plotted as a function the number of peptide units, n. The dependence of rate on length increases with increasing chain length, and for the longest peptides approaches $n^{-3/2}$.

What dependence is expected theoretically? Szabo, Schulten, and Schulten (SSS) [29, 30] calculated the end-to-end contact rate for an idealized chain with a Gaussian end-to-end distance distribution. They showed that when the contact distance a is small compared to the chain segment length, this rate is given by

$$(4) \qquad k_{D+}^{\text{uni}} = \frac{4\pi D^{\text{uni}} a}{(2\pi \langle r^2 \rangle / 3)^{3/2}},$$

where $\langle r^2 \rangle$ is the equilibrium mean-squared end-to-end distance. For a freely jointed (random walk) chain, $\langle r^2 \rangle \propto n$, so SSS theory predicts $n^{-3/2}$ dependence for the end-to-end contact rate, as observed for the longest peptides in fig. 6.

Thirumalai and coworkers have considered more realistic chains and predict a maximum in the rate of loop formation at about 10 peptide units [31-33]. The maximum occurs because of chain stiffness which makes loop formation more difficult for shorter chains. Chain stiffness is a probable origin for our observation of a smaller rate dependence for the shorter peptides (fig. 6). It will be interesting in future studies to search for a maximum in the end-to-end contact rate by making measurements on peptides without glycines and therefore with reduced flexibility.

4. – Mean-squared end-to-end distance and end-to-end diffusion constant

Assuming a Gaussian distribution of end-to-end distances, the equilibrium constant for contact formation is

$$K^{\text{uni}} = \frac{4/3 \pi a^3}{(2\pi \langle r^2 \rangle /3)^{3/2}} \tag{5}$$

(so the dissociation rate of the contact pair is simply $3D^{\text{uni}}/a^2$, as for the bimolecular case). The interesting aspect of the experiment using different quenchers is that it provides a novel way of determining $\langle r^2 \rangle$. Since $K^{\text{bi}} = 4/3 \pi a^3$, from eqs. (2) and (5) $K^{\text{bi}}/K^{\text{uni}} = (2\pi \langle r^2 \rangle /3)^{3/2} = 7\,\text{l/mol}$, and therefore $\langle r^2 \rangle = 2.4\,\text{nm}^2$.

This value of $\langle r^2 \rangle$ is similar to what is expected from theoretical calculations by Flory and coworkers. According to Flory [34], $\langle r^2 \rangle = C_n n l^2$, where l is the C_α-C_α distance ($= 0.38\,\text{nm}$) and C_n is the characteristic ratio. C_n is a measure of chain stiffness and is very close to the number of peptide units that correspond to a segment of the equivalent freely jointed (random walk) chain. The very few experimental determinations of end-to-end distances in polypeptides [35] indicate that for alanine and other amino acids with unbranched β-carbons $C_n \approx 9$ for $n > 100$, but is reduced to ~ 5 for $n = 10$. Flory also showed that the much more flexible glycine markedly reduces the value of C_n in copolypeptides. For an alanine/glycine copolypeptide with $\sim 30\%$ glycines (as in our peptide with $n = 10$) $C_n \approx 4$ for large n [34]. Naively, then, we can use Flory's theoretical results to estimate $\langle r^2 \rangle \approx (4/9) \times 5 \times 10 \times (0.38)^2 = 3.2\,\text{nm}^2$, in surprisingly good agreement with the value calculated from our experiments.

Determination of $\langle r^2 \rangle$ also allows us to estimate the end-to-end diffusion coefficient D^{uni} using SSS theory. With $a = 0.4\,\text{nm}$ and $\langle r^2 \rangle = 2.4\,\text{nm}^2$, eq. (4) gives $D^{\text{uni}} \approx 4 \times 10^{-7}\,\text{cm}^2/\text{s}$. Very similar values for D^{uni} have previously been reported for unfolded cytochrome c [22,23], and unfolded ribonuclease A [36]. The 16-fold smaller value compared to D^{bi} ($= 6.6 \times 10^{-6}\,\text{cm}^2/\text{s}$) presumably represents intrachain interactions not considered in SSS theory.

5. – Concluding remarks

It is clear that the technique we have described has considerable potential to explore the dynamical properties of polypeptides. Future studies will be directed toward determining how the contact rates depend on chain length, composition, and amino acid sequence. The time scale of these kinetics makes all-atom molecular dynamics calculations feasible, allowing the results to provide yet another experimental benchmark for testing the accuracy of computer simulations.

Studies of loop formation have immediate practical value for investigations of proteins. In order to understand the role of loops in protein stability and the mechanism of protein folding, there have been several studies in which protein engineering has been used to alter existing loops or to introduce new ones into proteins [37-41]. Quantitative interpretation of the experimental results has been hampered by the lack of a significant data base on isolated polypeptides, demonstrating the need for measurements of both the equilibrium and dynamical properties discussed here.

REFERENCES

[1] ONUCHIC J., LUTHEY-SCHULTEN Z. and WOLYNES P. G., *Ann. Rev. Phys. Chem.*, **48** (1997) 545.
[2] SHAKHNOVICH E. I., *Curr. Opin. Struct. Biol.*, **7** (1997) 29.
[3] PANDE V. S., GROSBERG A. Y., TANAKA T. and ROKHSAR D. S., *Curr. Opin. Struct. Biol.*, **8** (1997) 68.
[4] DOBSON C. M., SALI A. and KARPLUS M., *Angew. Chem. Int. Edit.*, **37** (1998) 868.
[5] CHAN H. S. and DILL K. A., *Proteins*, **30** (1998) 2.
[6] GAREL T., ORLAND H. and PITARD E., in *Spin Glasses and Random Fields*, edited by A. P. YOUNG (World Scientific, Singapore) 1998, pp. 387-443.
[7] THIRUMALAI D. and KLIMOV D., *Curr. Opin. Struct. Biol.*, **9** (1999) 197.
[8] FERSHT A., *Structure and Mechanism in Protein Science* (W. H. Freeman, New York) 1999.
[9] JONES C. M., HENRY E. R., HU Y., CHAN C.-K., LUCK S. D., BHUYAN A., RODER H., HOFRICHTER J. and EATON W. A., *Proc. Natl. Acad. Sci. USA*, **90** (1993) 11860.
[10] EATON W. A., MUÑOZ V., THOMPSON P. A., CHAN C.-K. and HOFRICHTER J., *Curr. Opin. Struct. Biol.*, **7** (1997) 10.
[11] CALLENDER R. H., DYER R. B., GILMANSHIN R. and WODDRUFF W. H., *Annu. Rev. Phys. Chem.*, **49** (1998) 173.
[12] GRUEBELE M., *Annu. Rev. Phys. Chem.*, **50** (1999) 485.
[13] EATON W. A., MUÑOZ V., HAGEN S. J., JAS G. S., LAPIDUS L., HENRY E. R. and HOFRICHTER J., *Ann. Rev. Biophys. Biomol. Struct.*, **29** (2000) 327.
[14] WILLIAMS K., CAUSGROVE T. P., GILMANSHIN R., FANG K. S., CALLENDER R. H., WOODRUFF W. H. and DYER R. B., *Biochemistry*, **35** (1996) 691.
[15] THOMPSON P. A., EATON W. A. and HOFRICHTER J., *Biochemistry*, **36** (1997) 9200.
[16] LEDNEV I. K., KARNOUP A. S., SPARRROW M. C., ASHER S. A., *J. Am. Chem. Soc.*, **121** (1999) 8074.
[17] THOMPSON P. A., MUÑOZ V., JAS G. S., HENRY E .R., EATON W. A. and HOFRICHTER J., *J. Phys. Chem.*, **104** (2000) 378.
[18] JAS G. S., EATON W. A. and HOFRICHTER J., *Phys. Chem. B*, **105** (2001) 261.
[19] MUÑOZ V., THOMPSON P. A., HOFRICHTER J., EATON W. A., *Nature*, **390** (1997) 196.
[20] MUÑOZ V., HENRY E. R., HOFRICHTER J., EATON W. A., *Proc. Natl. Acad. Sci. USA*, **95** (1998) 5872.
[21] DUAN Y. and KOLLMAN P. A., *Science*, **282** (1998) 740.
[22] HAGEN S. J., HOFRICHTER J., SZABO A. and EATON W. A, *Proc. Natl. Acad. Sci. USA*, **93** (1996) 11615.
[23] HAGEN S. J., HOFRICHTER J. and EATON W. A, *J. Phys. Chem.*, **101** (1997) 2352.
[24] BIERI O., WIRZ J., HELLRUNG B., SCHUTKOWSKI M., DREWELLO M. and KIEFHABER T., *Proc. Natl. Acad. Sci. USA*, **96** (1999) 9597.
[25] LAPIDUS L. J., EATON W. A. and EATON W. A., *Proc. Natl. Acad. Sci. USA*, **97** (2000) 7220.
[26] GONNELLI M. and STRAMBINI G. B., *Biochemistry*, **34** (1995) 13847.
[27] ZHOU H.-X. and SZABO A., *Biophys. J.*, **71** (1996) 2440.
[28] BENT D. V. and HAYON E., *J. Am. Chem. Soc.*, **97** (1975) 2612.

[29] SZABO A., SCHULTEN K. and SCHULTEN Z., *J. Chem. Phys.*, **72** (1980) 4350.
[30] PASTOR R. W., ZWANZIG R. and SZABO A., *J. Chem. Phys.*, **106** (1996) 3878.
[31] GUO Z. and THIRUMALAI D., *Biopolymers*, **36** (1995) 83.
[32] CAMACHO J. and THIRUMALAI D., *Proc. Natl. Acad. Sci. USA*, **92** (1995) 1277.
[33] THIRUMALAI D., *J. Phys. Chem. B*, **103** (1999) 608.
[34] FLORY P. J., *Statistical Mechanics of Chain Molecules* (Hanser/Gardner, Cincinnati) 1988.
[35] CANTOR C. R. and SCHIMMEL P. R., *Biophysical Chemistry, Part III, The Behavior of Biological Macromolcules* (W. H. Freeman, San Francisco) 1980.
[36] BUCKLER D. R., HAAS E. and SCHERAGA H. A., *Biochemistry*, **34** (1995) 15965.
[37] PACE C. N., GRIMSLEY G. R., THOMSON J. A. and BARNETT B. J., *J. Biol. Chem.*, **263** (1988) 11820.
[38] LADURNER A. G. and FERSHT A. R., *J. Mol. Biol.*, **273** (1997) 330.
[39] VIGUERA A.-R. and SERRANO L., *Nature Struct. Biol.*, **4** (1997) 939.
[40] ROBINSON C. R. and SAUER R. T., *Proc. Natl. Acad. Sci. USA*, **95** (1998) 5929.
[41] NAGI A. D., ANDERSON K. S. and REGAN L., *J. Mol. Biol.*, **286** (1999) 257.

Characterisation of the transition state for protein folding: Mutational analysis taken one step further

M. OLIVEBERG

Biochemistry, Umeå University - 901 87 Umeå, Sweden

1. – What is the folding transition state?

A characteristic feature of small proteins is that they fold and unfold in a highly concerted all-or-non process [1]. Half-denatured proteins in the test tube are thus not structural intermediates but represent ensemble averages of molecules which flick forwards and backwards between the unfolded (D) and fully native state (N). The behaviour tells that the structural intermediates are unstable and contribute to a free-energy barrier separating D and N. The highest point of this barrier, as probed by the folding and unfolding rates, is commonly defined as the protein folding transitions state ($\ddagger$) (fig. 1). Despite the high dimensionality of the folding reaction, the transition state very much behaves like a transition state for a simple low-dimensional reaction, and the folding process shows simple exponential time courses. It follows that the ratio of rate constants for folding (k_f) and unfolding (k_u) equals the equilibrium constant

$$K_{\text{D-N}} = \frac{[\text{D}]}{[\text{N}]} = \frac{k_u}{k_f}. \tag{1}$$

Proteins obeying this behaviour are referred to as two-state proteins. Although the meaning and structural composition of the transition state is unclear at the microscopic level, the treatment provides a strict and simple formalism for rationalising experimental

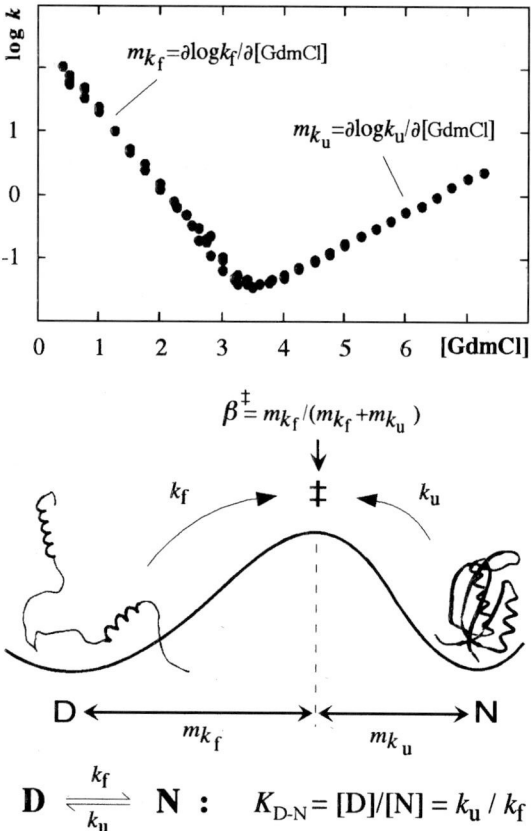

Fig. 1. – Top. Chevron plot of the split β-α-β protein S6. This type of v-shaped chevrons are typical for two-state proteins. The sensitivity of the ‡-stability relative to D and N ($\beta^\ddagger$) is derived by comparing the denaturant dependencies of the folding rate constant ($m_{k_f} = \partial \log k_f/\partial[\text{GdmCl}]$) and the unfolding rate constant ($m_{k_u} = \partial \log k_u/\partial[\text{GdmCl}]$). The parameter $\beta^\ddagger$, which in the present case is about 0.7, is often referred to as the position of the folding transition state. The linear limbs of the chevron plot indicate that $\beta^\ddagger$ is unaffected by denaturant. Upon certain mutations, however, S6 displays marked curvatures in chevron plot indicating changes of $\beta^\ddagger$ and the transition state structure. Adapted from [2]. Bottom. Two-state folding is modelled as a jump over a free-energy barrier of unstable conformations the top of which is the transition state ensemble (‡). From a microscopic perspective, the free-energy barrier, in turn, is the projection of the multidimensional folding energy landscape on the elusive experimental observable β, i.e. the relative sensitivity to denaturant (proportional to solvent exposure).

observations for comparison with theory. As long as the theorist knows how the results are derived, it is relatively simple to uncover the "raw" characteristics of the folding reaction and wrap them up in the context of a new model. The aim of this paper is to describe and discuss i) the experimental characterisation of the transition state structure by the commonly used phi-value method, ii) shifts of the transition state structure upon perturbation of the folding conditions, and iii) how these transition state shifts in combination with mutant analysis can be used to infere what happens before and after the snap-shot provided by the transition state structure.

2. – The gross behaviour of the transition state and the parameter $\beta^{\ddagger}$

For most proteins, the transition state ensemble is surprisingly robust and seems to maintain its structure as firmly as the native state. Most clearly, this is indicated by linear refolding and unfolding limbs in the so-called chevron plots, *i.e.* plots of $\log k_\mathrm{f}$ and $\log k_\mathrm{u}$ vs. GdmCl or urea concentration. The slopes of the chevron plot are defined as the kinetic m-values

$$\text{(2)} \qquad m_{k_\mathrm{f}} = \frac{\partial \log k_\mathrm{f}}{\partial [\mathrm{GdmCl}]} \qquad \text{and} \qquad m_{k_\mathrm{u}} = \frac{\partial \log k_\mathrm{u}}{\partial [\mathrm{GdmCl}]}$$

which, according to mass action and weak interaction models, reflect the changes in "solvent accessible surface area" in the D $\rightarrow$ ‡ and N $\rightarrow$ ‡ transitions. The higher the value of m_{k_f}, the larger the structural difference between D and ‡. The commonly

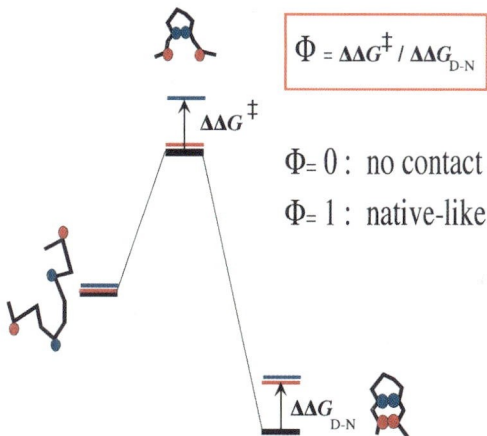

Fig. 2. – Phi-value analysis. Assume that a protein consists of two pairs of interactions, red and blue. Upon mutation of a red residue the native state is destabilised ($\Delta\Delta G_\mathrm{D\text{-}N} > 0$) but the free-energy level of the transition state remains unchanged ($\Delta\Delta G^{\ddagger} = 0$). The transition state is then said to be unfolded around the site of mutation and the phi-value for the muation is zero. Experimentally, the scenario is characterised by an increased rate constant for unfolding (k_u) and a rate constant for refolding (k_f) which is identical to that of the wild-type protein. Upon mutation of a blue residue, however, the transition state looses as much stability as the native state ($\Delta\Delta G^{\ddagger} = \Delta\Delta G_\mathrm{D\text{-}N}$) and the transition state structure is said to be native-like around the site of mutation; the phi-value is 1. In the latter case, k_u is unaffected by the mutation whereas k_f becomes slower. With real proteins, most mutations produce fractional phi-values, indicative of partially formed or weakened interactions in the transition state.

observed v-shaped chevron plots with constant values of m_{k_f} and m_{k_u}, are thus indicating that the folding and unfolding transitions are unaffected by denaturant and that the transition state stucture is the same in water as in 8 M of GdmCl. A crude measure of the transition state's strucure relative to D and N may then be inferred by the normalisation (fig. 1)

$$(3) \qquad \beta^{\ddagger} = \frac{m_{k_f}}{m_{k_f} + m_{k_u}}.$$

Depending on the protein, the value of $\beta^{\ddagger}$ is typically somewhere between 0.5 and 0.8 [3]. From a microscopic perspective, the ability of the transition state ensemble to resist even severe solvent perturbations is surprising and indicate that the barrier crossing is confined to a narrow rather than a broad pass in the energy landscape.

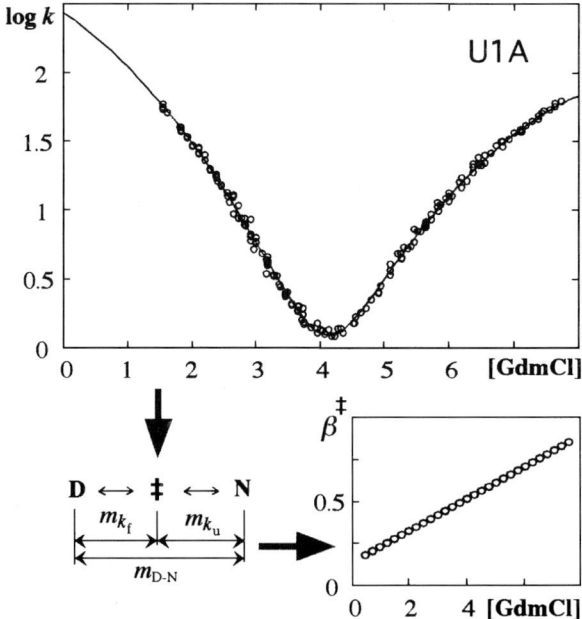

Fig. 3. – For the split β-α-β protein U1A the chevron plot is symmetrically curved. Consistent with two-state folding, the sum of the kinetic m-values $(m_{k_f} + m_{k_u})$ is nevertheless constant and matches precisely the m-value for the cooperative transition D to N $(m_{D\text{-}N})$. This suggests that the curvatures are coupled and result from changes of $\ddagger$ upon addition of GdmCl: $\beta^{\ddagger}$ moves progressively from 0.2 in pure water to 0.9 in 8 M GdmCl. The critical nucleus grows upon addition of denaturant. Adapted from [17].

3. – Phi-value analysis

To date, the most detailed information about the protein folding transition state has been derived from so-called phi-value studies [1]. Since the ‡ structure does not seem to reorganise significantly upon mutation, the contribution of the truncated side chain to the stability of the transition state may be deduced from changes in the folding kinetics. In essence, mutations which slow down the folding rate, *i.e.* raise the barrier, are considered to target interactions which are part of the transition state structure (fig. 2). If ‡ looses as much stability as N, $\Delta\Delta G^{\ddagger} = \Delta\Delta G_{\text{D-N}}$, the truncated moiety is said to experience a native-like environment in ‡ and yields a phi-value of $\Delta\Delta G^{\ddagger}/\Delta\Delta G_{\text{D-N}} = 1$. If $\Delta\Delta G^{\ddagger} = 0$, the side chain is said to experience a denatured-like environment in ‡ and phi $= \Delta\Delta G^{\ddagger}/\Delta\Delta G_{\text{D-N}} = 0$ [4].

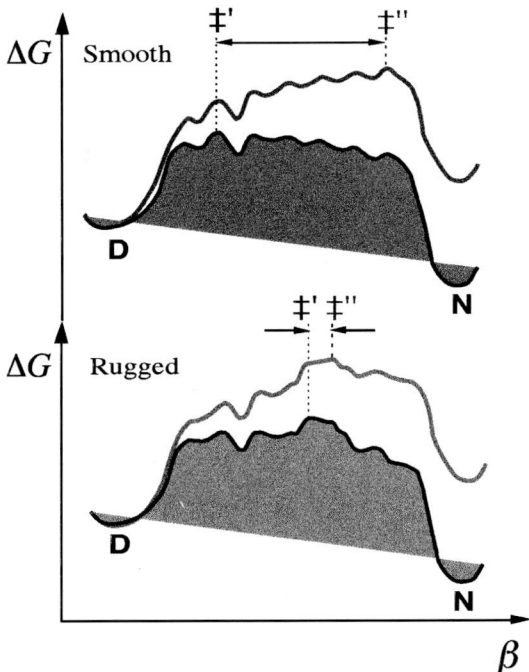

Fig. 4. – The broad barrier model. Transition state movements on rugged and smooth activation barriers. The difference in free energy between the grey and black barrier profiles is a linear ramp $\Delta\Delta G = \beta \times m_{\text{D-N}} \times [\text{GdmCl}]$, where β is the x-axis and $m_{\text{D-N}} \times [\text{GdmCl}]$ are constants; black illustrate the free-energy profile for folding under stabilising conditions and grey shows denaturing conditions. If the barrier is overall smooth, the transition state will describe large movements along the barrier profile (top), whereas if the transition state is confined to a pointed feature on top of the barrier its movements upon destabilisation will be small (bottom). In this way broad barriers unify a wide range of kinetic behaviours and account also for the high cooperativity in protein folding. Adapted from [19, 20].

Currently, there are several phi-value studies of small proteins where extensive sets of mutations have been analysed [5-14]. In most cases, the experiments yield a transition state structure with a large number of fractional phi-values diffusely centred around a small core of tertiary contacts with high phis. The observations indicate that folding is not as hierarchical as one might expect from the organisation of native protein structures: secondary and tertiary structures form concomitantly rather than consecutively. The

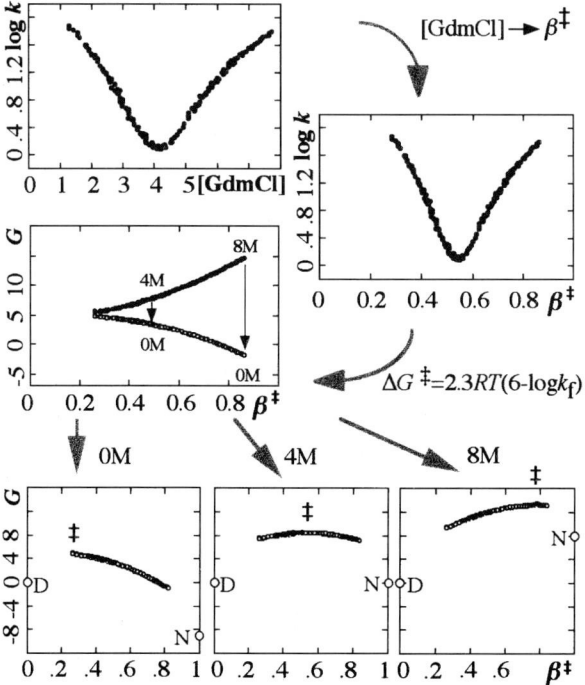

Fig. 5. – Derivation of barrier shape from chevron plot. Units are in s^{-1} and $2.3RT$. *Step 1.* To account for the curvatures, $\log k_f$ and $\log k_u$ were fitted by second-order polynomials the derivatives of which yield the kinetic m-values (eq. (2)). $\beta^\ddagger$ was then calculated from eq. (3), and used as x-axis for replotting the chevron data. *Step 2.* The barrier height (D → ‡) at each point of $\beta^\ddagger$ was approximated to $\Delta G^\ddagger = 2.3RT(6 - \log k_f)$, *i.e.* Eyrings rate equation with a pre-factor of 10^6 s^{-1}. Note that each point on this curve is obtained at a different [GdmCl]: early parts are linked to rate constants obtained at low [GdmCl] in the refolding region, whereas late parts are linked to high [GdmCl] in the unfolding region. A clearer representation of the free-energy profile is obtained if all points are extrapolated to a common [GdmCl]. This is done next by linear free-energy extrapolations. *Step 3.* As $\beta^\ddagger$ is simply the normalised sensitivity to GdmCl (eq. (3)), each value of $\Delta G^\ddagger$ can be extrapolated to a common [GdmCl] (Y) by standard linear free-energy relations: $\Delta G^\ddagger(\beta^\ddagger, [Y]) = \Delta G^\ddagger(\beta^\ddagger, [X]) + \beta^\ddagger m_{D-N} 2.3RT([Y] - [X])$, where [X] is the GdmCl concentration at which $\log k_f$ was measured. The lower panels show the resulting free-energy profiles at 0, 4 and 8 M GdmCl, and illustrate how ‡ moves along the top of a broad barrier as [GdmCl] is increased. The key assumption behind the barrier construction is thus that folding proceeds by the same events at all [GdmCl]. Adapted from [17].

behaviour is fingerprinted "nucleation condensation" [15] to underline the resemblance with nucleation mechanisms. But the question then arises, what happens before and after this diffuse snap-shot of the folding reaction?

4. – Transition state shifts and the broad barrier model

Although the folding and unfolding limbs of most chevrons are found to be linear, some proteins display distinct curvatures [16]. For example, for the split α-β-α protein U1A, the chevron is symmetrically "lily-shaped" showing that the m-values vary with [GdmCl]; m_{k_f} grows larger while m_{k_u} decreases (fig. 3). Notably, the increase in m_{k_u} is precisely coupled to the decrease in m_{k_u} maintaining a constant value of

(4)
$$-m_{k_f} + m_{k_u} = m_{\text{D-N}} = \frac{\partial \log K_{\text{D-N}}}{\partial [\text{GdmCl}]}.$$

This suggests that U1A displays two-state kinetics at all [GdmCl] (cf. eqs. (1), (2) and (4)), and that the curvatures result from changes of $\beta^{\ddagger}$ (eq. (3)). That is, the curvature results from changes of the U1A's transition state; at high concentrations of GdmCl, $\ddagger$ moves closer to N resulting in a smaller structural change upon activation N → $\ddagger$ and thus a lower value of m_{k_u} [17]. According to eq. (3), the $\ddagger$ movement of U1A ranges from ~ 0.2 at 0.45 M GdmCl to ~ 0.9 at 8 M GdmCl (fig. 3). Similar effects have been observed earlier for barnase and have been ascribed to Hammond postulate behaviour [18]. In the case of U1A, however, it is possible to further rationalise the phenomenon as transition state movements along the top of a broad free-energy barrier (fig. 4). The derivation of broad barrier profiles and transition state shifts from the curved chevron plot of U1A is given in fig. 5. Although $\beta^{\ddagger}$ is usually interpreted in terms of "relative solvent exposure" [1], it is helpful to note that $\beta^{\ddagger}$ is strictly an energetic parameter which quantifies the sensitivity of the transition state stability to denaturants: a species at $\beta^{\ddagger} = 0.5$ is destabilised half as much as N per every addition of GdmCl or

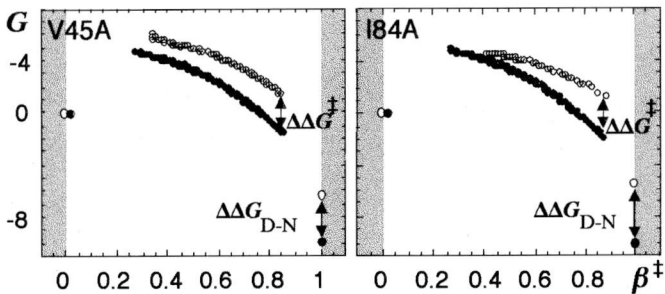

Fig. 6. – Examples of barrier profiles used for calculation of the phi-value graphs in fig. 7. (cf. fig. 2). The profiles are calculated for wild-type and mutant U1A according to fig. 5. Adapted from [17].

urea. This means that addition of denaturant may be said to tilt the free-energy profile, hence, providing an intuitive link between the extent of transition state movements and the shape of the activation barrier. The results have led to speculations that the barrier profile for two-state proteins may be generally broad, and that the commonly observed "fix" transition state in fact represents secondary features jutting up some distance from the rest of the barrier [16] (fig. 4). The idea is supported by experiments where curvatures have been induced in the normally v-shaped proteins CI2 and S6 by mutations that are expected to raise only the late part of the free-energy profile [19,20]. As will be described next, moving transition states in combination with phi-value analysis have been used to continuously scan the folding along the free-energy profile.

5. – From snap-shot to movie: scanning the activation barrier by a moving transition state

For U1A, phi-values have been calculated as a function of β by comparison of the barrier profiles for WT and mutant proteins (cf. figs. 2 and 6). The profiles in fig. 6 are calculated as in fig. 5. The continuous phi-values of U1A [17] provide a movie of the folding process which cover more than 60% of the coordinate (fig. 7). Or, more strictly, a movie of how the critical nucleus grows upon addition of denaturant. With reference to fig. 8, the U1A data is rationalised as follows. First contacts form between helix 1 and

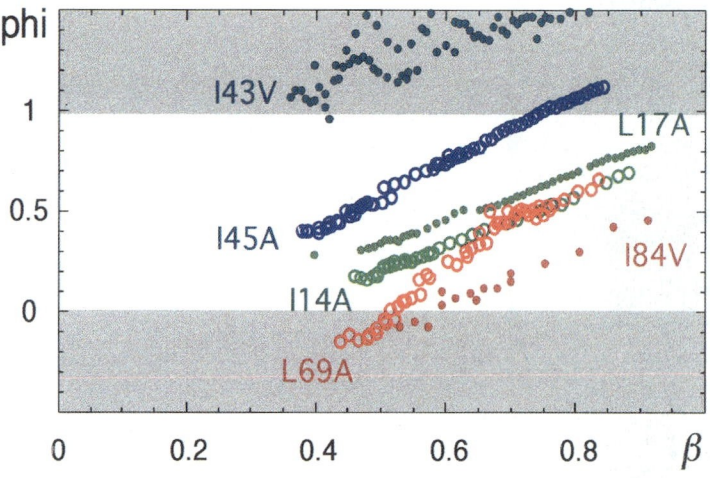

Fig. 7. – Representative examples of phi-value changes for the U1A protein as a function of transition state position ($\beta^\ddagger$). The data shows how the transition state structure becomes gradually more compact upon destabilisation by GdmCl, cf. smooth barrier movement in fig. 3. The colour coding refers to the structural interpretation in fig. 4. At $\beta^\ddagger = 0.6$, U1A shows a typical nucleation condensation pattern with diffuse fractional contacts (green and red) centred around a region of high phi-values (blue). The distribution of high and low phi values at each point of $\beta^\ddagger$ is given in fig. 9.

strands 2 and 3, which are all close in sequence (blue). The assembly is subsequently stabilised by strand 1 (green) and, finally, the structure is wrapped up by helix 2 and strand 4 at the C-terminal end (red). It is apparent that folding proceeds by joining structures which are close in sequence. However, the events are severely overlapping and triggered almost simultaneously throughout the structure, giving rise to an overall diffuse transition state structure resembling that of CI2 and the nucleation condensation behaviour (cf. [6]).

A related analysis of the analogous protein S6, which displays $\beta^{\ddagger}$ shifts upon most mutations [20], yields overall similar results but with one important difference. S6 forms an initial core between helix 1 and strand 1 that is subsequently stabilised by the flap of strands 2 and 3 (Daniel Otzen & M.O., unpublished) (fig. 9). Accordingly, S6 displays a shift of the folding nucleus towards the helix-2 side of the protein.

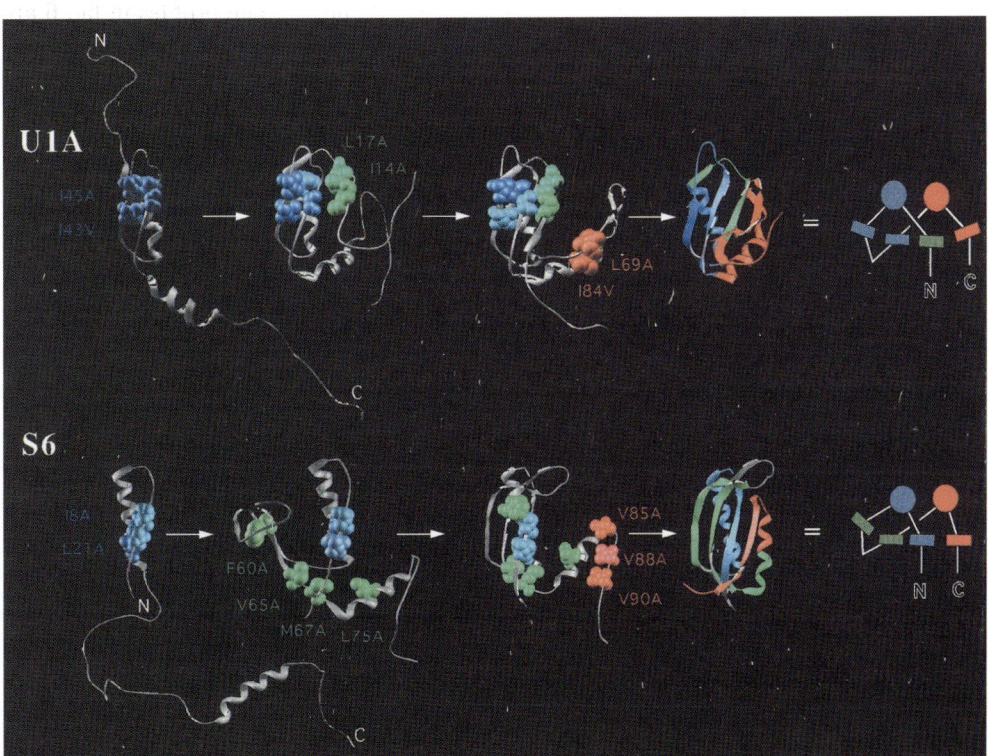

Fig. 8. – Schematic illustration of the folding reaction for the two split α-β-α proteins U1A and S6, as seen from analysis of transition state movements. Note that the figure exaggerates the order of folding events. In practice, the events are severely overlapping and triggered almost simultaneously throughout the structure giving rise to an overall diffuse transition state structure (cf. figs. 7 and 9).

6. – Determining the folding pathway directly from the native structures

The difference in nucleation between U1A and S6 may be understood from the detailed arrangement of the native structures. U1A comprises a series of good side chain contacts between helix 1 and strand 2 which are not present in S6. In S6, which has a straighter arrangement of the secondary structure elements, the centre of the hydrophobic core is formed by interactions between helix 1 and strand 1. The folding sequence for these proteins may then be inferred directly from a contact map with just a simple rule of thumb: join nearest elements with good contacts through the hydrophobic core. The method is even sufficient to predict the folding of S6 permutants where the sequence of the secondary structure is changed to shift the nucleus. Although, this crude, manual version of the calculations by Eaton and other groups [21-24] is sufficient to determine the folding events in the very clear cases of U1A and S6, problem arise when more elaborate trade-offs between contacts and entropy loss need to be made.

In summary, the continuous phi-value approach provides a new level of detail of the folding nucleation process and, as such, an improved handle for comparison with theory. This prompts us to be critical and explore further the mechanism behind "curved" chevron plots and possible pitfalls with the current interpretation models. For example, to what extent may ground-state changes, *i.e.* partial rupture of N in the dead-time of the unfolding experiment interfere with the transition state analysis? In this context, it would be nice to see more calculations on how the critical nucleus changes with stability. Such perturbation analysis of the transition state ensemble may help the development of new experiments to shed further light on the macro-micro relation and the poorly understood observable β.

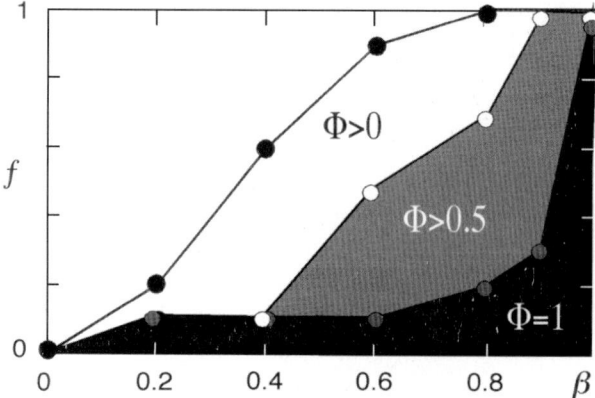

Fig. 9. – Development of structure in U1A's transition state *vs.* $\beta^{\ddagger}$. The graphs show the fraction of residues (f) with phi-values above 0, 0.5 and 1, respectively. Data are based on results from 16 mutations. It is apparent that full native contacts are formed only very late in folding, and that the transitions state overcomes the entropy cost of immobilising the chain by a diffuse set of fractional interactions. Perhaps this weak support from the side chain interactions causes the protein to gain structure where the entropy cost is minimal. Adapted from [17].

REFERENCES

[1] FERSHT A. R., *Structure and Mechanism in Protein Science: A Guide to Enzyme Catalysis and Protein Folding* (W. H. Freeman and Co., New York) 1999.
[2] OLIVEBERG M., *Curr. Opin.*, **11** (2001) 94.
[3] JACKSON S. E., *Fold. Des.*, **3** (1998) R81.
[4] MATOUSCHEK A., KELLIS J. T. JR., SERRANO L. and FERSHT A. R., *Nature*, **340** (1989) 122.
[5] JACKSON S. E., EL MASRY N. and FERSHT A. R., *Biochemistry*, **32** (1993) 11270.
[6] OTZEN D. E., ITZHAKI L. S., EL MASRY N. F., JACKSON S. E. and FERSHT A. R., *Proc. Natl. Acad. Sci. USA*, **91** (1994) 10422.
[7] BURTON R. E., MYERS J. K. and OAS T. G., *Biochemistry*, **37** (1998) 5337.
[8] KRAGELUND B. B., OSMARK P., NEERGAARD T. B., SCHIODT J., KRISTIANSEN K., KNUDSEN J. and POULSEN F. M., *Nature Struct. Biol.*, **6** (1999) 594.
[9] HAMILL S. J., STEWARD A. and CLARKE J., *J. Mol. Biol.*, **297** (2000) 165.
[10] MAIN E. R., FULTON K. F. and JACKSON S. E., *J. Mol. Biol.*, **291** (1999) 429.
[11] FULTON K. F., MAIN E. R., DAGGETT V. and JACKSON S. E., *J. Mol. Biol.*, **291** (1999) 445.
[12] VILLEGAS V., MARTINEZ J. C., AVILES F. X. and SERRANO L., *J. Mol. Biol.*, **283** (1998) 1027.
[13] TADDEI N., CHITI F., FIASCHI T., BUCCIANTINI M., CAPANNI C., STEFANI M., SERRANO L., DOBSON C. M. and RAMPONI G., *J. Mol. Biol.*, **300** (2000) 633.
[14] COTA E., HAMILL S. J., FOWLER S. B. and CLARKE J., *J. Mol. Biol.*, **302** (2000) 713.
[15] FERSHT A. R., *Proc. Natl. Acad. Sci. USA*, **92** (1995) 10869.
[16] OLIVEBERG M., *Acc. Chem. Res.*, **31** (1998) 765.
[17] TERNSTROM T., MAYOR U., AKKE M. and OLIVEBERG M., *Proc. Natl. Acad. Sci. USA*, **96** (1999) 14854.
[18] MATOUSCHEK A. and FERSHT A. R., *Proc. Natl. Acad. Sci. USA*, **90** (1993) 7814.
[19] OLIVEBERG M., TAN Y. J., SILOW M. and FERSHT A. R., *J. Mol. Biol.*, **277** (1998) 933.
[20] OTZEN D. E., KRISTENSEN O., PROCTOR M. and OLIVEBERG M., *Biochemistry*, **38** (1999) 6499.
[21] MUNOZ V. and EATON W. A., *Proc. Natl. Acad. Sci. USA*, **96** (1999) 11311.
[22] ALM E. and BAKER D., *Curr. Opin. Struct. Biol.*, **9** (1999) 189.
[23] GALZITSKAYA O. V. and FINKELSTEIN A. V., *Proc. Natl. Acad. Sci. USA*, **96** (1999) 11299.
[24] RIDDLE D. S., GRANTCHAROVA V. P., SANTIAGO J. V., ALM E., RUCZINSKI I. and BAKER D., *Nat. Struct. Biol.*, **6** (1999) 1016.

RSMR comparison of dynamic properties of various proteins

Yu. F. Krupyanskii and V. I. Goldanskii(*)

Institute of Chemical Physics, Russian Academy of Sciences
117977, ul. Kosygina 4, Moscow, Russia

1. – Introduction

The dynamical properties of proteins are determined by the energetic landscape of a macromolecule which is formed during the protein folding. There are two main hypotheses of folding within a post-translational folding mechanism. If the first one, the so-called thermodynamic hypothesis of folding [1], is valid, the protein native structure corresponds to the global minimum of energy. It means that the potential energy surface of the protein macromolecule is harmonic or quasiharmonic. The dynamic properties of all proteins should be very similar, as similar are the dynamic properties of different nanocrystalline particles, where harmonic or quasiharmonic vibrations are only possible. Naturally, short-lived fluctuations of a structure can also take place, as in any small (nanocrystalline) solids [2].

If the second hypothesis, the so-called kinetic hypothesis of folding [3], is valid, the protein native structure corresponds to the relatively deep and stable but local minimum of energy (see fig. 1). One can expect a great amount of conformational substates (CS) in a definite conformation of a protein macromolecule [4, 5]. It is possible to talk about the energetic landscape of a protein macromolecule. The dynamic properties of a protein may be much richer in this case than in the first one: protein specific motions namely transitions of an atomic group between different local minima of the type of bounded

---

(*) Deceased.

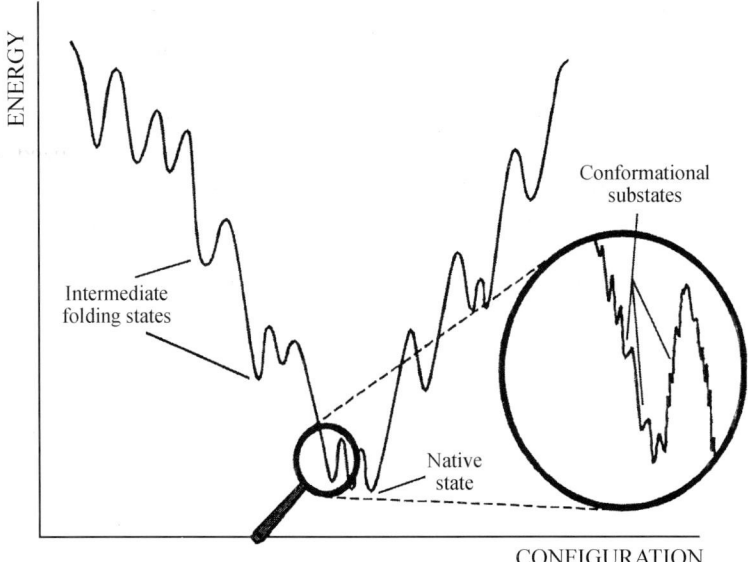

Fig. 1. – Folding funnel.

diffusion may appear (these motions are different from the usual motions in a solid). One can talk about liquid-like (on short times) motions [6]. Correspondingly, different proteins may have a strongly different energetic landscape and consequently strongly different dynamic properties. As one example of the proposed models of folding, we consider a sequential model proposed by Ptitsyn [7]. According to this model, protein folding goes

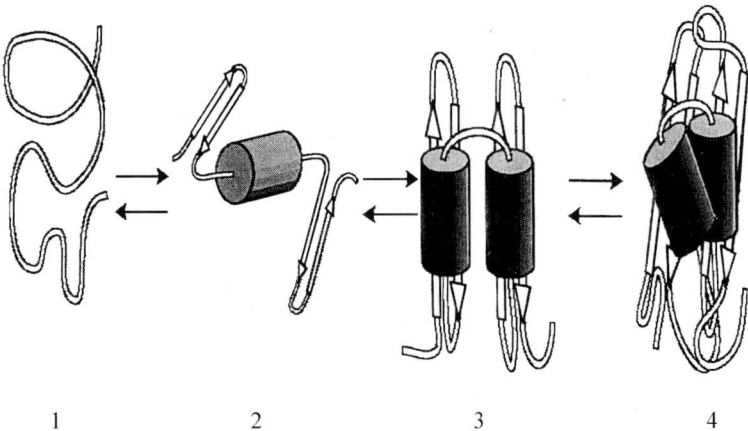

Fig. 2. – Sequential model of folding. 1) Unfolded chain (disordered coil), 2) fluctuating secondary structure, 3) intermediate (molten globular) structure, 4) native state.

at least via three stages (fig. 2): 1) the formation of the fluctuating secondary structure elements in an unfolded chain, stabilized by the peptide-hydrogen bonds; 2) the merging of pre-existing blocks with a secondary structure to an intermediate globular (molten globular) structure, stabilized by hydrophobic interactions, and 3) the adjustment of this intermediate structure to the final native tertiary structure, stabilized by van der Waals interactions. Different interactions "switch on" consecutively. The packing density increases at every step and the highest one is reached in the native state.

Evidently the dynamics of a condensed system is essentially connected with its packing density or free volume. Then an increase of density means the decrease of the mean square fluctuations of atoms and atomic groups. Therefore, the dynamic properties of the native state should be much less pronounced than in the molten (intermediate) state and, moreover, than in the fluctuating secondary structure or in the unfolded-coil state.

The observation of different dynamics for proteins with a different structural organization may give an additional argument in favor of the kinetic hypothesis of folding.

2. – Experimental part

Poly-L-glutamic acid (poly-Glu) is well known to be a classical model system: at $pH = 7.0$ poly-Glu is in a random coil state and at $pH = 2.0$ poly-Glu easily forms the α-helices [8]. Poly-Glu in its random coil state models an unfolded molecule. Myoglobin and human serum albumin (HSA) are well known as α-helical proteins, lysozyme as a partially β-sheet protein and α-lactalbumin at $pH = 2.0$ as a partially β-sheet protein in a molten globular state [9]. Among current physical methods for studying the dynamic properties of proteins, a notable role is played by the method of Rayleigh scattering of Mössbauer radiation (RSMR) [10]. The high-energy resolution of the method ($\sim 10^{-9}$ eV) accomplished by a combined use of a Mössbauer detector and source exceeds by several orders of magnitude the resolution provided by the most advanced neutron spectrometers (up to $\sim 10^{-6}$ eV), let alone that offered by X-ray dynamical analysis (~ 1 eV). It is precisely due to this fact that it is possible to detect comparatively slow motions with correlation times $\tau_c \sim 10^{-6}$ to 10^{-9} s from the broadening of energy spectrum lines, and motions with correlation times $\tau_c < 10^{-9}$ s from the decreased fraction of elastic scattering. The essential advantage of this method over Mössbauer absorption spectroscopy is its versatility, for the scatterer (the biopolymer under study) need not have Mössbauer nuclei. Due to this advantage, the scope of amenable biological objects can be substantially widened. There are several surveys dealing with RSMR published up to date, in which experimental and theoretical fundamentals of the method are considered in detail (see for example [10]). Shown in fig. 3 is the schematics of the RSMR experimental arrangement. Mössbauer radiation emitted by a ^{57}Co source mounted on a vibrator experiences Rayleigh scattering by electrons of a biopolymer B. The radiation scattered at an angle 2θ is measured by the detector D. In order to measure the elastic fraction f, an additional resonance absorber is implemented, which together with the detector makes up the so-called "Mössbauer" detector. By means of the resonance absorber, first the intensity of the incident beam, $I(o)$, is measured and then the intensity of

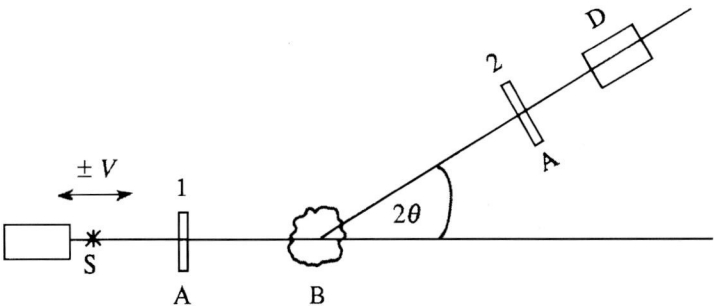

Fig. 3. – RSMR experimental set-up. S is the Mössbauer source, B is the protein or another biopolymer under investigation, D is the detector, and A is the resonant absorber which can alternatively be placed in position 1 and 2 to measure intensity $I(0)$ and $I(2\theta)$, respectively.

scattered radiation $I(2\theta)$. The elastic fraction is determined by the following expression:

(1) $$f = [I_{00}(2\theta) - I_0(2\theta)]/[I_{00}(0) - I_0(0)] = \gamma \exp\left[-Q^2 \langle x^2 \rangle\right],$$

where I_0 and I_{00} are measured under the resonance condition (*i.e.* with the source velocity $V = 0$) and far away from the resonance ($V = \infty$), respectively. Here $-\gamma$ is the ratio of the intensity of Rayleigh scattering to the intensity of Rayleigh and Compton scattering, $Q = 4\pi \sin\theta/\lambda$ is the momentum tranfer, $\lambda = 0.86$ Å is the wavelength of the incident radiation, and $\langle x^2 \rangle$ is the "overall" mean square displacement, which, to a first approximation, is identical for all atoms of the macromolecule. In measurements of the RSMR spectral lineshape,

$$\eta(V) = \frac{I(V = \infty) - I(V)}{I(V = \infty)},$$

the resonance absorber is to be placed in position 2 only. In the present paper we have no room for a detailed description of the treatment of RSMR spectra of proteins; such a treatment can be found elsewhere [10]. However, the digest of the results of such a treatment will be done below. A theoretical analysis shows the existence of three main types of movements in the protein: usual solid-state oscillations and two types of protein specific motions: a) large-scale (conformational) individual motions of small atomic groups of the protein globule and b) complex cooperative motion of large atomic groups and parts of the macromolecule (*i.e.* solid-state domains) proceeding by way of bounded diffusion and involving amplitudes up to 1 Å and correlation times $\tau_c \sim 10^{-7}$ to 10^{-8} s. Complex cooperative motions are usually described in terms of the bounded diffusion model or the model of a Brownian oscillator with strong damping [6, 10]. The first term of the spectral function, corresponding to this model ($n = 0$), contributes to the "narrow" component, all the other lines ($n = 1, \ldots, \infty$) produce a "wide" component with the effective linewidth Γ_w, so, the spectral function in such an approximation may

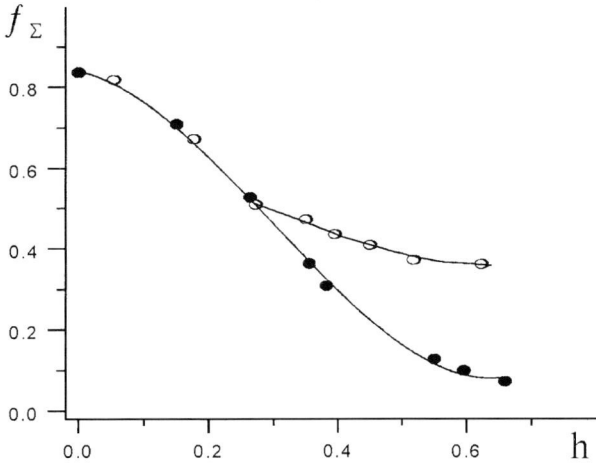

Fig. 4. – Hydration dependences of the elastic fraction for poly-Glu (○ $pH = 2.0$, ● $pH = 7.4$) at a scaterring angle $2\theta = (12 \pm 1.5)°$.

be represented as a weighted sum of the "narrow" and "wide" components.

$$(2) \quad g(\omega) = \exp\left[-Q^2\langle x^2\rangle\right]\Gamma/\left[2\pi\left((\Gamma/2)^2 + (\omega - \omega_0)^2\right)\right] + \\ + \left(1 - \exp\left[-Q^2\langle x^2\rangle\right]\right)\Gamma_w/\left[2\pi\left((\Gamma_w/2)^2 + (\omega - \omega_0)^2\right)\right].$$

This paper reports an experimental investigation of the dynamic properties of polyglutamic acid (poly-Glu) in the coil and α-helix state as well as the dynamic properties

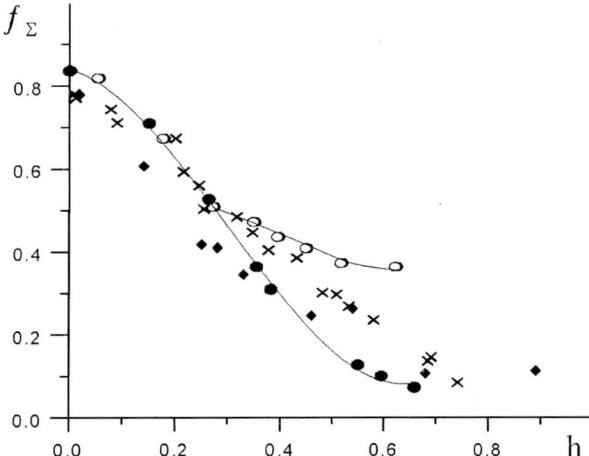

Fig. 5. – Hydration dependences of the elastic fraction for poly-Glu (○ $pH = 2.0$, ● $pH = 7.4$), HSA (×) and met-Mb (♦) at $2\theta = (12 \pm 1.5)°$.

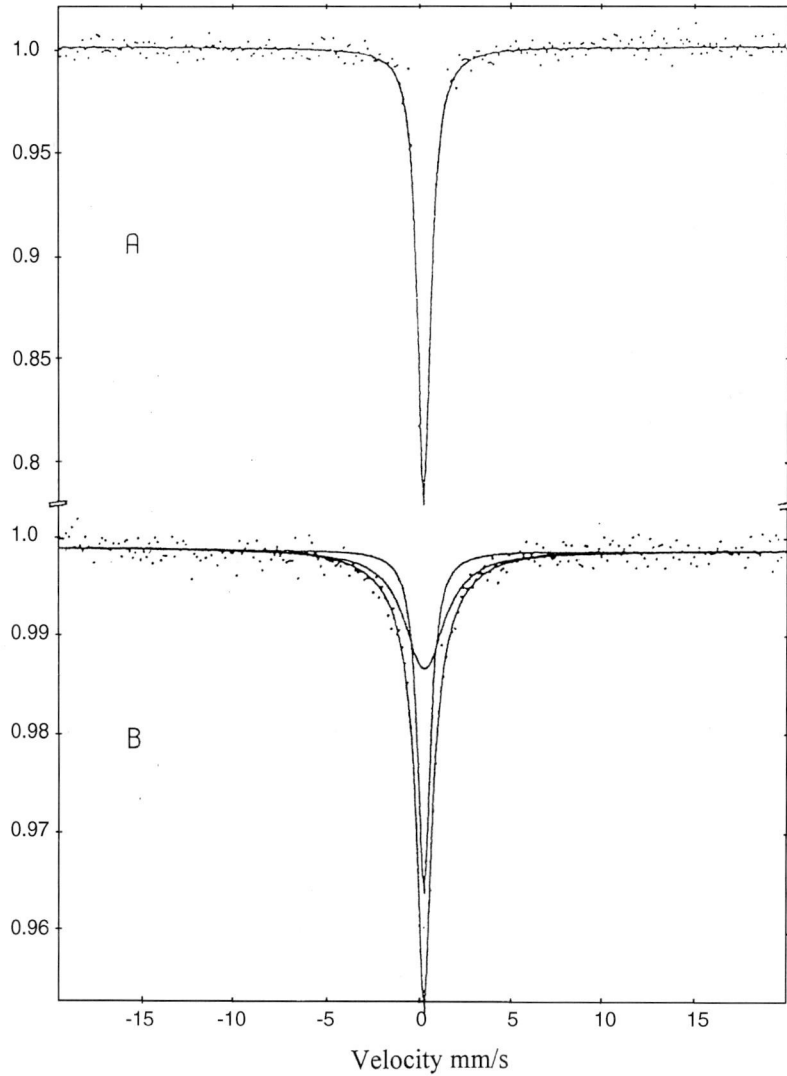

Fig. 6. – RSMR spectra for a HSA at $h = 0.77$ at different temperatures: A) $T = 100\,\mathrm{K}$; B) $T = 280\,\mathrm{K}$.

of the above-mentioned proteins by RSMR technique with soft collimating conditions. The dynamic properties were investigated at average scattering angles $2\theta = (12 \pm 1.5)°$, and $2\theta = (20 \pm 3)°$. The dynamic properties of poly-Glu in both α-helical and coil states are shown in fig. 4. We compare the dynamics of native α-helical proteins with the dynamics of poly-Glu as well (see fig. 5). Native α-helical proteins have dynamic properties intermediate between those of the random coil state and the helical state of poly-Glu, contrary to the expectation from the sequential folding model. By comparing

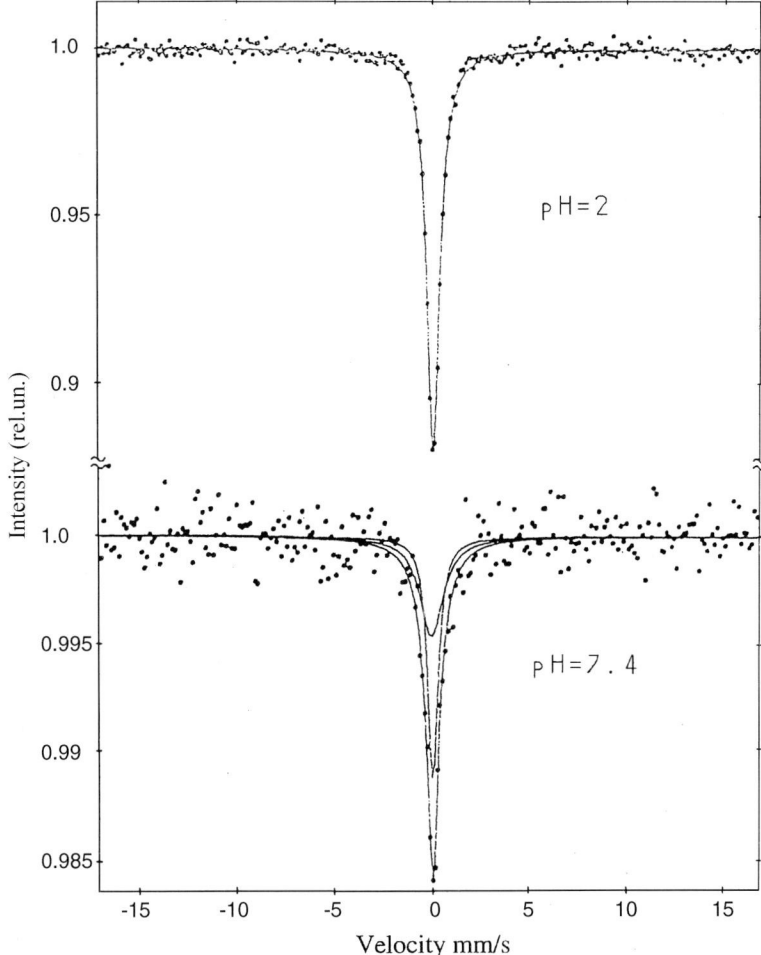

Fig. 7. – RSMR spectra of hydrated ($h = 0.6$) poly-Glu.

the energetic RSMR spectra for coil poly-Glu and for α-helical proteins (HSA and Mb) one can notice that these spectra at $h \geq 0.6$ are similar to one another in that they have both narrow and "wide" components. In other words, α-helical proteins, as well as poly-Glu in the coil state, have, in addition to high-frequency motions, low-frequency complex cooperative motion of large atomic groups with $\tau_c \sim 10^{-7}$ s (see figs. 6 and 7). RSMR spectroscopy demonstrates that α-helical proteins are between a random coil and a fluctuating secondary structure. Figure 8 demonstrates the comparison of the dynamics of poly-Glu states with the dynamic properties of proteins of another structure, a partially β-sheet structure (lysozyme and another protein homologous to lysozyme, *i.e.* α-lactalbumin). Data for the native state of the protein (lysozyme at $pH = 9.4$) and for the molten globular state (α-lactalbumin at $pH = 2.0$) are shown. The dynamic be-

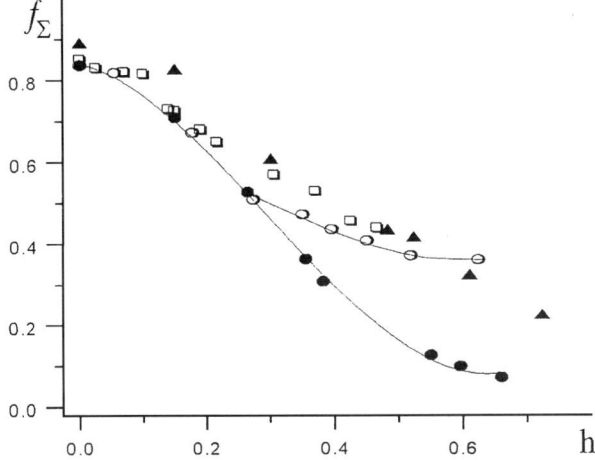

Fig. 8. – Hydration dependences of elastic fractions for poly-Glu (○ $pH = 2.0$, ● $pH = 7.4$), lysozyme (□) and α-lactalbumin (▲) at $2\theta = (12 \pm 1.5)°$.

haviors of lysozyme and α-lactalbumin are close to the dynamic behavior of poly-Glu in α-helix state, rather than to such behaviour of poly-Glu in a coil state. The RSMR energetic spectra for lysozyme, α-lactalbumin (fig. 9) and α-helix of poly-Glu (fig. 7) show no difference in the line shape for these biopolymers, indicating the presence of only a narrow component (the existence of high-frequency motions only). Thus we could (using the possibilities of the RSMR technique) but did not observe the noticeable difference in the dynamic behavior of the fluctuating secondary structure, the molten globular state and of the partially β-sheet (lysozyme) native protein state.

3. – Summary and discussion

RSMR experiments bring some surprising results:

1) The folding model of Ptitsyn suggests that all native proteins should have roughly equal high density of packing, as in any thermodynamic hypothesis of folding, the molten globular should have an intermediate, and the fluctuating secondary structure should have the smallest packing density. The dynamics of various native proteins, according to this, should be similar to one another, but less pronounced than in the molten globular state and, moreover, than in the random coil state.

2) Quite a different picture was observed: the dynamics of native α-helical proteins (HSA, Mb) are similar indeed but very different from a partially β-sheet protein (lysozyme, α-lactalbumin). The dynamics of native α-helical proteins are more pronounced than the dynamics of a molten globular state (α-lactalbumin) and, furthermore, than that of α-helical poly-Glu.

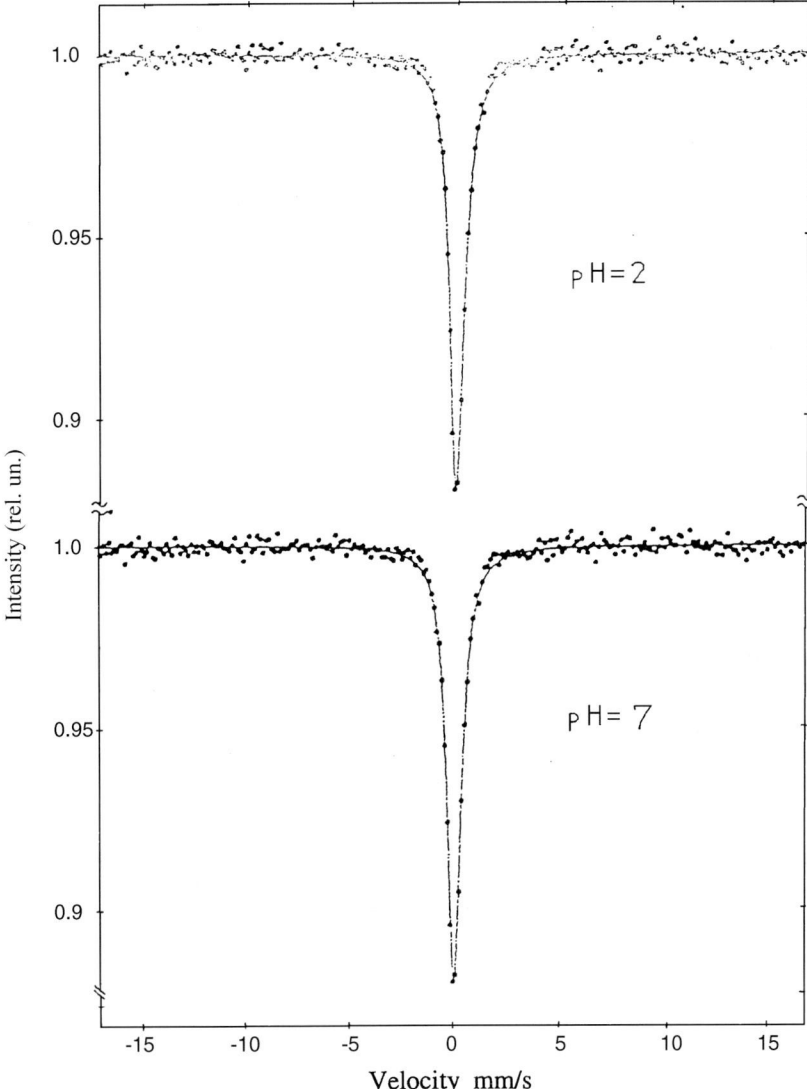

Fig. 9. – RSMR spectra of α-lactalbumin ($h = 0.6$) at native ($pH = 7$) and molten globular ($pH = 2$) states.

This fact, probably, is the result of the following reasons:

All the proteins studied are in a hydrated state and not in solution. Therefore, we have no direct evidences that α-lactalbumin in hydrated state represents a molten globular state as in solution as well. Probably due to this reason the comparison with a molten globular state is not correct.

The α-helix of poly-Glu could be not a very good model of the fluctuating secondary structure, since the persistent length of the α-helix of poly-Glu is much longer than those for α-helical proteins. Less ordered regions between α-helices in α-helical proteins give the ability for helices to move, to have low-frequency motions and to have large-scale fluctuations. The ability of moving for relatively large segments in model poly-Glu exists, probably, only in the random coil state, therefore the dynamics of native α-helical proteins is much closer to that of a random coil than to that of an α-helix.

3) Direct experimental observations show that the dynamics of native α-helical proteins (HSA, Mb) are similar indeed but very different from a partially β-sheet protein (lysozyme, α-lactalbumin). This fact is in favor of the kinetic hypothesis of folding, rather than of the thermodynamic hypothethis, since in the last case the dynamical properties of all proteins should be similar.

REFERENCES

[1] ANFINSEN C., *Science*, **181** (1973) 223.
[2] COOPER A., *Proc. Natl. Acad. Sci. USA*, **73** (1976) 2740.
[3] LEVINTHAL C. J., *Chim. Phys.*, **65** (1968) 44.
[4] FRAUENFELDER H., PETSCO G. A. and TSERNOGLOU D., *Nature*, **280** (1979) 558.
[5] GOLDANSKII V. I., KRUPYANSKII YU. F. and FLEUROV V. N., *Dokl. Akad. Nauk SSSR*, **272** (1983) 276.
[6] KRUPYANSKII YU. F., SHAITAN K. V., GOLDANSKII V. I., KURINOV I. V., RUBIN A. B. and SUZDALEV I. P., *Biofizika*, **32** (1987) 761 (translated from Russian).
[7] PTITSYN O. B., *J. Prot. Chem.*, **6** (1987) 273.
[8] LEHNINGER A. L., *Biochemistry* (Worth Publishers, Inc., New York) 1972.
[9] BYCHKOVA V. YE, private communication.
[10] GOLDANSKII V. I. and KRUPYANSKII YU. F., *Quart. Rev. Biophys.*, **22** (1989) 39.

PROTEIN FOLDING THEORY/SIMULATIONS

Protein folding: Matching theory and experiment

L. Mirny and E. I. Shakhnovich

Harvard University, Department of Chemistry and Chemical Biology
12 Oxford Street, Cambridge, MA 02138, USA

Abbreviations

CI2 Chymotripsin inhibitor 2
CoC Conservatism of conservatism
DCM Diffusion-collision model
HP Hydrophobic-polar (two-letter protein model)
MC Monte Carlo
MD Molecular dynamics
PCC Postcritical conformation
PE Protein engineering
RC Reaction coordinate
SFN Specific folding nucleus
TSE Transition state ensemble

1. – Introduction

Although no method exists that can reliably predict the native structure of a protein from its sequence, our understanding of the mechanisms that govern protein folding has progressed considerably.

In theory, major insights came from analytical studies of heteropolymer models and folding simulations of simplified lattice and off-lattice models. The relative simplicity and computational efficiency of these models allowed to characterize phase diagrams for these models and simulate thousands of folding-unfolding events to obtain a detailed statistical description of the folding process in model proteins. Furthermore it was also possible to model the evolution of proteins under various selective pressures. Although these models do not match the full complexity of the real protein architecture, they capture a core aspect of the physical protein folding problem: both real proteins and simplified lattice and off-lattice proteins find a conformation of the lowest energy out of the astronomically large number of possible conformations without prohibitively long exhaustive search. By simulating folding and evolution of simple model proteins, theoreticians gained insights into the general properties of the amino acid sequences that are required to provide stability and fast folding to model proteins. First, to exhibit cooperative folding transition and to fold fast, the native structure must be a pronounced energy minimum separated from the bulk of unfolded conformations by a large energy gap. Second, selected protein sequences fold by the nucleation mechanism whereby a small number of residues (folding nucleus) need to form their native contacts in order that the folding reaction proceed fast into the native state.

As always, important developments create new challenges for theoretical and experimental research. In particular substantial work is devoted by many groups to address a number of questions:

- Can we predict/rationalize which residues contribute mostly to the folding nucleus, given the sequence and the native structure of a protein?

- How to predict the stability and folding rate of a protein? At least, how to predict changes in stability and folding rate upon single mutation?

- What is the evolution of the folding nucleus and folding kinetics in general? Are nucleation residues under stronger evolutionary pressure than the rest of the protein core?

In this review we will survey many of those recent works and emphasize their strong and weak points.

The structure of this review is the following. First we introduce Basic Concepts essential for understanding of the further material. Next we turn to the discussion of recent theoretical studies and some experimental works. Finally we summarize the major conclusions and present an outlook for further research in the field of protein folding.

2. – Basic concepts

2`1. *Cooperativity: two-state thermodynamics and kinetics of protein folding.* – The concept of cooperativity is one of the most basic and fundamental for protein folding studies. Privalov and coworkers [1] applied the van't-Hoff criterion to evaluate the cooperativity of denaturational transition for several proteins. They found, with high degree of accuracy, that small-single-domain proteins fold like two-state systems with only folded and unfolded states being (meta)stable, while all intermediate—partly folded—states are unstable (see fig. 2). More recently Jackson and Fersht [2] showed that folding of a small protein, chymotripsin inhibitor 2 (CI2), can be considered also as a *kinetically* two-state process at which partly folded states are not significantly populated in the process of folding. Subsequently many more proteins were found to fold kinetically and thermodynamically as two-state systems. A list of such proteins as well as the data on their thermodynamics and kinetics are presented in an extensive review by Jackson [3].

The discovery of the remarkable cooperativity of protein folding (akin to first-order phase transition for a finite system [4]) inspired many theoretical and experimental studies aimed at explaining it. Earlier models (partly reviewed in [5]) considered factors such as polymer collapse [6], side-chain packing [4], directional or three-body interactions [7], or special folding pathway [8] as possible explanations.

More recently, cooperativity of protein folding received a detailed explanation within the analytical heteropolymer theory. Phenomenological [9] and microscopic statistical-mechanical theories [10-14] converged on the view that sequences that had undergone "evolutionary" selection for pronounced energy gap fold cooperatively, while random sequences do not.

Specifically, it was shown that "optimized" sequences that feature a large "stability gap" fold cooperatively, following the mechanism of the first-order transition (see fig. 1 in [11] and fig. 1 in [15] for a qualitative explanation of this result). The theoretical analysis [16, 17] revealed that sequence optimization for stability gap gives rise to folding cooperativity only for 3-dimensional polypeptides while two-dimensional models lack cooperative behavior [16-18] (a qualitative explanation of this result is given in [17]). Furthermore, not any type of sequences can be optimized to provide a large stability gap and cooperative folding transition: it was argued in [19, 20] that the maximal stability gap depends on the number of amino acid types in the protein "alphabet". Specifically, no sequences composed of two types of amino acids were predicted to have a sufficient energy gap for cooperative folding [18-21]. Several simulation studies support this view [15, 21-23].

An extended analysis of the thermodynamics of designed sequences was performed recenly by Wilder and Shakhnovich [24]. It differed from an initial analysis of Ramanathan and Shakhnovich [12] and that of Pande et al. [14] in that it extended the consideration beyond the pure mean-field approach taking into account higher-order fluctuations in microphase separation order parameters as well as the possibility of a two-step replica symmetry breaking (RSB) in the overlap order parameter. It was established that the one-step RSB is still a stable solution to the problem and a new phase diagram for the model was presented. It differs slightly from the original one developed in [12] due to a more accurate approximation, however qualitatively it is similar to the earlier version [12] and also predicts cooperative, first-order phase transition between the native and disordered states for designed sequences and absence of cooperative transition for random sequences.

The results of the analysis are summarized in the phase diagram shown in fig. 1.

In addition to the phase diagram analysis of the states with designed sequences, Wilder and Shakhnovich analyzed the nonrandomness of the sequence distribution in designed sequences. They found an explicit expression for correlation amino acids in different locations on the chain (a model was used in [24] where two types of monomers

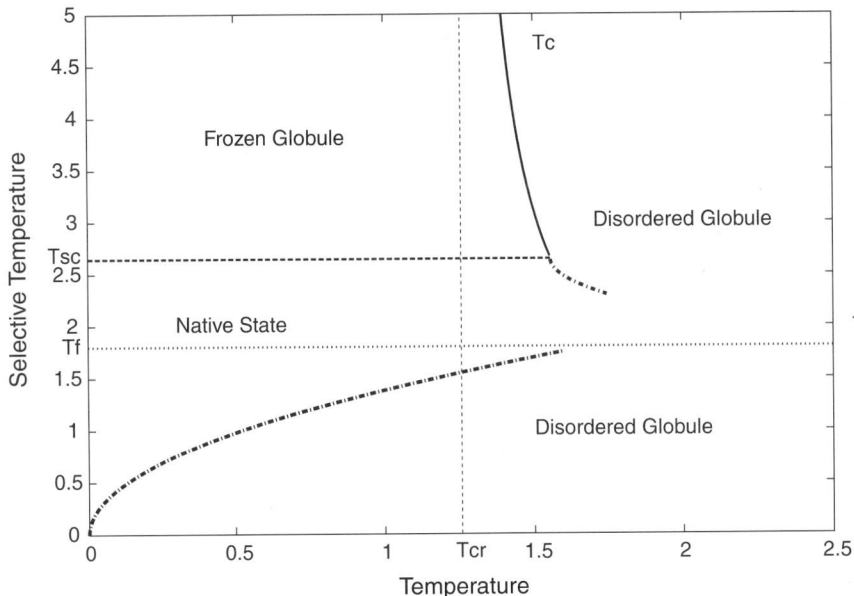

Fig. 1. – Phase diagram for a heteropolymeric chain with a selected sequence with the real temperature and selective or polymerization temperatures parameters in arbitrary units. The selective temperature represents a measure of the evolutionary optimization of the protein sequences.

were allowed: "hydrophobic" and polar):

$$\langle \sigma_l \sigma_{l+k} \rangle_{\mathrm{av}} = -\frac{2\chi}{T_{\mathrm{s}}} \delta(\mathbf{r}_l^0 - \mathbf{r}_{l+k}^0). \tag{1}$$

The average sign $\langle \rangle$ means averaging over all sequences that fold into the same protein structure that is characterized by coordinates of all its atoms $\{\mathbf{r}_l^0\}$. The parameter χ characterizes the interaction strength between "hydrophobic" and "polar" amino acids. σ_i marks the type of aminoacid i: $\sigma_i = 1$, if the residue i is "hydrophobic" and -1, otherwise. T_{s} is the "selective" temperature [11, 12] that serves as a measure of the degree of design of the sequences: sequences designed at low T_{s} feature a greater energy gap and stability in the native state, while $T_{\mathrm{s}} = \infty$ corresponds to random sequences. The result indicates that there is a correlation between the monomers that form a native contact, which is proportional to the strength of interaction and inversely proportional to the degree of sequence optimization (stability) expressed in terms of "temperature in sequence space" T_{s}.

In recent papers [25, 26] Chan and coworkers analyzed cooperativity in various types of lattice models and found that out of all the studied sequence models, only 3-dimensional models with twenty types of amino acids (Chan studied the same 3-dimensional 36-mer sequence as was used in [27]) exhibit a folding cooperativity comparable to that of natural proteins. Two-dimensional models and HP models were found to be far less cooperative than real proteins. These results support earlier detailed analyses [12, 14, 17, 18] of the origin of protein cooperativity.

The requirement of protein-like cooperativity narrows down the selection of viable models to study folding kinetics to those that feature protein-like cooperativity in thermodynamics. Such model proteins are capable to reproduce a range of kinetic behaviors including two-state kinetics that was observed for many small proteins [2, 3].

2˙2. *Transition state ensemble and folding nucleus*. – Consider folding of a protein that has a two-state kinetics. In this case folding proceeds through a single *free*-energy barrier. The height of the transition state $\Delta G_{\ddagger-D}$ controls the rate of folding:

$$k_\mathrm{f} = C \exp\left[-\frac{\Delta G_{\ddagger-D}}{RT}\right],$$

where C is a constant. The approximate best fit experimental value of C is $10^6\,\mathrm{s}^{-1}$ [28].

The transition state is not a single conformation of a protein. Rather, it is an *ensemble* of conformations (transition state ensemble, or TSE [18]). Correspondingly, it is the *free* energy of the TSE $\Delta G_{\ddagger-D}$ that determines the folding rate, *i.e.* the number of conformations constituting the TSE is as important in determining the protein folding rate as is the energy of the TSE.

The concept of TSE is a natural generalization of the concept of transition state in chemical kinetics to protein folding with the important consideration that folding occurs in a highly multidimensional space. A simple chemical reaction can be characterized by one (or very few) *reaction coordinates* (RC). The unique role of reaction coordinates in chemical kinetics is that it provides a relation between the structure of the reagents (coordinates of the nuclei) and the time course of the reaction. Specifically the value of RC can serve as a predictor of the *transmission coefficient* for the reaction: the top of the barrier separates the region where flux is towards products from the region where flux is towards reactants. In a simple chemical reaction described by one-dimensional RC the separation occurs at one particular value of RC which is the transition state. In the case of complex protein folding "reaction" no single RC is found at this point [29] (see below). Nevertheless, the concept of TSE, which is more general than RC, still applies. Indeed the signature of a transition state in chemical kinetics is that the transmission coefficient for reactions that originate from TS is 1/2, *i.e.* if one imagines a statistical ensemble of reacting molecules, each starting a reaction from the transition state configuration, half of the molecules in the ensemble will transform fast to products and half of them will transform back to reactants. This view can be generalized to multidimensional protein folding reactions: the TSE can be defined as such set of conformations that folding trajectories starting from each of them have probability $p_\mathrm{fold} = 1/2$ to reach the *fast and downhill* folded state before unfolding, and $p_\mathrm{unfold} = 1/2$ to reach the unfolded state before folding. Apparently the TSE separates the basin of attraction for the folded state from the unfolded basin of attraction. The folded basin of attraction (called "postcritical" in [30]) consists of conformations that are committed to fast folding: every folding trajectory starting from any conformation belonging to the postcritical ensemble reaches the folded state *fast* prior to unfolding via directed "downhill" motion in the configurational space. It is very important to note that the folding dynamics from postcritical conformations (PCCs) is *qualitatively* different from that starting from any conformation *before* the transition state: in the former case a steady "descent" to the native state occurs (see fig. 2), while in the latter case the dynamics features a two-state character: most of the time the chain appears to be making stochastic fluctuations with no apparent progress towards the folded state until at some point it passes one of the conformations belonging to TSE and then rapidly "jumps" to the native state. On the time scale of total folding simulation, the descent from a TSE conformation to the native state looks like a jump, while on a finer time scale it represents gradual biased descent, see fig. 2.

The nucleation concepts of protein folding kinetics were proposed and tested in the context of lattice models, in [30]. The postcritical set of conformations for a simple 36-mer protein was determined and it was directly verified that simulations that start from any of the conformations from this postcritical set indeed reach the native state via directed dynamics that no longer involves crossing of a major free-energy barrier (see fig. 2). Furthermore, it was shown in [15], for the same model, that the dynamics starting from any PCC remains fast even at very low temperature, in contrast to the dynamics that starts from an arbitrary non-PCC. This is a most direct verification that the dynamics from post-critical conformations no longer involves major barrier crossing, which is energetic at low temperatures [15].

Which features distinguish conformations belonging to TSE and postcritical conformations from all the other conformations? In most of the studied proteins and protein models conformations of the TSE, especially PCCs feature a certain set of native interactions between specific residues—a *folding nucleus*. Hence a conformation with assembled nucleus may look like a distorted native conformation sometimes with large unstructured loops [31] (see fig. 2). Some proteins, however, have a nucleus with partially formed secondary structure [32]. In these cases, along with long-range interactions, short-range interactions that stabilize the formed secondary structures contribute to the stability of the nucleus. The nucleation mechanism does not require secondary structure elements to be formed before the transition state is reached. In contrast to diffusion-collision and hierarchical folding models, the secondary structure may be formed simultaneously with the formation of the tertiary structure.

Several mechanisms alternative to nucleation were discussed in recent literature.

2˙3. *Folding funnels.* – Originally "folding funnels" were introduced in [33] to indicate a special requirement of kinetic accessibility for few viable native structures in the 27-mer lattice model. However, subsequent detailed lattice simulations did not support the view proposed in [33] that only special structures are kinetically accessible (simulations presented in [33] in support of this proposal appeared to be insufficient: only two structures were compared and kinetic accessibility was determined for each of them from

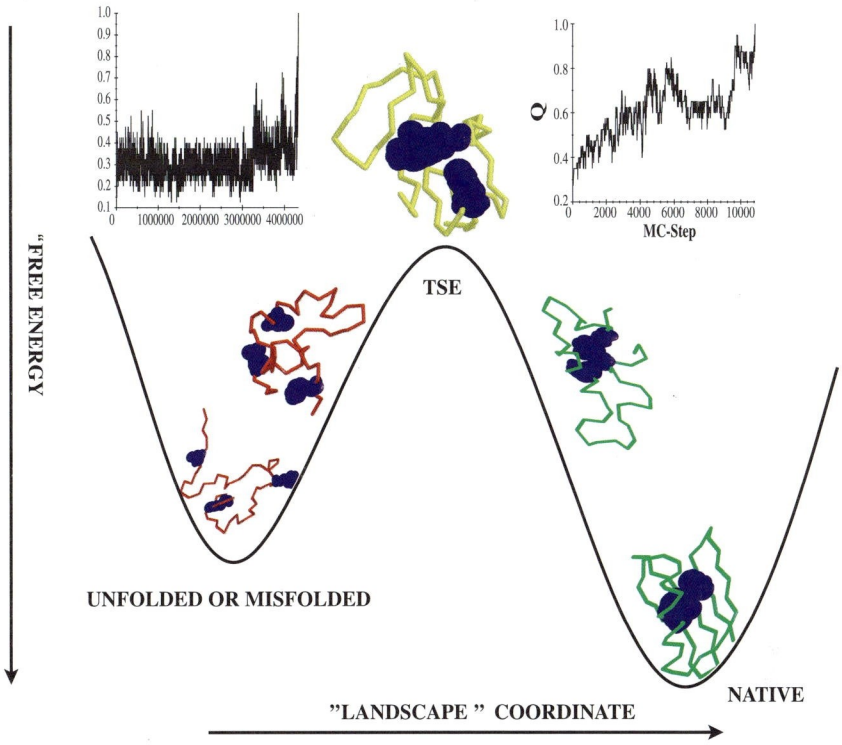

Fig. 2. – Specific-nucleus mechanism of protein folding. The free-energy landscape is shown schematically as one-dimensional. Nucleus residues are shown schematically as blue space-filling residues. Conformations that do not have the nucleus assembled are precritical (shown in red), they are committed to unfold. Folding-Folding trajectories starting from precritical conformations are committed to unfolding and long fluctuations in the unfolded state prior to the folding event (upper left panel). Conformations that are past the free-energy barrier have the folding nucleus assembled. Those are PCCs that are committed to rapid continuous folding without further major free-energy barriers (right upper panel). The separating set of conformations corresponding to the top of the free-energy barrier is TSE whereby the nucleus is "almost formed". One of the conformations of the TSE ensemble is shown schematically in yellow. All representations in this figure are schematic. Each conformation shown (except the native one) is a member of the corresponding broad ensemble.

only ten runs). Correspondingly, the concept of "folding funnel" has transformed into an intuitive reflection of the fact that the native state of a protein represents a deep energy minimum, *i.e.* most contacts in the native state are stabilizing. This view makes it natural to explain the solution of the search problem by energetic bias that rewards every move towards the native state (hence energetic "funnels"). This view is pictorially and metaphorically represented in "landscape pictures" [34]. While such pictures have some appeal due to their simplicity and artistic beauty, the funnel concept and related pictorial landscapes may be misleading. In fact the "funnel" picture presents the folding process as a multitude of pathways each representing a directed descent to the native state exemplified in the "downhill skier" analogy suggested by Chan and Dill [34]). As noted before, this view directly contradicts the results of lattice simulations that show that directed descent to the native state occurs only at the very latest (in time) stage of a folding trajectory when one of the conformations from the TSE is passed. Furthermore, it was pointed out that the "descending skier" picture is inconsistent with the experimentally observed exponential folding kinetics [35].

The major weakness of the "funnel" concept is that it downplays the importance of conformational entropy in the folding kinetics. In fact the free-energy, not energy, minimization determines the direction and the time course of the protein folding reaction. Conformational entropy is a crucial component of the free-energy balance in protein folding that metaphoric "landscape" pictures fail to take into account properly.

2˙4. *Diffusion-collision mechanism.* – A more substantial possible kinetic mechanism alternative to nucleation is suggested by framework or "diffusion-collision" models (DCM). The hierarchical models stipulate that folding starts from the formation of local stable structural elements, that serve as pre-formed units for subsequent stages of folding [36, 37].

The key difference between nucleation mechanism and "hierarchical" or DCM process is that they provide qualitatively different predictions concerning the effect of mutations on the folding rate. The DCM predicts that the stabilization of any local secondary-structures element (an alpha-helix or beta-strand) will *always* lead to the acceleration of folding. In contrast, the nucleation mechanism predicts that the strength of *tertiary* contacts formed in the transition state may be primary determinants of the folding rate [38, 18]. According to the nucleation mechanism, the stabilization of the local structure accelerates the folding rates only if a particular element of secondary structure is formed in the TSE.

The relative importance of secondary *vs.* tertiary interactions in determining the folding rates was studied experimentally for a number of proteins: the activation domain of carboxypeptydase A (CPA) [39], CheY [39], the helical coiled-coil GCN4 [40-43], the GB1 domain [44], its structural homolog protein L [45], the lambda-repressor [46], and the ribosomal protein L9 [47]. It was shown *in all cases* that secondary interactions play no role, or only a minor one, in determining the folding kinetics. Instead, strong and *specific* hydrophobic contacts play a dominant role. These experimental findings rule out a necessary kinetic role for preformed secondary-structure elements in all studied proteins and by implication rule out the DCM as a folding mechanism for these proteins. Specifically, it was shown that strengthening both native helices in CheY and helix 1 in CPA decelerates folding [39], while weakening of the native helix in L protein and GCN4 does not significantly affect the folding rate [45, 40, 41]. Furthermore, strengthening the hydrophobic contacts in GCN4 at the expense of helical stability raises the folding rate by two orders of magnitude, in sharp contrast to the order-of-magnitude deceleration predicted by the DCM [42, 41]. Perhaps the most spectacular demonstration of the irrelevance of the preformed secondary structure for folding kinetics is provided in a recent study by Luis Serrano's group [44] where the authors replaced the alpha-helix of the GB1 domain by sequences with strong [Beta]-hairpin propensities. While the mutant proteins were destabilized (and up to 80% of the destabilization could be accounted for by lost native helical propensity in the alpha-helix), the folding rates remained largely unaffected by such dramatic replacements!

The authors of most of the cited experimental papers note that their results are not consistent with DCM and related hierarchical models for the studied proteins [41, 45, 44, 39]. An opposite view has been put forth by Oas and his coworkers, who proposed that precollision helix formation is consistent with the observed rates of folding of GCN4 [43]. However, as was pointed out by Sosnick and his coworkers in [41], Oas' analysis was based on an incorrect assumption of uniform helical propensity that grossly overestimated the length of helical stretches in isolated monomers and grossly underestimated the effects of mutations in regions of high helical propensity. Similarly, Oas and coworkers asserted that the DCM model quantitatively explains their experiments of lambda-repressor folding. However, their own data contradict this assertion. Indeed the only mutations that significantly affect the folding rate are M15 and M66. The DCM predictions are in qualitative disagreement with experimental data on those mutations (see table 2 in [46]).

Using a C_α off-lattice model with a Gō potential Zhou and Karplus observed that for a three-helix bundle the "fast track" folding pathway, where no kinetic intermediates/traps were encountered, was the diffusion-collision pathway. Perhaps this result is not surprising given the conditions at which simulations were carried out. The three-helix bundle was being folded by Zhou and Karplus at extremely low temperature: for a bias gap -1.3 (native interactions -1, non-native interactions $+0.3$) the protein was being folded at a temperature that is 0.25 of T_f—folding transition temperature. If we estimate T_f for actual proteins to be roughly 350 K, this implies that the folding simulations of Zhou and Karplus were carried out at less than 100 K. Folding a similar three-helix bundle using Langevin dynamics, Berriz and Shakhnovich (unpublished) found that the diffusion-collision behavior was observed at very low temperature but it disappeared as the temperature was raised.

Since in most experimentally studied cases isolated helices are hardly marginally stable [39, 48-50], it remains to be seen whether the DCM can be observed in real proteins.

Experimental results along with a number of computational studies for a variety of models (see below) establish nucleation as a plausible kinetic folding mechanism for small proteins—consistent with the cooperative character of their thermodynamics [30, 51]. (This does not exclude possible hierarchical mechanisms for higher-order structural organization of proteins, such as quaternary structure, or multidomain arrangement [38, 52].)

2˙5. *Determining* TSE *in experiment: protein engineering analysis.* – Although several methods can be used to characterize the transition state ensemble in computer simulations, the major experimental method is protein engineering analysis. This method allows to quantify the contribution of each residue (or even specific atom/group) to stability of the transition state ensemble, and hence, their contribution to the folding nucleus.

In the ϕ-value analysis one mutates a particular residue in a protein and determines the influence of this mutation on protein stability $\Delta\Delta G_{N-D}$ and folding kinetics $\Delta\Delta G_{\ddagger-D}$, then

$$\phi = \frac{\Delta G_{\ddagger-D}}{\Delta\Delta G_{N-D}},$$

where $\Delta\Delta G_{\ddagger-D} = RT\log(k_f^m/k_f^{wt})$ is computed from the rates of folding of the mutant k_f^m and the wild type k_f^{wt}.

The value $\phi = 0$ indicates that the mutant folds as fast as the wild type and hence the mutated residue is not involved in stabilizing the transition state, *i.e.* it does not belong to the nucleus. On the opposite, $\phi = 1$ means that mutation (de)stabilized transition and native states equally. This indicates that the mutated residue has the same network of interactions in the transition state as it has in the native one.

It is more difficult to interpret fractional ϕ-values. There are several possibilities. 1) Fractional $0 < \phi < 1$ arises when a residue (or group removed by mutation) is involved in more stabilizing interactions in the native state than in the transition state. Clearly, in such a case mutation destabilizes the native state more than the transition state and leads to fractional ϕ. Importantly, this does not mean that the residue with fractional $\phi < 1$ contributes to the stability of the folding nucleus less than a residue with $\phi = 1$! ϕ measures the *relative* (not the absolute) contribution to the stability of the nucleus. 2) Fractional ϕ can be also coming from a mixture of states, some with fully formed interactions that involve a mutated group and some with fully broken ones. A kinetic test has been suggested to discriminate between these two possibilities [53]. 3) Another possibility is a numerical artifact. When $\Delta\Delta G_{N-D}$ is small, even small inaccuracy in measuring $\Delta\Delta G_{N-D}$ leads to large errors in the value of ϕ. Hence, one must be very careful considering ϕ values that were obtained with small $\Delta\Delta G_{N-D}$. 4) Experimental artifacts such as changes in the native structure upon mutation, reorganization of the transition state ensemble, mobile transition state, non-native interactions of the mutant, etc. lead to fractional ϕ-values. The usual recipe is to make multiple mutations at the same site. Inconsistent ϕ-values of multiple mutations are an indicator of such artifacts.

A negative ϕ-value indicates an opposite effect of mutation on the native and transition states. $\phi > 1$ indicates that mutation affects the transition state more than the native one. This can happen if the mutated residue is involved in certain interactions that stabilize the transition state and that are not present in the native state, *i.e.* the transition state is stabilized by some non-native interactions. Below we discuss experimental and computational studies supporting this view.

Experimentally the ϕ-values were determined for a number of small and medium-size proteins: all-beta SH3 domains [54,55], all-beta Ig-fold domains (cd2: [56]), tenascin [57], alpha-helical acyl-coenzyme A-binding proteins (ACBPs) [58], Rossman fold protein CheY [59], Alpha-Beta plait activator domain of procarboxypeptidase A ADA2h [60] and its structural homolog muscle acylphosphatase, AcP [61], alpha/beta protein FKBP [62], small alpha/beta domain CI2 (the classic work from A. Fersht's lab [38]) and a larger protein, villin, that features two hydrophobic cores [63].

As expected, the ϕ-values for most studied proteins were distributed in a broad range with some of them exhibiting anomalous values less than zero or greater than one. A naive interpretation of this observation would suggest that no specific features of TSE ensemble can be discerned ("multiple nuclei model" of Wolynes [64] and coworkers or Multiple Folding Nuclei model of Thirumalai [65]). It is not surprising that mutational analysis reveals mostly fractional ϕ-values: most amino acid residues form many interactions with other amino acids both in the native state and in TSE. Not surprisingly, mutations change these interactions hence intermediate ϕ-values for most residues. Nevertheless, a more detailed analysis of the experimental data allows to distinguish between possible scenarios. Serrano and coworkers [54] analyzed possible models for TSE and came to the conclusion that their data (on spectrin SH3 domains) is consistent with SFN mechanisms and practically rules out two other alternative scenarios that they considered (non-specific nucleus and multiple folding nuclei). These authors pointed out that a clear manifestation of the SFN mechanism in protein engineering studies is clustering of residues that have high ϕ-values in 3D structure of a protein [38] and robustness of ϕ-values with respect to changes in the environment [54]. It is remarkable that in *all* studied proteins, the residues that have high ϕ-values tend to form tight clusters in structure forming some kind of nucleation minicores, consistent with the SFN model.

It is also important to note that the experimental analysis of nucleation in proteins goes beyond just identifying residues with high ϕ-values. In several cases an apparent ϕ-value obtained in a single-mutant experiment may be low, especially if the corresponding $\Delta\Delta G$ value is small, like I57 in CI2 [38]. However, careful analysis with multiple mutations can still reveal that the residue in question may be an integral part of the folding nucleus. It appears reasonable to follow the notation for the nucleus as identified by the experimental group that studies a given protein (see below for more details).

3. – Simple lattice and topology-based models

Experimental results [38, 54, 55, 57, 63, 66] along with a number of computational studies for a variety of models (see below) establish nucleation as a plausible kinetic folding mechanism for small proteins—consistent with the cooperative character of their thermodynamics [11, 14, 21, 30, 51]. (This does not exclude possible hierarchical mechanisms for the higher-order structural organization of proteins, such as the quaternary structure, or multidomain arrangement [38, 52].)

While the general validity of the nucleation mechanism of protein folding can be considered established for many small proteins, several crucial details remain unclear. In particular, it is very important to understand what determines the folding nucleus— protein topology or sequences? This question is closely related to the problem of protein evolution [67, 68]. While there are certain indications from simulations and experiment that protein structure may be a strong determinant of nucleus location [30, 54, 61, 67, 69], a complete understanding of what determines the spatial location of the folding nucleus has not yet been reached. Another important question concerns the relative importance of short- vs. long-range interactions in nucleation. Recent interesting observations by Plaxco and coworkers concerning correlation between contact order and folding rate [70] may provide a clue for more in-depth theoretical analysis of this issue. These questions are beginning to evolve as central to the whole field of protein folding [44, 67-69, 71]. Addressing them has been the focus of the theoretical studies of folding mechanisms carried out by several groups over the last few years.

3˙1. *Determining* TSE *in computer simulations of simple models*. – One of the key issues in computational studies of protein folding kinetics is the determination of TSE. Several authors took an approach to derive TSE from equilibrium [64,72] or umbrella [73] sampling by determining a "one-dimensional" free energy profile for a certain order parameter $\mathcal{A}$. The free energy is defined as

$$\mathcal{F}(\mathcal{A}) = -kT \log f(\mathcal{A}), \tag{2}$$

where $f(\mathcal{A})$ is the frequency of observing the value of order parameter $\mathcal{A}$ in the range $(\mathcal{A}, \mathcal{A} + \Delta \mathcal{A})$.

The maximum on the plot $\mathcal{F}(\mathcal{A})$ at some value of $\mathcal{A}^*$ is identified as transition state. Presumably, the TSE consists of conformations with $\mathcal{A} = \mathcal{A}^*$.

This approach was first used in the context of protein folding by Sali *et al.* [72] to determine the TSE for lattice 27-mer model [72]. Q, a normalized number of native contacts, was chosen as an order parameter sampling in [72]. This choice was motivated by earlier analytical studies of random and designed heteropolymers [10,12] where Q, an overlap with the native state, was shown to be a key order parameter in the thermodynamic analysis.

Later Onuchic and coauthors [64, 74] used the same approach of equilibrium sampling and the same order parameter Q to derive TSE for a similar 27-mer lattice model. These authors proposed in [64] a "multiple delocalized nuclei" model. It is not entirely clear what is meant by that ("delocalized nuclei" sounds like a contradiction in terms). However, this model is proposed as an alternative to SFN mechanism of [30]. The arguments against SFN in [64] were based on the observation that in the ensemble of conformations with $Q = Q^*$ identified as TSE in [64] no specific contacts or interactions were clearly dominant.

However, the approach to determine the TSE from the equilibrium sampling of any order parameter may be problematic [75, 18]. Du *et al.* directly evaluated the correlation between order parameter Q and transmission coefficient p_{fold} for various lattice models and found no correlation between the two [29]. More specifically, the distribution of p_{fold} in the ensemble of conformations with $Q = Q^*$ was found very broad ranging from 0 to values close to 1. This analysis suggests that Q is not a viable RC for protein folding. Further studies by H. Angerman and ES (unpublished results) and by Li *et al.* [76] supported this conclusion showing again no correlation between Q and p_{fold} (*i.e.* between Q and TSE) for various models. This analysis clearly shows that it may be not correct to determine the TSE from equilibrium sampling [64, 77, 73]. The reason for the failure of the equilibrium or umbrella sampling methods to provide an adequate "low-dimensional" description of kinetics and TSE was explained in more detail in [18].

The TSE for complex systems, like protein folding models, can be determined only directly from kinetics in the cases when RC is not known [30]. A possible method to determine TSE from kinetics was proposed in [30]. This study focused on contacts that appeared on a "steep" part of folding trajectories (see fig. 2) just preceding *in time* a folding event. Special attention was paid to check explicitly that conformations identified that way are indeed postcritical, *i.e.* that they feature $p_{\text{fold}} = 1$. The importance of such check was highlighted in [78].

A related approach was taken in [76] to study the folding nucleus in a more complex model, cubic lattice with side chains introduced by Klimov and Thirumalai [79]. Conformations that were close in time to the native state yet having low-structural similarity to it ($Q \approx 0.40$) were identified as *putative* PCCs. Then each of them was subjected to the test of running simulations that start from these conformations and only a small fraction of putative PCCs turned out to be the real ones, committed to fast folding. This simulation allowed to identify a folding nucleus in the lattice model with side chains. Interestingly, a few non-native interactions were found to play an important role in determining folding nucleus in this model. This finding helped to rationalize some experimental observations with SH3 domains [55, 54].

Another approach to determine TSE in lattice simulations was proposed in [27]. A virtual protein engineering (PE) study was carried out for lattice model protein where each residue was mutated to all 19 possible alternatives. Folding and unfolding rates of the mutants were determined and ϕ-values were obtained using a standard definition [38]. In comparison with real experiments, the simulations have two major advantages: i) the TSE and the intermediates can be evaluated independently of simulations by other methods (see above) without resorting to the PE analysis. This provides a valuable reference point to evaluate strengths and weaknesses of the PE method; ii) artificial "mutations" that change the energy of certain particular contacts are possible in simulations. This provides a way to evaluate the degree of formation of specific (native and non-native) contacts as reported by the PE analysis, without being obscured by the fact that real mutations change a multitude of contacts simultaneously. The results of the study [27] provide detailed guidelines for interpreting PE experiments. They support the view that PE approach may be a good way to evaluate TSE, provided that multiple mutation scans are made. On the other hand, the analysis in [27] pointed out to certain limitations of the PE method. The most important of them is that that ϕ-values become unreliable when mutations result in small change in stability $\Delta\Delta G$. The main reason is not an obvious increase in error bars when the denominator gets smaller but rather a possible peculiar "compensation effect" of competing strong interactions of various magnitude and sign (see details in [27]).

Furthermore, the temperature dependence of the TS ensemble was studied by the virtual PE method in [27]. This analysis provided, for the first time, a microscopic interpretation of pronounced temperature shifts in ϕ-values that were observed in CI2 by Fersht and coworkers [80].

Several kinetic methods of analysis applied to lattice models of various degree of complexity provided a consistent view on the character of PCCs that share a common folding nucleus. However, this conclusion was questioned by Klimov and Thirumalai [65] who studied similar lattice models but failed to detect a specific nucleus. The reason for this discrepancy was explained in [81]: Klimov and Thirumalai collected and analyzed conformations that appear prior in time to reaching the native state (*i.e.* they collected *putative* PCCs). But these authors did not check which of the putative PCCs are actual postcritical ones, *i.e.* they did not run a test of starting simulations from putative PCCs. This test is a crucial part of the search for PCCs: as shown in [76] only a small fraction (less than 20%) of putative PCCs are actual ones, *i.e.* are committed to fast folding. Thus the analysis of Klimov and Thirumalai has a technical problem that prevented them from correct determination of the actual set of PCCs.

The kinetic analysis of simulations, when properly applied to lattice model Monte Carlo simulations, points out to the specific nucleus as defining feature of the PCCs.

3˙1.1. Nucleation in off-lattice models. How universal is this conclusion? Is it transferable to a more realistic, off-lattice model and/or another dynamic simulation algorithm? This question was addressed in a recent publication [82] where off-lattice folding was simulated using Discrete Molecular Dynamics [83, 84]. The authors used Gō model [85] to study the folding of a small (56 residues) notional off-lattice protein. The thermodynamics of this model was presented in detail in an earlier publication [84] where folding transition was shown to be cooperative. The search for folding nucleus in the off-lattice model [82] consisted of an analysis of the conformations obtained in deep equilibrium folding and unfolding fluctuations. The idea of this analysis is that partly unfolded conformations that are committed to immediately return back to the folded state are the ones that retain the folding nucleus, while fluctuations from partly folded conformations that return back to unfolded state fast represent conformations that have not formed the folding nucleus. Comparison of such "Folded-Folded" fluctuations with "Unfolded-Unfolded" ones allowed to identify the folding nucleus in the off-lattice Gō model of a protein. Further test showed that fixation of nucleus contacts (like introducing virtual disulfide bonds between nucleus residues) eliminated the barrier between folded and unfolded conformations, while fixation of the same number of control non-nucleus contacts did not change the landscape qualitatively retaining the barrier (fig. 3).

3˙2. *Topology may define folding nucleus. A key finding from simulations and experiment and its evolutionary implications.* – The analysis of lattice and simple off-lattice simulations resulted in the remarkable conclusion that the location of the folding nucleus may be determined to a greater extent by the native structure than by the details of the sequence that fold into that structure. This finding was first made in [30]. It was found that the folding nucleus was identical for 50 sequences designed to fold into the same lattice conformation despite the fact that the sequences were quite different. It was concluded in [30] that the nucleus location is determined primarily by the native

The CoC analysis can provide very specific indications of the folding nucleus, in contrast to topology-based oversimplified models that can show some correlation with experimental ϕ-values but are not accurate enough to predict specific nucleation residues. However, the CoC analysis is very data-demanding and not all proteins have many known analogs and homologs to carry out the CoC analysis. To this end attempts are made to determine the folding transition state(s) at the atomic level of resolution from molecular-dynamics simulations of individual proteins.

5. – Molecular dynamics: folding pathways inferred from unfolding simulations

Molecular dynamics (MD) allows to study proteins at very high space and time resolution: an all-atom protein model moves according to Newtonian laws in a liquid of explicit water molecules. However, a high price should be paid for such resolution. Simulations not longer than 10–50 ns can be performed at current computer power (with the exception of a single study that reached 1000 ns landmark [105]). Although 10 ns time may be sufficient to observe hydrophobic effects, electrostatic screening and friction, it is clearly not enough to simulate either folding or unfolding under natural or mildly denaturing conditions. To speed molecular events, extreme conditions of more than 500–1000 K and high pressure are applied. Clearly under such condition (close to the conditions in a gun shell during firing) only *unfolding* of a protein can be observed. Then to infer folding from unfolding trajectories one has to rely on microscopic reversibility and reverse sequence of events observed in unfolding in the hope that the folding mechanism at 500–1000 K is the same as at room temperature.

In spite of these problems and the fundamental ambiguity of force fields, MD simulations of high-temperature unfolding were able to recover the coarse-grained sequence of folding events consistent with experimental data on CI2 [106, 107], SH3 [108], barnase [109], lysozyme [110], segment B1 from protein G [73] and other proteins (also see [111, 112] for reviews). Unfortunately, most of these studies consider sequence of (un)folding events on a scale of formation of secondary-structure elements, melting of domains or melting of the whole hydrophobic core etc. Hence the contribution of an individual residue into kinetics of folding can hardly be evaluated directly.

Li and Daggett analyzed a possible transition state for CI2 unfolding and suggested mutations that can stabilize the transition state and hence speed up folding. Ladurner *et al.* tested these predictions with impressive results of substantial (up to 40 times!) increase in the rate of folding for the mutants [113]. Importantly, most other mutations in CI2 slow folding-down. To complete the cycle the authors performed MD simulations of the mutants and identified their transition states.

Another MD study of CI2 was performed by Lazaridis and Karplus who simulated 24 unfolding trajectories using implicit solvent model. While emphasizing the diversity of unfolding trajectories, they found, in accord with experiment, the disruption of tertiary interactions between the helix and a two-stranded portion of the beta sheet as the primary unfolding event. It is important to note that these authors used experimental information about ϕ-values in CI2 in order to determine the transition state in their simulations.

Thus it is difficult to say whether the observed correlation with experiment is a mere consequence of this fact or if simulations provided some non-trivial results.

A different approach to the analysis of the transition state ensemble is suggested by Pande and Rokhsar [114]. They mapped the transition states between different metastable states for small 16-residue β-hairpin peptide using a p_{fold} analysis similar to [29]. This approach is elegant but it is not clear whether such method can be used to determine TSE between globally folded and unfolded states in MD simulations.

Success in reconstructing a folding picture consistent with experiment indicates that MD simulations can be helpful in interpreting experimentally observed ϕ-values and, perhaps, for their predictions in the future. Combination of experiment with MD simulation seems to be a promising approach as MD may complement an experimental picture of the transition state with atomic details and short-time (on the scale of folding) dynamics.

One, however, needs to be very careful inferring the sequence of folding events or the structure of transition state from very few high-temperature unfolding trajectories. As more studies are being done, another major limitation of MD becomes apparent. A limited number of MD trajectories (< 25) precludes consistent comparison with experiment where all observations are averaged over a huge ensemble of folding proteins. Recently Kazmirski *et al.* studied several unfolding trajectories of BPTI, CI2 and barnase [115]. These authors developed a variety of techniques to compare trajectories. The main conclusion of this study is that unfolding trajectories do not follow a narrow path from the native state, but are rather diverse. However, when structures sampled in unfolding trajectories were characterized by a small (< 10) number of properties (*e.g.* radius of gyration, helix content etc.) a common path can be observed for some proteins. In BPTI, however, even the order of secondary-structure melting was different in different MD simulations requiring to decrease the resolution of analysis to fewer coarse-grained properties. The only scenario common to all three proteins was that unfolding started from the expansion of the core followed by fraying of secondary-structure elements leading to the transition state, in accord with earlier predictions from analytical theory [4]. Afterwards the transition state trajectories diverged very fast sharing no commonality even in such a coarse-grained picture. This is consistent with the (inverse) description of the folding process that emerged in lattice simulation: stochastic search prior to the transition state (nucleus) and directed pathway past the nucleus [30]. Another lesson of the study [115] is the danger of projecting high-dimensional trajectories on a small number of dimensions [116, 117]. While projections may appear similar, the actual trajectories can be very different and vice versa. The apparent similarity of trajectories can be a result of our selection of projection axis that may not adequately describe the changes within the system.

In summary, MD operates on a very high resolution protein model at a cost of the necessity to simulate unfolding at extreme conditions. However, in order to have a consistent interpretation of these results from multiple trajectories, one needs to sacrifice high resolution and get back to a low-resolution picture. This problem suggests that intermediate-resolution models that allow to study hundreds of folding trajectories at normal conditions may be promising.

6. – All-atom Monte Carlo simulations: possible folding mechanism at an atomic level of detail

An intermediate-resolution model has been developed recently by Shimada, Kussell and ES (unpublished). In this approach all heavy atoms are represented as interacting hard spheres of various sizes corresponding to their van der Waals radii. This model includes all degrees of freedom relevant to folding—all side chain and backbone torsions— and uses a Gō atom-atom potential that makes two atoms attract each other (when they are within certain atom-type specific cut-off distance) if they were neighbors in the native state and repel each other otherwise. This energy function strongly favors the native state making it the global energy minimum.

Folding dynamics is simulated using Monte Carlo technique whereby at each step a window of six consecutive residues is selected. Then a move is made in which up to three (Φ, ψ) pairs out of this window are changed and each (Φ, ψ) move is accompanied by 10 moves of sidechain dihedrals ξ. Importantly, the move amplitudes are drawn from Gaussian distribution with zero mean and variance of 2 degrees for mainchain and 10 degrees for sidechain moves. This move set provides an optimal combination of long-range moves at the initial stage of folding and small moves when the chain is already compact enough.

Using a small protein, crambin, 165 folding transitions from random coil to compact conformation differing less than 1 Å rms. with the native state were recorded. The disulfide bonds were treated as normal Gō interactions, $i.e.$ at the beginning the protein was fully unfolded. By recording many folding events over a wide range of temperatures a possible folding mechanism for crambin is obtained. Folding occurs via a cooperative first-order process and many folding pathways to the native state exist. One particular sequence of events constitutes a "fast-folding" pathway where kinetics traps (due to chain misfolding and side-chain mispacking) are avoided. This pathway includes formation of alpha-helical hairpin followed by a rate-limiting step of nucleation of beta-sheet propagating to the full native structure with concurrent side-chain packing. These results present a "proof-of-principle" for the possibility of a solution of protein folding problem at an all-atom level, provided that one has a realistic all-atom potential energy function that correctly favors the native state, and which does not use information about the native state.

Several other small proteins (SH3 domains IgG binding domain) have also been folded using this approach. Of those, the most interesting are simulations of IgG binding domain (protein G) folding for which extensive experimental data are available [118]. Protein G appears to fold slower than crambin (1.5 billion steps in average) so that a smaller number of fodling vents was collected (50). Nevertheless, the dominant pathway appears to be intersting with flickering hairpin 2 forming first. Next an intermediate complex between hairpin I and alpha-helix is formed cooperatively. This complex remains stable until it associates with flickering hairpin 2 completing the folding event. Interestingly, it was shown in a separate study (N. Ingolia, Jun Shimada and E. Shakhnovich, unpublished) of folding of isolated hairpin 1 and hairpin 2 of protein G (in the same Gō model) that hairpin 1 is less stable and folds cooperatively at low temperature, while hairpin 2 is less cooperative and folds/unfolds at higher temperature. It will be interesting to analyze further how folding thermodynamics and kinetics of individual hairpins affects the folding scenario of intact protein G, especially in the light of finding that its structural homolog, protein L, features an inverse sequence of folding events (the first hairpin 1 is formed first in protein L in contrast to protein G [119]).

The all-atom Monte Carlo simulations may provide, for the first time, an insight into folding mechanisms of small proteins at the atomic level of resolution.

7. – Concluding remarks

The protein folding field has undergone an interesting evolution since its inception in the sixties. First, "biochemical" thinking dominated, that viewed each protein as a unique system which requires a full atomic-level description of its folding pathway. The folding pathway itself was viewed as a sequence of microscopically well-defined events akin to a simple chemical reaction.

This view changed in the early nineties when simplified analytical and lattice models demonstrated their power to explain several key aspects of protein folding such as cooperativity, nucleation, resolving the earlier paradigm known as "Levinthal paradox". A "new view" (term suggested by R. Baldwin [120] in appreciation of the lattice model simulations reported in [72]) on protein folding as a statistical process emerged. This makes folding akin to phase transition—an analogy explored by many researchers both from thermodynamic and kinetic perspectives [4, 6, 11, 18, 51, 75, 121-125]. The "phase transition" analogy shifts the focus in addressing folding pathways from a microscopic, step-by-step description to the analysis of essential milestones on the pathway—unfolded state, transition state, native state and possible intermediates—viewing each of them as dynamic ensembles of conformations corresponding to local minima or saddle point(s) in the free-energy landscape. Those developments brought experiment and theory closer together [38, 126]. A key lesson from studies of simple models is the appreciation of certain aspects of universality in protein folding. This means that, at a coarse-grained level of description, there is a small number of major scenarios and many proteins fall into one of them [127].

However, most recently models were developed that allowed to simulate folding of small proteins at atomic or near atomic level of detail. Success of those simulations partly came from the enhanced power of computers and, to a great extent, from better understanding of general principles of protein folding. The latter played a key role in formulating the right questions and carrying out appropriate analysis of all-atom simulations. Probably, we are entering a new era of folding studies which will elevate our understanding of this uniquely important phenomenon to the atomic level of detail and will result finally in a search strategy and energy function that are powerful enough to predict structure from sequence at an atomic resolution of 1–2 Å. This may finally render theoretical protein folding useful for application in functional genomics and drug design.

REFERENCES

[1] PRIVALOV P. L., *Annu. Rev. Biophys. Biophys. Chem.*, **18** (1989) 47.
[2] JACKSON S. E. and FERSHT A. R., *Folding of chimotrypsin inhibitor 2. 1. evidence for a two-state transition*, Biochemistry, **30** (1991) 10428.
[3] JACKSON S., *How do small single-domain proteins fold?*, Folding & Design, **3** (1998) R81.
[4] SHAKHNOVICH E. I. and FINKELSTEIN A. V., *Theory of cooperative transitions in protein molecules. I. Why denaturation of globular protein is a first-order phase transition* Biopolymers, **28** (1989) 1667.
[5] KARPLUS M. and SHAKHNOVICH E. I., *Protein Folding* (W. H. Freeman and Company, New York) 1992, Chapt. 4, p. 127.
[6] PTITSYN O. B., KRON A. and EIZNER YU. YE., *Simple statistical theory of self-organization of protein molecules*, J. Polym. Sci. Pt. C, **16** (1968) 3509.
[7] HAO M. H. and SCHERAGA H., *Molecular mechanisms for cooperative folding of proteins*, J. Mol. Biol., **277** (1998) 973.
[8] DILL K., *Theory for the folding and stability of globular proteins*, Biochemistry, **24** (1985) 1501.
[9] BRYNGELSON J. D. and WOLYNES P. G., *Spin glasses and the statistical mechanics of protein folding*, Proc. Natl. Acad. Sci. USA, **84** (1987) 7524.
[10] SHAKHNOVICH E. and GUTIN A., *Formation of unique structure in polypeptide chains. Theoretical investigation with the aid of replica approach*, Biophys. Chem., **34** (1989) 187.
[11] SHAKHNOVICH E. I. and GUTIN A., *Engineering of stable and fast-folding sequences of model proteins*, Proc. Natl. Acad. Sci. USA, **90** (1993) 7195.
[12] RAMANATHAN S. and SHAKHNOVICH E., *Statistical mechanics of proteins with "evolutionary selected" sequences*, Phys. Rev. E, **50** (1994) 1303.
[13] SFATOS C., GUTIN A. M. and SHAKHNOVICH E. I., *Phase diagram of random copolymers*, Phys. Rev. E, **48** (1993) 465.
[14] PANDE V., GROSBERG A. YU. and TANAKA T., *Freezing transition of random heteropolymers consisting of arbitrary sets of monomers*, Phys. Rev. E, **51** (1995) 3381.
[15] ABKEVICH V., GUTIN A. and SHAKHNOVICH E., *Free energy landscape for protein folding kinetics. Intermediates, traps and multiple pathways in theory and lattice model simulations*, J. Chem. Phys., **101** (1994) 6052.
[16] SHAKHNOVICH E. I. and GUTIN A. M., *Frozen states of disordered globular heteropolymers*, J. Phys. A, **22** (1989) 1647.
[17] ABKEVICH V., GUTIN A. and SHAKHNOVICH E., *Impact of local and non-local interactions on thermodynamics and kinetics of protein folding*, J. Mol. Biol., **252** (1995) 460.
[18] SHAKHNOVICH E. I., *Theoretical studies of protein-folding thermodynamics and kinetics*, Curr. Opin. Struct. Biol., **7** (1997) 29.
[19] YUE K., FIEDIG K., THOMAS P., CHAN H. S., SHAKHNOVICH E. and DILL K., *A test of lattice protein folding algorithms*, Proc. Natl. Acad. Sci. USA, **92** (1995) 325.
[20] SHAKHNOVICH E., *Protein design: a perspective from simple tractable models*, Folding & Design, **3** (1998) R45.
[21] SHAKHNOVICH E. I., *Proteins with selected sequences fold to their unique native conformation*, Phys. Rev. Lett., **72** (1994) 3907.
[22] HAO M. H. and SCHERAGA H., *Monte Carlo simulation of a first order transition for protein folding*, J. Phys. Chem., **98** (1994) 4940.
[23] HAO M. H. and SCHERAGA H., *Statistical thermodynamics of protein folding: Sequence dependence*, J. Phys. Chem., **98** (1994) 9882.
[24] WILDER J. and SHAKHNOVICH E., *Proteins with selected sequences. A heteropolymeric study*, Phys. Rev. E, **62** (2000) 7100.

[25] CHAN H. S., *Modelling protein density of states: additive hydrophobic effects are insufficient for calorimetric two-state cooperativity*, Proteins: Struct. Function, Genetics, **40** (2000) 543.

[26] KAYA H. and CHAN H. S., *Polymer principles of protein calorimetric two-state cooperativity*, Proteins: Struct. Function, Genetics, **40** (2000) 637.

[27] GUTIN A., ABKEVICH V. and SHAKHNOVICH E., *A protein engineering analysis of the transition state for protein folding: simulation in the lattice model*, Folding & Design, **3** (1998) 183.

[28] FERSHT A. R., *Structure and Mechanism in Protein Science: A Guide to Enzyme Catalysis and Protein Folding* (W. H. Freeman & Co, San Francisco) 1999.

[29] DU R., PANDE V. S., GROSBERG A. YU., TANAKA T. and SHAKHNOVICH E. I., *On the transition coordinate for protein folding*, J. Chem. Phys., **108** (1998) 334.

[30] ABKEVICH V. I., GUTIN A. M. and SHAKHNOVICH E. I., *Specific nucleus as the transition state for protein folding: Evidence from the lattice model*, Biochemistry, **33** (1994) 10026.

[31] FERSHT A. R., *Transition-state structure as a unifying basis in protein-folding mechanisms: Contact order, chain topology, stability, and the extended nucleus mechanism*, Proc. Natl. Acad. Sci. USA, **97** (2000) 1525.

[32] PRIETO J. and SERRANO L., *C-capping and helix stability: The pro c-capping motif*, J. Mol. Biol., **274** (1997) 276.

[33] LEOPOLD P., MONTAL M. and ONUCHIC J., *Protein folding funnels: A kinetic approach to the structure-sequence relationship*, Proc. Natl. Acad. Sci. USA, **89** (1992) 8721.

[34] CHAN H. S. and DILL K., *From levinthal to pathways to funnels*, Nature Struct. Biol., **4** (1997) 10.

[35] BICOUT D. and SZABO A., *Entropic barriers, transition states, funnels, and exponential protein folding kinetics: A simple model*, Protein Sci., **9** (2000) 452.

[36] KARPLUS M. and WEAVER D., *Protein-folding dynamics*, Nature, **160** (1976) 404.

[37] BALDWIN R. L. and ROSE G. D., *Is protein folding hierarchic? II. Folding intermediates and transition states*, Trends Biochem. Sci., **24** (1999) 77.

[38] ITZHAKI L., OTZEN D. and FERSHT A., *The structure of the transition state for folding of chymotrypsin inhibitor 2 analyzed by protein engineering methods: Evidence for a nucleation-condensation mechanism for protein folding*, J. Mol. Biol., **254** (1995) 260.

[39] MUNOZ V. and SERRANO L., *Local versus nonlocal interactions in protein folding and stability—an experimentalist's point of view*, Folding & Design, **1** (1996) R71.

[40] SOSNICK T. R., JACKSON S., ENGLANDER S. W. and DEGRADO W., *The role of helix formation in the folding of a fully α-helical coiled-coil*, Proteins, **24** (1996) 427.

[41] MORAN L., SCHNEIDER J., KENTISIS A., REDDY G. and SOSNICK T. R., *Transition state heterogeneity in gcn4 coiled coil folding studied by using multisite mutations and crosslinking*, Proc. Natl. Acad. Sci. USA, **96** (1999) 10699.

[42] DURR E., JELESAROV I. and BOSSARD H. R., *Extremely fast folding of a very stable leucine zipper with a strengthened hydrophobic core and lacking electrostatic interactions between helices*, Biochemistry, **38** (1999) 870.

[43] MYERS J. K. and OAS T. G., *Reinterpretation of gcn4-p1 folding kinetics: Partial helix formation precedes dimerization in coiled coil folding*, J. Mol. Biol., **289** (1999) 205.

[44] CREGUT D., CIVERA C., MACIAS M., WALLON G. and SERRANO L., *A tale of two secondary structure elements: when a [beta] hairpin becomes an alpha-helix*, J. Mol. Biol., **292** (1999) 389.

[45] KIM D. E., YI Q., GLADWIN S. T., GOLDBERG J. M. and BAKER D., *The single helix in protein l is largely disrupted at the rate-limiting step in folding*, J. Mol. Biol., **284** (1998) 807.

[46] BURTON R. E., MYERS J. K. and OAS T. G., *Protein folding dynamics: Quantitative comparison between theory and experiment*, Biochemistry, **37** (1998) 5337.

[47] LUISI D. L., KUHLMAN B., SIDERAS K., EVANS P. A. and RALEIGH D. P., *Effects of varying the local propensity to form secondary structure on the stability and folding kinetics of a rapid folding mixed alpha/beta protein: Characterization of a truncation mutant of the n-terminal domain of the ribosomal protein l9*, J. Mol. Biol., **289** (1999) 167.

[48] DYSON J. et al., J. Mol. Biol., **201** (1988) 161.

[49] DYSON J. et al., J. Mol. Biol., **201** (1988) 201.

[50] BAI Y. W., KARIMI A., DYSON H. J. and WRIGHT P. E., *Absence of a stable intermediate on the folding pathway of protein a*, Protein Sci., **6** (1998) 1449.

[51] BRYNGELSON J. D. and WOLYNES P. G., *A simple statistical field theory of heteropolymer collapse with application to protein folding*, Biopolymers, **30** (1990) 177.

[52] MATEU M., SANCHEZ DEL PINO M. and FERSHT A., *Mechanism of folding and assembly of a small tetrameric protein domain from tumour suppressor p53*, Nature Struct. Biol., **6** (1999) 190.

[53] OLTZEN D., ITZHAKI L., ELMASRY N. F., JACKSON S. E. and FERSHT A. R., *Structure if the transition-state for the foldin/unfolding of the barley chymotrypsin inhibitor-2 and its implications for mechanisms of protein folding*, Proc. Natl. Acad. Sci. USA, **91** (1994) 10422.

[54] MARTINEZ J., PISSABARRO T. and SERRANO L., *Obligatory steps in protein folding and the conformational diversity of the transition state*, Nature Struct. Biol., **5** (1998) 721.

[55] GRANTCHANOVA V., RIDDLE D., SANTIAGO J. and BAKER D., *Important role of hydrogen bonds in the structurally polarized transition state for folding of the src sh3 domain*, Nature Struct. Biol., **5** (1998) 714.

[56] LORCH M., MASON J. M., CLARKE A. R. and PARKER M. J., *Effects of core mutations on the folding of a beta-sheet protein: implications for backbone organization in the i-state*, Biochemistry, **38** (1999) 1377.

[57] HAMILL S. J., STEWARD A. and CLARKE J., *The folding of an immunoglobulin-like greek key protein is defined by a common-core nucleus and regions constrained by topology*, J. Mol. Biol., **297** (2000) 165.

[58] KRAGELUND B. B., OSMARK P., NEERGAARD T. B, SCHIODT J., KRISTIANSEN K., KNUDSEN J. and POULSEN F. M., *The formation of a native-like structure containing eight conserved hydrophobic residues is rate limiting in two-state protein folding of acbp*, Nature Struct. Biol., **6** (1999) 594.

[59] LOPEZ-HERNANDEZ E. and SERRANO L., *Structure of the transition state for folding of the 129 aa protein chey resembles that of a smaller protein, ci2*, Folding & Design, **1** (1996) 43.

[60] VILEGAS V., MARTINEZ J., AVILEZ F. and SERRANO L., *Structure of the transition state in the folding process of human procarboxypeptidase a2 activation domain*, J. Mol. Biol., **283** (1998) 1027.

[61] CHITI F., TADDEI N., WHITE P., BUCCIANTINI M., MAGHERINI F., STEFANI M. and DOBSON C., *Mutational analysis of acylphosphatase suggests the importance of topology and contact order in protein folding*, Nature Struct. Biol., **6** (1999) 1005.

[62] MAIN E. R. G., FULTON K. F. and JACKSON S. E., *Folding pathway of fkbp12 and characterisation of the transition state*, J. Mol. Biol., **291** (1999) 429.

[63] CHOE S., LI L., MATSUDAIRA P., WAGNER G. and SHAKHNOVICH E., *Differential stabilization of two hydrophobic cores in the transition state of the villin 14t folding reaction*, J. Mol. Biol., **304** (2000) 99.

[64] ONUCHIC J., SOCCI N., LUTHEY-SCHULTEN Z. and WOLYNES P., *Protein folding funnels: The nature of the transition state ensemble*, Folding & Design, **1** (1996) 441.
[65] KLIMOV D. and THIRUMALAI D., *Lattice models for proteins reveal multiple folding nuclei for nucleation-collapse mechanism*, J. Mol. Biol., **282** (1998) 471.
[66] MICHNICK S., ROSEN M., WANDLESS T., KARPLUS M. and SCHREIBER S., Science, **252** (1991) 836.
[67] MIRNY L., ABKEVICH V. and SHAKHNOVICH E., *How evolution makes proteins fold quickly*, Proc. Natl. Acad. Sci. USA, **95** (1998) 4976.
[68] MIRNY L. and SHAKHNOVICH E., *Universally conserved residues in protein folds. Reading evolutionary signals about protein function, stability and folding kinetics*, J. Mol. Biol., **291** (1999) 177.
[69] CLARKE J., COTA E., FOWLER S. and HAMILL S. J., *Folding studies of immunoglobulin-like β-sandwich proteins suggest that they share a common folding pathway*, Structure with Folding & Design, **7** (1999) 1145.
[70] PLAXCO K., SIMONS K. T. and BAKER D., *Contact order, transition state placement and the refolding rates of single domain proteins*, J. Mol. Biol., **277** (1998) 985.
[71] GALZITSKAYA O. and FINKELSTEIN A. V., *A theoretical search for folding/unfolding nuclei in three-dimensional protein structures*, Proc. Natl. Acad. Sci. USA, **96** (1999) 11299.
[72] SALI A., SHAKHNOVICH E. and KARPLUS M., *How does a protein fold?*, Nature, **369** (1994) 248.
[73] SHEINERMAN F. and BROOKS C., *Calculations on folding of segment b1 of streptococcal protein g*, J. Mol. Biol., **278** (1998) 439.
[74] SOCCI N. D., ONUCHIC J. N. and WOLYNES P., *Diffusive dynamics of the reaction coordinate for protein folding funnels*, J. Chem. Phys., **104** (1996) 5860.
[75] PANDE V. S., GROSBERG A. YU., ROKSHAR D. and TANAKA T., *Pathways for protein folding: is a "new view" needed?*, Curr. Opin. Struct. Biol., **8** (1998) 68.
[76] LI L., MIRNY L. and SHAKHNOVICH E. I., *Kinetics, thermodynamics and evolution of non-native interactions in protein folding nucleus*, Nature Struct. Biol., **7** (2000) 336.
[77] BOCZKO E. and BROOKS C., *First-principle calculation of the folding free energy of a three-helix bundle protein*, Science, **269** (1995) 393.
[78] SHAKHNOVICH E., *Folding nucleus: specific or multiple insights from lattice models and experiments*, Folding & Design, **3** (1999) R108.
[79] KLIMOV D. and THIRUMALAI D., *Cooperativity in protein folding: from lattice models with sidechains to real proteins*, Folding & Design, **3** (1998) 127.
[80] OLIVEBERG M., TAN Y., SILOW M. and FERSHT A., *The changing structure of the protein folding transition state: implications for the shape of the free-energy profile for folding*, J. Mol. Biol., **277** (1998) 933.
[81] SHAKHNOVICH E. I., *Folding nucleus: specific of multiple? Insights from simulations and comparison with experiment*, Folding & Design, **3** (1998) R108.
[82] DOKHOLYAN N., BULDYREV S., STANLEY H. E. and SHAKHNOVICH E., *Identifying the protein folding nucleus using molecular dynamics*, J. Mol. Biol., **296** (2000) 1183.
[83] ZHOU Y. and KARPLUS M., *Folding thermodynamics of a model three-helix-bundle protein*, Proc. Natl. Acad. Sci. USA, **94** (1997) 14429.
[84] DOKHOLYAN N., BULDYREV S., STANLEY H. E. and SHAKHNOVICH E., *Discrete molecular dynamics studies of the folding of a protein-like model*, Folding & Design, **3** (1998) 577.
[85] UEDA Y., TAKETOMI H. and GO N., *Studies on protein folding, unfolding and fluctuations by computer simulation*, Int. J. Peptide Prot. Res., **7** (1975) 445.
[86] SHAKHNOVICH E. I., ABKEVICH V. I. and PTITSYN O. B., *Conserved residues and the mechanism of protein folding*, Nature, **379** (1996) 96.

[87] CLEMENTI C., NYMEYER J. and ONUCHIC J., *Topological and energetic factors: What determines the structural details of the transition state ensemble and en-route intermediates for protein folding? An investigation for small globular proteins*, J. Mol. Biol., **298** (2000) 937.

[88] PORTMAN J., TAKADA S. and WOLYNES P. G., *Variational theory for site resolved protein folding free energy surfaces*, Phys. Rev. Lett., **81** (1997) 5237.

[89] SHOEMAKER B., WANG J. and WOLYNES P. G., *Exploring structures in protein folding funnels with free energy functionals: The transition state ensemble*, J. Mol. Biol., **287** (1999) 675.

[90] ALM E. and BAKER D., *Prediction of protein-folding mechanisms from free-energy landscapes derived from native structures*, Proc. Natl. Acad. Sci. USA, **96** (1999) 11305.

[91] MUNOZ V. and EATON W., *A simple model for calculating the kinetics of protein folding from three-dimensional structures*, Proc. Natl. Acad. Sci. USA, **96** (1999) 11311.

[92] GALZITSKAYA O. and FINKELSTEIN A., *A theoretical search for folding/unfolding nuclei in three-dimensional protein structures*, Proc. Natl. Acad. Sci. USA, **96** (1999) 11299.

[93] BAKER D., *A surprising simplicity to protein folding*, Nature, **405** (2000) 39.

[94] DOKHOLYAN N., LI L. and SHAKHNOVICH E., *Structural determinants of folding transition state in proteins*, to be published in Protein Str. Funct. Gen.

[95] FLORY P., *Principles of Polymer Chemistry* (Cornell University Press, Ithaca) 1953.

[96] GUTIN A. and SHAKHNOVICH E., *Statistical mechanics of polymers with distance constraints*, J. Chem. Phys., **100** (1994) 5290.

[97] DODGE C., SCHNEIDER R. and SANDER C., *The hssp database of protein structure-sequence alignments and family profiles*, Nucleic Acids Res., **26** (1998) 313.

[98] PTITSYN O., *Protein folding and protein evolution: Common folding nucleus in different subfamilies of c-type cytochromes?*, J. Mol. Biol., **278** (1998) 655.

[99] PTITSYN O. and TING KLH, *Non-functional conserved residues in globins and their possible role as a folding nucleus*, J. Mol. Biol., **291** (1999) 671.

[100] MICHNICK S. and SHAKHNOVICH E., *A strategy for detecting the conservation of folding-nucleus residues in protein superfamilies*, Folding & Design, **3** (1998) 239.

[101] SHAKHNOVICH E. and GUTIN A., *A novel approach to design of stable proteins*, Protein Engin., **6** (1993) 793.

[102] KUHLMAN B. and BAKER D., *Native protein sequences are close to optimal for their structures*, Proc. Natl. Acad. Sci. USA, **97** (2000) 10383.

[103] KIM D., GU H. and BAKER D., *The sequences of small proteins are not extensively optimized for rapid folding by natural selection*, Proc. Natl. Acad. Sci. USA, **95** (1998) 4982.

[104] PLAXCO K., LARSON S., RUCZINSKI I., RIDDLE D., BUCHWITZ B., DAVIDSON A. and BAKER D., *Evolutionary conservation in protein folding kinetics*, J. Mol. Biol., **298** (2000) 303.

[105] DUAN Y. and KOLLMAN P., *Pathways to a protein folding intermediate observed in a 1-microsecond simulation in aqueous solution*, Science, **282** (1998) 740.

[106] LI A. J. and DAGGETT V., *Characterization of the transition state of protein unfolding by use of molecular dynamics - chymostrypsin inhibitor-2*, Proc. Natl. Acad. Sci. USA, **91** (1994) 10430.

[107] LAZARIDIS T. and KARPLUS M., *"New view" of protein folding reconciled with the old through multiple unfolding simulations*, Science, **278** (1997) 1928.

[108] TSAI J., LEVITT M. and BAKER D., *Hierarchy of structure loss in md simulations of src sh3 domain unfolding*, J. Mol. Biol., **291** (1999) 215.

[109] DAGGETT V., LI A. and FERSHT A., *Combined molecular dynamics and phi-value analysis of structure-reactivity relationships in the transition state and unfolding pathway of barnase: Structural basis of hammond and anti-hammond effects*, J. Amer. Chem. Soc., **120** (1998) 12740.

[110] KAZMIRSKI S. and DAGGETT V., *Non-native interactions in protein folding intermediates: Molecular dynamics simulations of hen lysozyme*, J. Mol. Biol., **284** (1998) 793.

[111] DAGGETT V., *Long timescale simulations*, Curr. Opin. Struct. Biol., **10** (1998) 160.

[112] BROOKS C., *Simulations of protein folding and unfolding*, Curr. Opin. Struct. Biol., **8** (1998) 222.

[113] LADURNER A., ITZHAKI L., DAGGETT V. and FERSHT A., *Synergy between simulation and experiment in describing the energy landscape of protein folding*, Proc. Natl. Acad. Sci. USA, **95** (1998) 8473.

[114] PANDE V. and ROKHSAR D., *Molecular dynamics simulations of unfolding and refolding of a beta-hairpin fragment of protein g*, Proc. Natl. Acad. Sci. USA, **96** (1999) 9062.

[115] KAZMIRSKI S., LI A. and DAGGETT V., *Analysis methods for comparison of multiple molecular dynamics trajectories: Applications to protein unfolding pathways and denatured ensembles*, J. Mol. Biol., **290** (1999) 283.

[116] DINNER A. and KARPLUS M., *Is protein unfolding the reverse of protein folding? A lattice simulation analysis*, J. Mol. Biol., **292** (1999) 403.

[117] DINNER A., SALI A., SMITH L., DOBSON C. and KARPLUS M., *Understanding protein folding via free-energy surfaces from theory and experiment*, Trends Biochem. Sci., **25** (2000) 331.

[118] MCCALLISTER E., ALM E. and BAKER D., *Critical role of beta-hairpin formation in protein g folding*, Nature Struct. Biol., **7** (2000) 669.

[119] KIM D. E., FISHER C. and BAKER D., *A breakdown of symmetry in the folding transition state of protein l*, J. Mol. Biol., **298** (2000) 971.

[120] BALDWIN R., *Protein folding - matching speed and stability*, Nature, **369** (1994) 183.

[121] LIFSHITZ I. M., GROSBERG A. Y. and KHOHLOV A. R., *Some problems of statistical physics of polymers with volume interactions*, Rev. Mod. Phys., **50** (1978) 683.

[122] SHAKHNOVICH E. I. and FINKELSTEIN A. V., *On the theory of cooperative transitions in proteins*, Dokl. Acad. Nauk SSSR, **243** (1982) 1247.

[123] FINKELSTEIN A. V. and SHAKHNOVICH E. I., *Theory of cooperative transitions in protein molecules. II. Phase diagram for a protein molecule in solution*, Biopolymers, **28** (1989) 1668.

[124] KLIMOV D. and THIRUMALAI D., *A criterion which determines foldability of proteins*, Phys. Rev. Lett., **76** (1996) 4070.

[125] PLOTKIN S. and ONUCHIC J., *Investigation of routes and funnels in protein folding by free energy functional methods*, Proc. Natl. Acad. Sci. USA, **97** (2000) 6509.

[126] FERSHT A. R., *Optimization of rates of protein folding: The nucleation-condensation mechanism and its implications*, Proc. Natl. Acad. Sci. USA, **92** (1995) 10869.

[127] MIRNY L., ABKEVICH V. and SHAKHNOVICH E., *Universality and diversity of the protein folding scenarios: A comprehensive analysis with the aid of lattice model*, Folding & Design, **1** (1996) 103.

Mechanism of folding and aggregation of proteins

R. A. Broglia

Dipartimento di Fisica, Università di Milano - Via Celoria 16, I-20133 Milano, Italy
INFN, Sezione di Milano - Via Celoria 16, I-20133 Milano, Italy
The Niels Bohr Institute, University of Copenhagen - 2100 Copenhagen, Denmark

G. Tiana

Dipartimento di Fisica, Università di Milano - Via Celoria 16, I-20133 Milano, Italy
INFN, Sezione di Milano - Via Celoria 16, I-20133 Milano, Italy

1. – Introduction

Although all the information required for the folding of a protein chain is contained in its amino acid sequence [1], one has not yet learnt how to read this information so as to be able to predict the detailed three-dimensional structure of a real protein whose sequence is known [2-4]. This is the protein folding problem (one of the great unsolved problems of science, cf. ref. [4], p. 508), a problem which have been deemed too difficult to be solved even within the framework of simple models. This is the reason why the inverse folding strategy was developed [5]. Within this scenario, the protein folding problem is turned upside down into the quest for the sequences which fold on short call into a selected native conformation. This problem has, at least for small, single-domain proteins, a simple solution: good folder sequences are characterized by a large gap δ (compared to the standard deviation σ of the contact energies) between the energy E_n of the sequence in the native conformation and the lowest-energy E_c of the conformations structurally dissimilar to the native conformation [6-10], E_c being a quantity which is solely determined by the composition of the protein. Good folders are obtained by minimizing the energy of the chain in the native conformation with respect to amino acid sequence for fixed composition.

Making use of the insight provided by these results and studying, through Monte Carlo simulations, the dynamical evolution of designed sequences, from the random coil to the native conformation, and how this process is affected by mutations, we have identified [11-16] a hierarchy of specific elementary phenomena which control the way single domain (lattice-model) proteins fold: a) formation of few, local elementary structures, b) creation of the (post-critical) folding nucleus through the assemblage of these local elementary structures, c) relaxation of the remaining amino acids to the native conformation. The single, most important feature common to all designed sequences folding to the same native conformation is the presence of few, conserved, strongly interacting amino acids [11] which stabilize the local elementary structures and which are buried inside the folding nucleus of the protein in its native conformation.

Making use of these results we have derived a strategy [16] which allows one to predict the native conformation of the protein from its primary structure. Three are the main steps to be taken to achieve this goal: 1) to start from a notional sequence and make use of the contact energies employed in its design find the local elementary structures, 2) to determine the possible folding cores by allowing the local elementary structures to interact among them, 3) to relax the position of the remaining amino acids and determine the corresponding energies. The (single) compact structure which displays an energy smaller than E_c is the native conformation of the protein.

To the extent that lattice models capture some of the basic features of real proteins there is thus essentially a single requirement to be fulfilled in trying to generalize the 1D → 3D strategy to the case of real proteins: to calculate accurately, combining *ab initio* and classical methods (cf., *e.g.*, [17] and references therein), the potential acting between the (few) strongly interacting amino acids of the real protein, taking into account the solvent.

In these lectures we shall try to convey some of the details and most of the excitement connected with our finding of a solution of the protein folding problem within the framework of lattice models.

In sect. **2** we present the model and characterize, through mutations, the role different amino acids of a notional protein play in the folding process. Also discussed in this section are questions connected with the stability and designability of the protein. In sect. **3** we present evidence for the role local elementary structures play in the phenomenon of protein folding. Some of the consequences of the presence of local elementary structures in the aggregation of proteins are also discussed. In sect. **4** examples of the strategy devised to read the three-dimensional structure of a notional protein from its amino acid sequence are discussed. The success of this strategy vindicates the correctness of the picture of protein folding developed in sects. **2** and **3**. The conclusions of the work which is at the basis of these lectures are presented in sect. **5**, together with the perspectives opened by it: a) lattice model studies of aggregation of single-domain and of two-native-state (prion-like) proteins, as well as of folding of the homo- and hetero-oligomers; b) *ab initio* strategies for the calculation of the potential acting among selected amino acids.

2. – The model

A useful theoretical approach to study protein folding is a simplified lattice model, where the protein is a string of beads that is arranged on a cubic lattice. The configurational energy of a chain of N monomers is given by

$$(1) \qquad E = \frac{1}{2} \sum_{i,j}^{N} U_{m(i),m(j)} \Delta(|\vec{r}_i - \vec{r}_j|),$$

where $U_{m(i),m(j)}$ is the effective interaction potential between monomers $m(i)$ and $m(j)$, $\vec{r}_i$ and $\vec{r}_j$ denote their lattice positions and $\Delta(x)$ is the contact function. In eq. (1) the pairwise interaction is different from zero when i and j occupy nearest-neighbour sites, i.e., $\Delta(a) = 1$ and $\Delta(na) = 0$ for $n \geq 2$, where a indicates the step length of the lattice. In addition to these interactions, it is assumed that on-site repulsive forces prevent two amino acids to occupy the same site simultaneously, so that $\Delta(0) = \infty$ (excluded-volume ansatz). The folding of the chain is simulated by Monte Carlo (MC) methods. We shall consider throughout a 20-letters representation of protein sequence where U is a 20×20 matrix. A possible realization of this matrix (cf. fig. 1) is given in ref. [18] (table VI), where it was derived from frequencies of contacts between different amino acids in protein structures. The model we study here is a generic heteropolymer model which has been shown to reproduce important generic features of protein folding thermodynamics and kinetics, independent of the particular potential chosen [10,19]. This is achieved by using the same potential to design sequences and to simulate folding. However, in using such an approach, one should keep in mind that the labelling of amino acids (spherical beads all of the same size and with no side chain) is generic too and may be no obvious relation between those labels and labels for real amino acids.

	A	C	D	E	F	G	H	I	K	L	M	N	P	Q	R	S	T	V	W	Y
A	-0,13	0	0,12	0,26	0,03	-0,07	0,34	-0,22	0,14	-0,01	0,25	0,28	0,1	0,08	0,43	-0,06	-0,09	-0,1	-0,09	0,09
C	0	-1,06	0,03	0,69	-0,23	-0,08	-0,19	0,16	0,71	-0,08	0,19	0,13	0	0,05	0,24	-0,02	0,19	0,06	0,08	0,04
D	0,12	0,03	0,04	-0,15	0,39	-0,22	-0,39	0,59	-0,76	0,67	0,65	-0,3	0,04	-0,17	-0,72	-0,31	-0,29	0,58	0,24	0
E	0,26	0,69	-0,15	-0,03	0,27	0,25	-0,45	0,35	-0,97	0,43	0,44	-0,32	-0,1	-0,17	-0,74	-0,26	0	0,34	0,29	-0,1
F	0,03	-0,23	0,39	0,27	-0,44	-0,38	-0,16	-0,19	0,44	-0,3	-0,42	0,18	0,2	-0,29	0,41	0,29	0,31	-0,22	-0,16	0
G	-0,07	-0,08	-0,22	0,25	-0,38	-0,38	0,2	0,25	0,11	0,23	0,19	-0,14	-0,11	-0,06	-0,04	-0,16	-0,26	0,16	0,18	0,14
H	0,34	-0,19	-0,39	-0,45	-0,16	0,2	-0,29	0,49	0,22	0,16	0,99	-0,24	-0,21	-0,02	-0,12	-0,05	-0,19	0,19	-0,12	-0,34
I	-0,22	0,16	0,59	0,35	-0,19	0,25	0,49	-0,22	0,36	-0,41	-0,28	0,53	0,25	0,36	0,42	0,21	0,14	-0,25	0,02	0,11
K	0,14	0,71	-0,76	-0,97	0,44	0,11	0,22	0,36	0,25	0,19	0	-0,33	0,11	-0,38	0,75	-0,13	-0,09	0,44	0,22	-0,21
L	-0,01	-0,08	0,67	0,43	-0,3	0,23	0,16	-0,41	0,19	-0,27	-0,2	0,3	0,42	0,26	0,35	0,25	0,2	-0,29	-0,09	0,24
M	0,25	0,19	0,65	0,44	-0,42	0,19	0,99	-0,28	0	-0,2	0,04	0,08	-0,34	0,46	0,31	0,14	0,19	-0,14	-0,67	-0,13
N	0,28	0,13	-0,3	-0,32	0,18	-0,14	-0,24	0,53	-0,33	0,3	0,08	-0,53	-0,18	-0,25	-0,14	-0,14	-0,11	0,5	0,06	-0,2
P	0,1	0	0,04	-0,1	0,2	-0,11	-0,21	0,25	0,11	0,42	-0,34	-0,18	0,26	-0,42	-0,38	0,01	-0,07	0,09	-0,28	-0,33
Q	0,08	0,05	-0,17	-0,17	-0,29	-0,06	-0,02	0,36	-0,38	0,26	0,46	-0,25	-0,42	0,29	-0,52	-0,14	-0,14	0,24	0,08	-0,2
R	0,43	0,24	-0,72	-0,74	0,41	-0,04	-0,12	0,42	0,75	0,35	0,31	-0,14	-0,38	-0,52	0,11	0,17	-0,35	0,3	-0,16	-0,25
S	-0,06	-0,02	-0,31	-0,26	0,29	-0,16	-0,05	0,21	-0,13	0,25	0,14	-0,14	0,01	-0,14	0,17	-0,2	-0,08	0,18	0,34	0,09
T	-0,09	0,19	-0,29	0	0,31	-0,26	-0,19	0,14	-0,09	0,2	0,19	-0,11	-0,07	-0,14	-0,35	-0,08	0,03	0,25	0,22	0,13
V	-0,1	0,06	0,58	0,34	-0,22	0,16	0,19	-0,25	0,44	-0,29	-0,14	0,5	0,09	0,24	0,3	0,18	0,25	-0,29	-0,07	0,02
W	-0,09	0,08	0,24	0,29	-0,16	0,18	-0,12	0,02	0,22	-0,09	-0,67	0,06	-0,28	0,08	-0,16	0,34	0,22	-0,07	-0,12	-0,04
Y	0,09	0,04	0	-0,1	0	0,14	-0,34	0,11	-0,21	0,24	-0,13	-0,2	-0,33	-0,2	-0,25	0,09	0,13	0,02	-0,04	0,06

Fig. 1. – The interaction matrix derived by Miyazawa and Jernigan (ref. [18], table VI).

Sequences are designed by minimizing, for fixed amino acid concentration, the energy of the native conformation with respect to the amino acid sequence. Good-folder sequences are characterized by a large fap $\delta = E_c - E_n$ (compared to the standard deviation σ of the contact energies) between the energy of the sequence in the native conformation E_n, and the lowest energy (threshold energy) of the conformations structurally dissimilar to the native conformation [6, 7]. In other words, good folders are associated with an normalized gap $\xi = \delta/\sigma \gg 1$, quantity closely related to the z-score [20]. Furthermore,

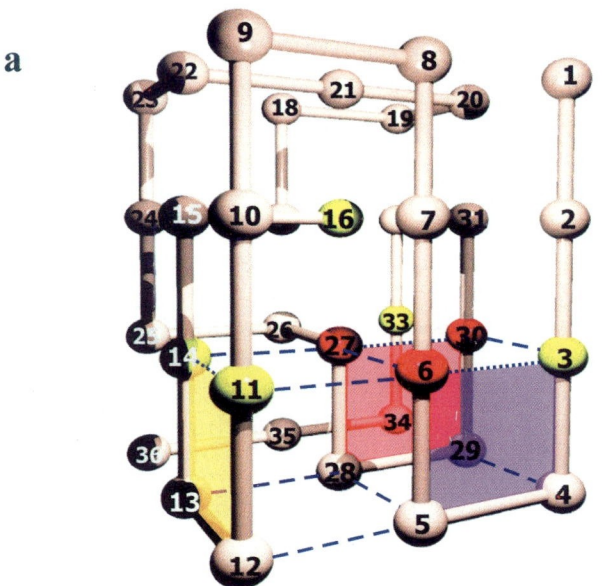

b SQKWLERGATRIADGDLPVNGTYFSCKIMENVHPLA

c YPDLTKWHAMEAGKIRFSVPDACLNGEGIRQVTLSN

Fig. 2. – (a) The conformation of the 36-mer chosen as the native state in the design procedure. Each amino-acid residue is represented as a bead occupying a lattice site. The design tends to place the most strongly interacting amino acids in the interior of the protein where they can form most contacts. The strongest interactions are between groups D, E and K (cf. (b) and fig. 1), the last one being buried deep in the protein (amino acid in site 27). Amino acids occupying "hot" sites (sites 6, 27, 30, cf. table I) have been represented by red beads, those occupying "warm" sites (sites 3, 5, 11, 14, 16 and 28) by yellow beads and those occupying cold sites by light brown beads. The local elementary structures (LES) formed by the amino acid sequences $S_4^1 \equiv (3, 4, 5, 6)$, $S_4^2 \equiv (27, 28, 29, 30)$ and $S_4^3 \equiv (11, 12, 13, 14)$ and stabilized by the contacts 3–6, 27–30 and 11–14 (drawn by dotted blue lines) are explicitly shown making use of light violet, light red and light yellow shades. The contacts between the LES are shown by dashed blue lines. (b) Designed amino acid sequence S_{36}. (c) Designed sequence S'_{36} corresponding to $E_n = -17.13$.

starting from a designed sequence which displays a large gap, all mutated sequences which preserve (to some extent) the gap fold into the native conformation [13].

2˙1. Role of the different amino acids in the folding process. – We shall now discuss some of the results of a Monte Carlo simulation study of the dynamics of a 36-monomer chain characterized by a polymer sequence, denoted S_{36} (cf. fig. 2(b)), designed by minimizing the energy in the target (native) conformation shown in fig. 2(a). In the units we are considering [18] ($RT_{room} = 0.6$ kcal/mol), the energy of S_{36} is $E_{nat} = -16.5$. While this is not the sequence of lowest energy, in particular the sequence displayed in fig. 2(c) has an energy in the native conformation of -17.13, S_{36} has a sufficiently low energy and a large value of ξ ($= 8.33$) so that it can encode for a "wild type" protein.

Monte Carlo simulations of folding performed on S_{36} at $T = 0.20$ (in our units) and using a standard algorithm described extensively in the literature [21, 22], in which, at each MC step, a monomer is picked up at random and end and crankshaft moves as well as corner flips are considered, indicate that this designed chain folds in a rather short "time" of $8 \cdot 10^6$ MC steps, and that at $T = 0.28$ the folding time is even shorter, $6.5 \cdot 10^5$ MC steps. The fractional population of the native state corresponding to these two temperatures is 91% and 10%, respectively, to be compared with a population of 0.5 and of 10^{-5} for the heteropolymer folding temperatures of $T = 0.25$ (temperature at which the probability for folding as well as for unfolding is 1/2) and $T = 0.40$ (temperature at which bonds break essentially as fast as they are formed due to thermal fluctuations), respectively. All the calculations discussed below were carried out (unless otherwise stated) at the temperature $T = 0.28$, optimal from the point of view of allowing for the accumulation of representative samples of the different simulations, and at the same time leading to a consistent population of the native conformation.

Mutations in the designed sequence are introduced by replacing a single monomer in S_{36} by a monomer of a different type. In our case, there exist 19 such possible substitutions for each monomer in S_{36}, implying $36 \times 19 = 684$ different sequences to study. Actually, a systematic behavior of folding can be deduced by analyzing a relatively small number of such altered chains. To show this, we started our analysis by choosing a monomer of S_{36} arbitrarily, say the 17th monomer, and studied the folding dynamics of the corresponding 19 altered chains. The MC simulations were performed up to a maximum number of $15 \cdot 10^6$ steps, averaging over 12 different starting random coil configurations for each altered sequence. By repeating the same procedure for other few selected sites along the original chain, it is possible to conclude that the behavior of altered chains can be generally classified into three categories:

1) chains which still fold to the native structure,
2) chains which fold to a unique compact structure, but different than the native one, and
3) chains which, although becoming compact, do not fold to a unique structure at all during simulations.

To characterize quantitatively the above three different behaviors, we find that the quantity ΔE_{loc} defined as the difference between the energies of the altered and the intact ("wild type") chain, both calculated in the native configuration, plays a key role. More precisely, such local energy difference, $\Delta E_{\mathrm{loc}}[m'(i) \to m(i)]$, for a mutation at site i, where the monomer $m(i)$ in S_{36} is replaced by a monomer $m'(i) \neq m(i)$ is given by

$$\text{(2)} \qquad \Delta E_{\mathrm{loc}}[m'(i) \longrightarrow m(i)] = \sum_{j \neq i} \left(U_{m'(i),m(j)} - U_{m(i),m(j)} \right) \Delta(|\mathbf{r}_i - \mathbf{r}_j|).$$

We have calculated all the 684 values of $\Delta E_{\mathrm{loc}}[m' \to m]$, and found that they fall in the range $0 \leq \Delta E_{\mathrm{loc}}[m' \to m] < 5.66$, a number to be compared with the energy gap $\delta = 2.5$ associated with S_{36} in the native conformation. We classify the impact of mutation by the ability of the mutated sequence to fold into or close to the native conformation. We define the degree of folding (similarity parameter) q [23] as the fraction of correctly formed contacts ($q = 1$ corresponds to the native state and $q \ll 1$ corresponds to misfolded states).

The following rules are obtained from the results of the calculations:

1) $\Delta E_{\mathrm{loc}}[m' \to m] < 1$: the altered chain always folds to the native structure ($q = 1$).

2) $1 < \Delta E_{\mathrm{loc}}[m' \to m] < 2.5$: the altered chain folds to a unique structure, sometimes different than the native one, with q being smaller than but close to one. In some cases, however, folding to the native structure may still occur ($q = 1$).

3) $2.5 < \Delta E_{\mathrm{loc}}[m' \to m] < 4$: twilight zone: for some mutations, chains fold into near native structure with $q \sim 1$, other mutations lead to misfolding with $q \ll 1$.

4) $\Delta E_{\mathrm{loc}}[m' \to m] > 4$: the altered chain does not fold to a unique structure at all during the simulation time, and now $q \ll 1$.

For a given site i, we can therefore classify the 19 possible mutations according to the rules (1) and (3), (4). For rule (2), additional information about the dynamical behavior of the chain is required. We find that 73.675% mutations fall into the first class $\Delta E_{\mathrm{loc}}[m' \to m] < 1$, 26.3% into the second class $1 < \Delta E_{\mathrm{loc}}[m' \to m] < 4$, and the rest 0.015% into the class (4). Thus, a relatively large fraction of mutations yield altered chains still folding into the native structure, some mutations lead to limited misfolding ($q \approx 1$) and only a small fraction of mutations leads to complete misfolding.

Mutations at a given site may yield values $\Delta E_{\mathrm{loc}}[m' \to m]$ which do not correspond to a single class, *i.e.* sometimes values smaller and larger than one occur at the same site. However, assuming that mutations are not selective, *i.e.* that they all occur with equal probability at a single site, an approximate scheme can be envisaged to classify the different sites according to the average magnitude of the damage (impact energy) caused by mutations. This is done by calculating the average value of $\Delta E_{\mathrm{loc}}[m' \to m]$ for each site i as

$$\text{(3)} \qquad \Delta \bar{E}_{\mathrm{loc}}(i) = \frac{1}{19} \sum_{m'} \Delta E_{\mathrm{loc}}[m'(i) \longrightarrow m(i)].$$

In this way, to each site i is associated a mean value $\Delta \bar{E}_{\mathrm{loc}}(i)$ and the following simple scheme emerges (remember that $\delta = 2.5$ for S_{36}):

TABLE I. – *The average values of $\Delta \bar{E}_{\text{loc}}$ for each site of S_{36} (sequence (B) in fig. 2). Bold numbers indicate the hot sites.*

Site	$\Delta \bar{E}_{\text{loc}}$	Site	$\Delta \bar{E}_{\text{loc}}$	Site	$\Delta \bar{E}_{\text{loc}}$	Site	$\Delta \bar{E}_{\text{loc}}$
1	0.24	10	0.64	19	0.38	28	1.31
2	0.73	11	1.46	20	0.63	29	0.95
3	1.94	12	0.50	21	0.65	30	**2.48**
4	0.75	13	0.55	22	0.40	31	0.99
5	1.09	14	1.85	23	0.28	32	0.78
6	**2.79**	15	0.88	24	0.77	33	0.68
7	0.77	16	1.79	25	0.44	34	0.42
8	0.27	17	0.80	26	0.54	35	0.66
9	0.09	18	0.38	27	**3.46**	36	0.26

1) when $\Delta \bar{E}_{\text{loc}}(i) < 1$, chains having mutations at site i are likely to belong to the first category, *i.e.* on average they fold to the native structure and we denote i as a "cold" site,

2) when $1 < \Delta \bar{E}_{\text{loc}}(i) < 2$ they behave on average as in the second category mentioned above, and i is denoted as a "warm" site. Finally,

3) when $\Delta \bar{E}_{\text{loc}}(i) > 2$ the resulting altered chains are likely to yield unfolded structures, and the site is denoted as a "hot" site.

We have classified the 36 monomers of S_{36} according to this scheme as shown schematically in fig. 2(a). We find that 27 sites can be considered as cold sites (light brown beads), 6 as warm sites (yellow beads) and only 3 (red beads) as hot ones (see table I). Thus, about 75% of the heteropolymer chain admits single error substitutions in the correct amino acid sequence yielding altered chains still folding to the native structure. Only in a relatively small fraction of the chain (about 10%, *i.e.* on the hot sites), mutations have catastrophic effects leading to complete misfolding. Additional simulations have confirmed the general trend predicted by this empirical scheme.

TABLE II. – *The number of mutated sequences S'_{36} which fold into the native conformation shown in fig 2(a). In column one the number of mutations m is shown. Columns 2 and 3 are associated with composition conserving results (c.), while columns 4 and 5 correspond to pointlike mutations (n.c.). Columns 2 and 4 display the number of sequences associated with a change in energy ΔE smaller than the gap δ, while columns 3 and 5 display the total number of sequences associated with the number of mutations m.*

m	$\Delta E < \delta$ (c.)	Tot. (c.)	$\Delta E < \delta$ (n.c.)	Tot. (n.c.)
1			613	684
2	447	630	$1.59 \cdot 10^5$	$2.27 \cdot 10^5$
3	6518	14280	$2.30 \cdot 10^7$	$4.89 \cdot 10^7$
4	$1.37 \cdot 10^5$	$5.30 \cdot 10^5$	$1.99 \cdot 10^9$	$7.68 \cdot 10^9$
5	$2.10 \cdot 10^6$	$1.66 \cdot 10^7$		
6	$2.53 \cdot 10^7$	$5.16 \cdot 10^8$		
7	$2.78 \cdot 10^8$	$1.55 \cdot 10^{10}$		

2˙2. *How many mutations can a designed protein tolerate?* – A complete enumeration of all sequences S'_{36} obtained by introducing mutations in S_{36} have been carried out up to seven mutations keeping fixed the amino acid composition of the chain (swappings), and up to four mutations without this constraint (pointlike). Carrying out Monte Carlo simulations of the dynamics of the mutated sequences with the same composition of S_{36} (all sequences S'_{36} with two and three swappings and 10^3 randomly chosen sequences among those obtained from S_{36} by carrying out up to seven swappings of amino acids) and $\Delta E < \delta$, it turns out that, in 98% of the cases, they can reach the native conformation in a time comparable to the folding time of S_{36}. Repeating the same analysis on 10^3 sequences with up to four pointlike mutations, we observed that only in 57 cases did the chain find conformations dissimilar from the native one, with lower energy, and it

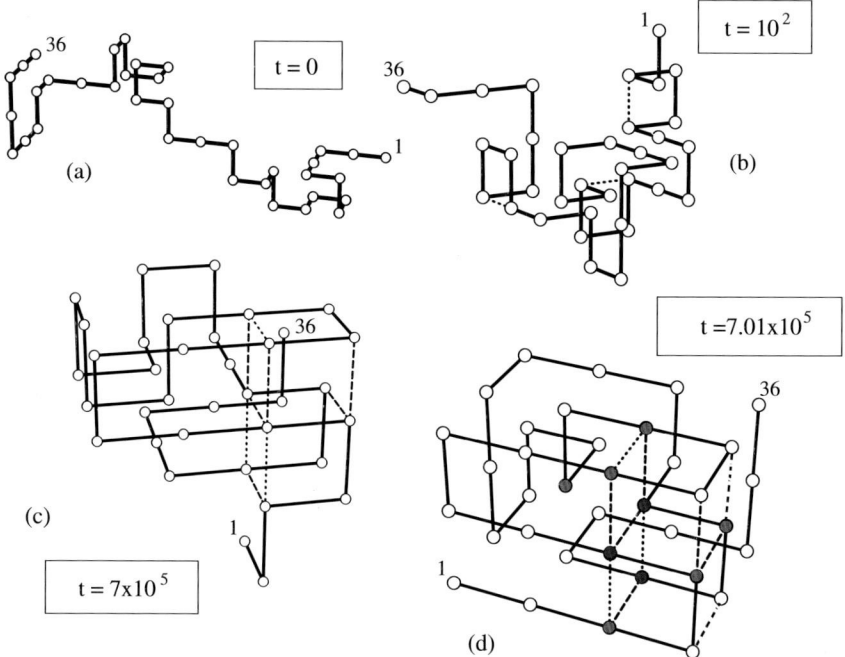

Fig. 3. – Snapshots of the folding of the sequence S'_{36} (cf. fig. 2(c)), whose energy in the native conformation is $E_n = -17.13$. Starting from a random conformation (a), the system forms after $\approx 10^2$ MC steps partially folded intermediates (b), involving three sets of four amino acids (3–6, 11–14, 17–30), whose stability is provided by the bonding indicated by dotted lines. When the partially folded intermediates come together to form the folding core (indicated by dotted and dashed lines) after $7 \cdot 10^5$ MC steps (c), the system folds to the native conformation after only 10^3 MC steps (d). The amino acids participating in the bonding of the partially folded intermediates (dotted lines) are among some of the most strongly interacting amino acids, which occupy, in the native conformation (d), "hot" and "warm" sites [3] indicated by dark- and light-gray beads, respectively (red and yellow beads in fig. 2). The monomers number 36 of the sequence S'_{36} are indicated for each conformation.

was not able to find the native conformation within the simulation time. The numbers shown in table II, in particular that there are $2.78 \cdot 10^8$ sequences, *i.e.*, 1.9% of all the sequences generated by seven composition conserving mutations which preserve (to any extent) the gap, provide a lower limit to the total number of sequences folding into the same structure. These results demonstrate that good folders are highly designable.

Summing up, the stability of a designed sequence to mutations, and thus the number of mutations which preserve folding to the same native structure, is controlled by the energy gap δ. Neutral mutations are those which lead to sequences which preserve (to any extent) the gap, a result which is tantamount to saying that the amino acids occupying the hot sites of the native conformation are conserved. This result sets the stage for the next section.

3. – How does a notional protein fold?

Let us study the time evolution of the different native contacts of the designed sequence. A surprisingly simple movie runs under our eyes [15]. Starting from a random coil (fig. 3(a)), few local elementary structures (LES) built out of the monomer sequences $S_4^1 \equiv (3, 4, 5, 6)$, $S_4^2 \equiv (27, 28, 29, 30)$ and $S_4^3 \equiv (11, 12, 13, 14)$ are formed only after $\approx 10^2$ steps of the MC simulation (fig. 3(b)). They are controlled by the local contacts 3–6, 27–30 and 11–14, involving some of the most strongly interacting amino acids, in any case all those occupying the hot sites and some of those occupying warm sites in the native conformation. These contacts achieve stability ($> 80\%$) after 10^5 MC steps and when they assemble together after $\approx 10^6$ MC steps (fig. 3(c)) they form the (post-critical) folding nucleus [24, 25] of the notional protein, from which it reaches the native conformation (fig. 3(d)) in a short time ($\approx 10^3$ MC steps), provided the energy of the system is lower than E_c, where it has not to compete with the bulk of misfolded conformations (cf. also fig. 4). The local elementary structures S_4^1, S_4^2 and S_4^3, and not the individual monomers, thus take care, through local guidance and non-local long range correlations([1]) (bonding between local elementary structures), of the process of protein folding([2]), and can be viewed as the local "bricks" of a dynamical LEGO kit to model protein folding.

([1]) Within the present scenario of protein folding the question of the relative importance of local- *versus* non-local contacts is somewhat trascended. In fact, they become complementary expressions of the same reality and, in the same way that nothing is gained in expressing a quantum phenomenon in terms of particles, that cannot equally well be described in terms of waves, the paradigm of non-local contact description of protein folding, the formation of the folding core, can be described in terms of local elementary structures, entities which epitomize the local-contact picture. Vice versa, the local elementary structures we have observed in the very last steps of the unfolding of notional proteins, can be described as arising from stabilizing bondings between individual, strongly interacting amino acids wich lie well protected inside the protein and form the folding nucleus of the native conformation (amino acids occupying "hot" and "warm" sites of the protein, cf. subsect. 2˙1).
([2]) This result is also supported by the disruptive effect mutations which affect the stability of these structures have on the folding ability of the design sequence.

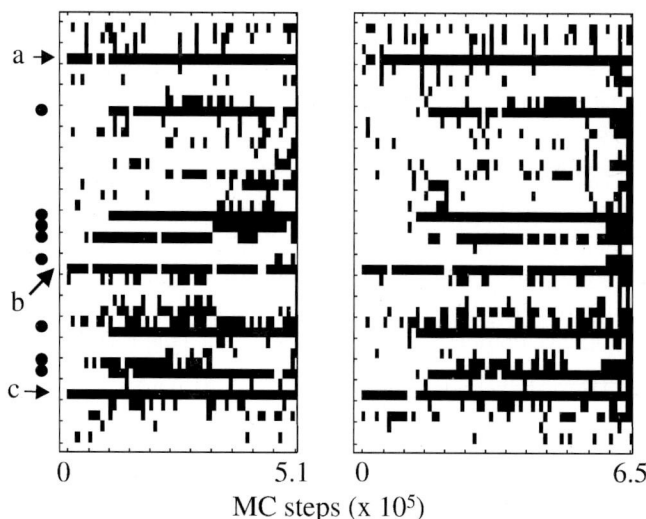

Fig. 4. – Dynamics of contact formation for two MC simulations of the folding of the S_{36} sequence. With a, b, c we have labeled the contact 27–30, 11–14 and 3–6, stabilizing the partially folded intermediates (cf. figs. 2 and 3). With a solid dot along the vertical axis we label (from top to bottom) the contacts: 5–28, 3–30, 14–27, 6–11, 13–28, 6–27, 12–5 and 4–29 forming, together with contacts 27–30, 11–14 and 3–6 (i.e., a, b, c) the folding core. In the simulation leading to the results displayed in the left panel, the protein folds in $5.1 \cdot 10^5$ MC steps, while in that associated with the right pannel it folds in $6.5 \cdot 10^5$ MC steps.

The fast formation of few local elementary structures, and of their bonding, reduces, in a conspicuous way, the number of conformations that need to be searched (in the case of the chain S_{36} to 10^{12} as compared to 10^{24} for the random-coil), leading to the resolution of the Levinthal paradox (cf. fig. 5). It is also a very efficient way to squeeze entropy out from the system ($\approx 50\%$) at the very early stages of the folding process,

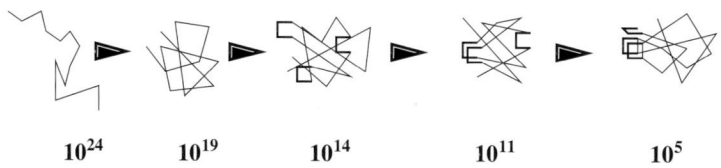

Fig. 5. – Schematic representation of the mechanism according to which the chain squeezes out its entropy. Starting from a random conformation (there are 10^{24} such conformations), it gets compact (10^{19} conformations) and the local elementary structures are formed (10^{14} conformations). The binding of the first two elementary structures reduces the number of conformations to 10^{11}, while the completion of the folding nucleus further decreases them to 10^5 (cf. also appendix B).

and to repeat this feat when the partially folded intermediates come together to form the folding nucleus (cf. also ref. [12]), at which stage the integrated decrease of entropy amounts to a large fraction of the original random-coil value (of the order of 80% in the case of the process shown in fig. 3).

To state that the ability a notional protein has to fold is connected with the presence of a small number of conserved contacts or, equivalently, of conserved amino acids [26], is tantamount to saying that foldability is connected with the presence of a small number of local elementary structures. In fact, although most of the conserved contacts found in the folding nucleus of ref. [26] are non-local, few of them are local. These few contacts stabilize the local elementary structures already at the initial stage of the folding process (cf. figs. 3 and 4). It is then natural that the non-local contacts of the folding nucleus arise from the assembling together of the local elementary structures. Because these local structures are both few and strongly interacting as they are mediated, by the few, strongly interacting amino acids occupying "hot" sites in the protein (cf. fig. 2), they can come together both fast ($\approx 10^5$–10^6 MC steps) and in a unique fashion, to form the (post-critical) folding core of the protein. Consequently, the findings displayed in fig. 3 agree in detail with the results of ref. [26] providing it a microscopic picture(^{3}).

We have found similar results from systematic studies of folding dynamics of representative members of essentially all the classes of lattice designed sequences available in the literature: 27 mers [28-31], 36 mers [11-13, 27, 32-34], 48 mers [24] and 80 mers [35]. Even the 36 mers sequence designed in ref. [36] under the constraint that the corresponding folding core of the protein contained as few local contacts as possible, folded through local elementary structures. Of course in this case, as well as in the case of longer chains, the local elementary structures are different from those found in the present analysis (cf., e.g., fig. 12(b)), although they are equally unique and stable. Consequently, the study which is at the basis of the results presented here [15] is deemed sufficiently comprehensive and systematic so as to allow us to conclude that the presence of local elementary structures in the folding process of designed sequences is a universal feature.

In order to investigate the dynamical behaviour of designed proteins with energy $E_n \leq E < E_c$ (that is sequences which can also be marginally stable), a database of (composition conserving) sequences of specified energy has been created making use of a Monte Carlo algorithm. The database is divided into 6 groups whose elements have energy

---

(^{3}) It also agrees with the findings of ref. [27] which, from an analysis of the folding of a 27 mer and a 36 mer concludes that there are multiple (transition states) folding nuclei. In fact, while the stability of the local elementary structures is not 100%, the corresponding contacts are operative with an incidence much higher than that associated with non-conserved native contacts (cf. fig. 4 as well as fig. 5 of ref. [27]). In keeping with the definition of transition states (which in the case of the 36 mer under discussion are $\approx 10^4$) as those in which the protein has equal probability to proceed to the native conformation as it has to unfold, not all the conserved contacts are, in these 10^4 states, operative (different transition states). On the other hand, any good folder passes, with probability 1, through the (post-critical) folding nucleus (fig. 3(c)) *en route* to the native structure.

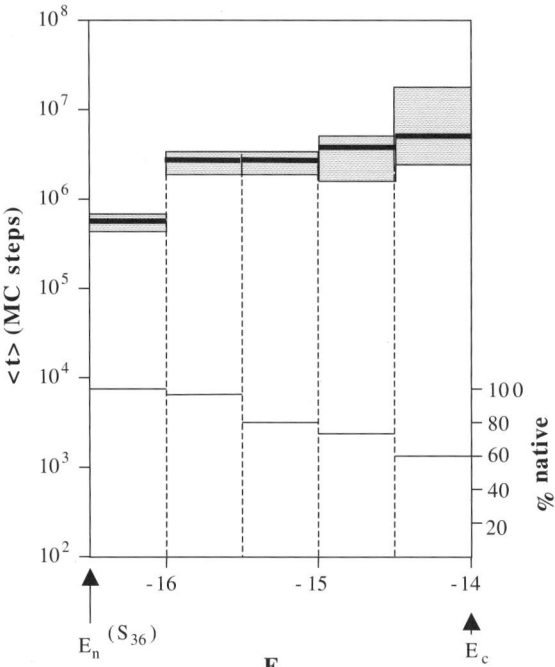

Fig. 6. – Average first passage time $\langle t \rangle$ needed to reach the ground-state conformation, associated with the different energy bins of the database groups (scale to the left). The associated standard deviations are reported (grey areas around the mean), as well as the percentage of sequences folding to the exact target structure (scale to the right). The energy of the sequence S_{36} in the native conformation ($E_n^{S_{36}}$) and of the threshold energy E_c are also indicated.

$-17.00 < E < -16.50$, $-16.50 < E < -15.00, \ldots, -14.50 < E < -14.00$, each group containing 500 sequences. For each group the Monte Carlo simulation has been performed at a temperature such that the average energy lies in the associated energy interval.

Folding simulations of the sequences of the database, performed at the same temperature ($T = 0.28$), show that essentially all sequences (92%) with $E < E_c$ fold in rather homogeneous times (cf. fig. 6), a time which is much shorter than that associated with a random search in the space of compact conformations ($\sim 10^{12}$), let alone the full space of conformations ($\sim 10^{24}$, cf. fig. 5). The sequences fold either to the native structure or to a unique structure with a value of the similarity parameter $q > 0.6$. This process takes always place through partially folded intermediates, a result which seems to find strong experimental support (cf., *e.g.*, [37] and references therein). Note that for a given native structure, all designed sequences are characterized by a very limited choice of local elementary structures [14]. For example, in the case of the native conformation chosen for the analysis (fig. 2(a)), these local substructures involve monomers 3–6, 11–14 and 27–30 [12] (the only other choice for partially folded intermediates involves monomers can take place in a number of different ways, and not only in the ones which mimic the disposition of the "bricks" in the native core configuration, in keeping with the LEGO analogy. Because of the strongly interacting character of the amino acids occupying sites 27, 30 and 6, aggregation is, for all purposes, an irreversible process under native-like conditions, as testified by the results of simulations leading to aggregation which have been followed over 10^8 MC steps.

To check whether the local structures are an artifact or not of the conspicuous dispersion displayed by the contact energies used in the calculations (cf. fig. 1), we have repeated the simulations making use of the Go model [44-47]. We have found that the presence of the local elementary structures is, if anything, better defined in this case as compared to the case discussed previously, and that their role in the aggregation process is again essential. In particular, due to the fact that now all these local structures have equal energy content, in most of the events leading to aggregation, all three local structures of one chain find a local structure partner belonging to the other chain with which they interact. We have also repeated the calculations making use of the contact energies $U_{m(i),m'(j)}$ (cf. eq. (1)) reported in table V of ref. [18], and obtained results very similar to the ones discussed above.

Making again use of the contact energies of ref. [18] (table VI), we have found that the rate of aggregation can increase in a significant manner, by introducing "cold" (neutral) mutations, if the chosen mutations are able to affect in a significant way the stability of one of the local structures, without much changing the ability the resulting isolated sequence S'_{36} have to fold on short call to the native conformation. In particular, substituting the amino acid R at position 11 of the designed sequence, by amino acid A (cf. figs. 2(a) and 2(b)), the rate with which aggregation takes place increases by 70% (i.e. from 22% to a 37% at a distance $d = 4$, where d represents the initial distance, in units of lattice spacing, between monomers number 18 of each of the chains). The reason for this increase lies in the fact that it takes 0.6×10^6 MC steps for the pair of monomers 11–14 of the mutated sequence S'_{36} to establish a stable contact (as compared to 0.25×10^6 MC steps for S_{36}). Consequently, the other two local structures (associated with the monomer groups S^1_4 and S^2_4) have more time and thus a better chance to interact with the homologous structures of the other chain, than in the case of the simultaneous folding of two S_{36} sequences. Similar results have been obtained by performing single and multiple mutations in other "cold" and "warm" sites of the native conformation. Because 75% of all sites are "cold" sites, and thus associated with neutral mutations [11], there is a large number of mutations which, while destabilizing the elementary structures, and thus increasing the rate of aggregation, do not affect in an important way the stability of the protein. These results are consistent with a number of observations, in particular those carried out in the study of the amyloid-forming system, transthyretin. When altered by any of 50 different mutations, this protein, which normally occurs in the blood plasma, deposits in the heart, lungs and gut, causing a lethal disease called familial amyloidotic polyneuropathy (FAP) [48,49]. These mutations do not alter normal folding of the protein but do destabilize the protein structure, facilitating the formation of partially folded intermediates that readily aggregate to one another [50, 51].

From the discussion carried out in this section it emerges that a given protein will have a (small) number of local elementary structures (which in some cases may also show as local partially folded intermediates) which controls both protein folding and aggregation. Within the model of designed proteins these are the elementary structures which build the folding nucleus. Consequently, most of the aggregates of this protein as well as of the sequences homologous to it will display similar native-like structures, independent of the nature of the effect triggering the aggregation.

3˙2. *Statistical analysis of contact formation.* – Local elementary structures are built of residues which are very close along the chain (typically 3–5 beads distant). Consequently, they can find their native conformation in a very short time (small number of MC steps), since they do not need to explore a large conformational space. For example, an elementary structure built out of four monomers has only 9 non-native conformations, so it can find its native state through a random search in ≈ 10 elementary steps. Furthermore, local elementary structures contain some of the most strongly interacting amino acids, resulting in rather stable structures.

Elementary structures are then small parts of the chain with a high probability of being locally in their native state at any time of the dynamics. As already discussed, elementary structures build the folding nucleus when they assemble together and, once this is done, the chain finds its native conformation almost immediately, due to the reduced size of the conformational space (fig. 4). Why does the protein need elementary structures to build the nucleus fast or, in other words, why is the nucleus not composed by single isolated residues, independent of each other? This is because elementary structures enhance the formation of the folding nucleus in a variety of ways. First of all, they effectively reduce the length of the chain. Being remarkably stable, they behave as single entities and freeze some degrees of freedom. For example, in the case of the sequence S_{36} which has 3 elementary structures each built out of 4 monomers, the chain becomes built out of 27 elementary blocks. Accordingly, the number of conformations available to the chain are reduced from γ^{35} to γ^{26} (if $\gamma = 3.45$ cf. appendix B, this means from 10^{19} to 10^{14}).

Another property of the local elementary structures is that each of them interacts with the rest of the chain as a whole. If, *e.g.*, two sets of four residues interact strongly (*e.g.*, through the largest, most attractive, contact energies([6]) $U_{\min}$.) but independently of each other, the stability of the resulting composite structure is proportional to $\exp[-U_{\min}/T]$. On the other hand, if the four monomers interact cooperatively, the interaction energies sum together, and the stability is proportional to $\exp[-4U_{\min}/T]$. Furthermore, this very fact also makes it more unlikely for an elementary structure to match a non-native structure and build strong attractive contacts with it than for a single amino acid to establish a non-local contact. Note that in some cases elementary structures are "degenerate", in the sense that they do not have internal interactions like, for example, the case

---

([6]) Note that $U_{\min} \approx -3\sigma \approx -3T$.

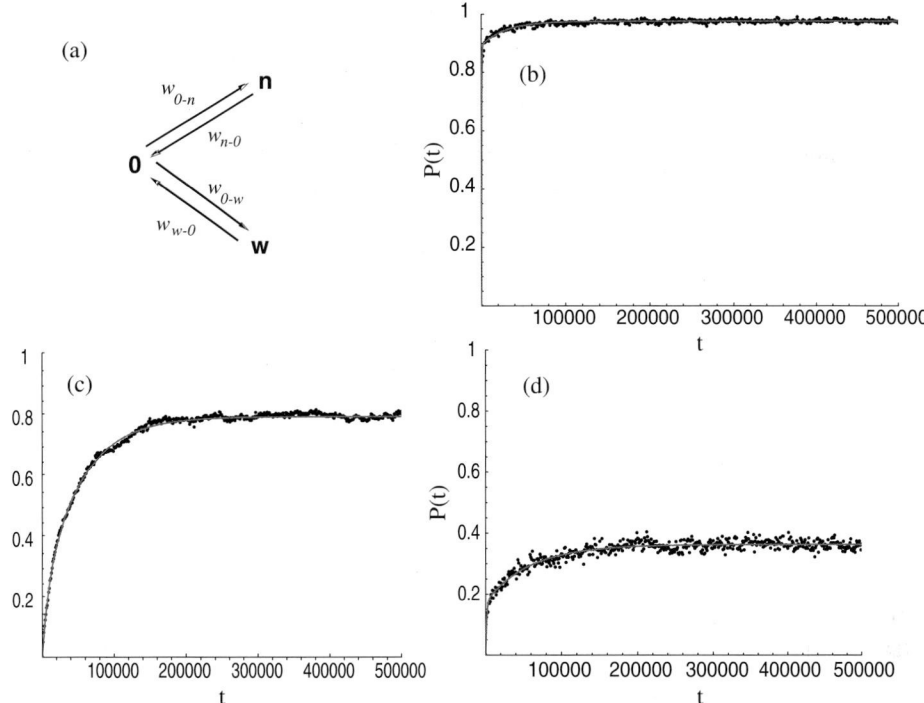

Fig. 9. – (a) Each native contact can be thought to be in its native state ("n"), in an unbound state ("0") or "wrongly" bound to some non-native monomer ("w") (cf. eq. (5)). (b) The probability $P_{i-j}(t)$ for the bond 3-6 (cf. eq. (4)) associated with the kinetic evolution of the sequence S'_{36}. To the curve obtained by simulations is superimposed the least-squares fit done making use of eq. (6). (c) The same for bond 5-28. (d) The same for bond 15-24.

of two consecutive residues (cf. fig. 12(b)). In this case, the elementary structure is kept together only by the polymeric bond.

A statistical characterization of bonds can be achieved studying the quantity

(4) $$P_{i-j}(t) = \langle \Delta(|r_i - r_j|) \rangle,$$

which is the probability that the bond between the i-th and the j-th monomer is formed as a function of time, averaged over independent dynamical runs at fixed temperature. The plots of the function associated with a bond within elementary structures (elementary structure S_4^1, i.e. bond 3-6), a bond among elementary structures (S_4^1 and S_4^2, i.e. bond 5-28) and a bond between two monomers not belonging to any of the elementary structures S_4^1, S_4^2 or S_4^3 (15-24) are displayed in fig. 9(b), (c) and (d), respectively. In keeping with the fact that the three bonds differ substantially both concerning the formation times and their stability, it is possible to give a precise characterization of them with the help of a simple model, in which it is assumed that each bond $i - j$ can be found in one of three possible states (cf. fig. 9(a)): native state (which we will denote "N"), unbound

("0") or bound to some non-native residues ("W"). We also assume that the system can transit the states "0" and "N", "0" and "W", but not the states "N" and "W".

According to this model, it is possible to write, for each bond, master equations:

(5)
$$\dot{P}_n(t) = w_{0 \to n} P_0(t) - w_{n \to 0} P_n(t),$$
$$\dot{P}_w(t) = w_{0 \to w} P_0(t) - w_{w \to 0} P_w(t),$$
$$P_0(t) = 1 - P_n(t) - P_w(t),$$

where $P_n(t)$, $P_w(t)$ and $P_0(t)$ are the probabilities that the bond of the chain S_{36} under discussion is in its native state, in a non-native state or in a unbound state, respectively. The initial condition for the folding calculations is that bonds start in an unbound state, so that $P_0(0) = 1$. Assuming the transition rates $w_{0 \to n}$, $w_{n \to 0}$, $w_{0 \to w}$ and $w_{w \to 0}$ to be constant, the probability that a native bond is formed is given by

(6)
$$P_n(t) = \alpha(1 - \exp[-\lambda t]) + \beta(1 - \exp[-\mu t]),$$

with the coefficients reported in appendix A. For every native bond of the sequence S_{36}, a least-square fit to the results of the MC simulation has been carried out making use of the above expression. From the resulting values of the parameters λ, μ, α and β the associate transition probabilities were derived (cf. appendix A).

The bonds within elementary structures (cf., e.g., fig. 9(b)) are formed remarkably fast (in a time of the order of $(\alpha\lambda)^{-1} + (\beta\mu)^{-1} \sim 10$ steps). The bonding free energy ΔF can be found from $w_{n \to 0} = w_0 \exp[\Delta F/T]$, w_0 being the inverse of the time step. Since at each MC step (before folding) every monomer has the same probability of being the one which is selected for the move, it is reasonable to choose $w_0 = 2/N$, N being the length of the chain, the factor 2 taking into account the fact that it is possible to create or to break a bond by moving one of the two monomers which form it. From the fitting analysis one finds that the bonding formation free energy is $\Delta F = -1.05$ for bond 3–6 (similar values of ΔF are found for the bonds associated with the other elementray structures S_4^2 (27–30) and S_4^3 (11–14)). The bonding free energy is $\Delta F = F_n - F_0 = E_n - E_0 - TS_n + TS_0$, with $E_n - E_0 = -0.97$ (i.e. for bond 3–6, and similar values for the other fast bonds) and $S_n = 0$ by definition, leading to $S_0 \approx 0$. Consequently, there is in average only one non-bound state available to the system, due to the excluded-volume constrain (note that, in the present simulations, the chain starts from a random generated conformation, which is usually rather compact). If the chain was non-compact, there would be 9 possible unbound states for any contact separated by four monomers, in particular for the contact 3–6, leading to an entropy equal to $S_0 = 2.19$ and, consequently, to an overall stability of $\alpha + \beta = 0.35$ (instead of ≈ 1). According to the present model, a collapse of the chain to a compact conformation is necessary before the fastest native bonds are stabilized([7]).

([7]) This seems not to be true if the interaction matrix is changed to a new one whose elements are, in average, repulsive.

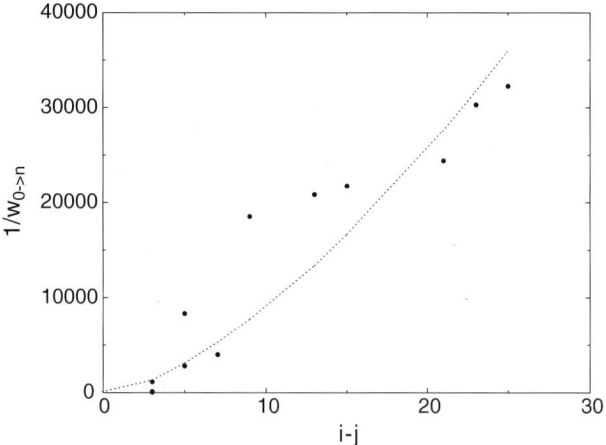

Fig. 10. – The inverse of the parameter $w_{0\to n}$ associated with the kinetic evolution of the sequence S'_{36} (cf. fig. 2(c)) is plotted with respect to the distance $|i-j|$ of the monomers involved in the bond, measured along the chain (circles), compared to the probability that the two monomers come close during a random search (dotted line).

The free energy associated with bonds among elementary structures ranges from -1.57 to -2.43. The bond $28-5$ between S_4^1 and S_4^2, for example, has $\Delta F = -2.61$ and the monomers interact with a contact energy -0.41. Assuming again that the native state is unique, this bond should satisfy the relation

$$-0.41 + TS_0 = -2.61. \tag{7}$$

This is not possible, being the entropy of the unbound state S_0 a positive quantity. On the other hand, we can consider that the monomers 3, 4, 5, 6 (stabilized by the contact 3–6) form a rigid plaquette (S_4^1) which interact, as a whole, with a plaquette (S_4^2) built out of the monomers 27, 28, 29, 30 (the bond 27–30 being also very stable). In this case, the total plaquette (LES) interaction energy is the sum of the energies of the four bonds involved, that is $E = -3.02$. The energy balance becomes now

$$-3.02 + TS_0 = -2.61, \tag{8}$$

a relation which is satisfied if $S_0 = 1.46$, implying the existence of approximately 4 non-bound states for monomers 28 and 5. The same argument applies to the other elementary structure (S_4^3), built out of monomers 11, 12, 13, 14 and stabilized by the contact 11–14.

An interesting fact is that the transition rate $w_{0\to n}$ is, in average, determined by the distance $j-i$ along the chain of the two monomers involved in the bond. The results shown in fig. 10 display the behaviour of the inverse rate with respect to the distance $j-i$ (solid dots) and the inverse of the probability that the two monomers are in contact, calculated just as the number of conformation which display the bond $j - i$ over the total number of conformations (dotted line). Such probability is simply the number of conformations $\gamma^{N-2}(j-i)^{-1.68}$, γ being the effective coordination number of the cubic lattice [52], in which the i-th monomer is nearest neigbour of the j-th monomer divided by the total number of conformations γ^{N-1} of the chain. The above expression for the number of conformations with a bond between the i-th and j-th monomer is given by the product of terms associated with the three parts which compose each conformation, namely two branches of length i and $N - j + 1$, respectively, and a loop of length $j - i$. The first two terms give γ^{i-1} and γ^{N-j}, respectively. We have calculated the term associated to the number of loops through a fit of the number of conformations obtained by a complete enumeration up to([8]) $j - i = 16$ [54].

4. – Predicting the 3D structure of a notional protein from its amino acid sequence

With the help of the results discussed above, we have developed a strategy which allows to predict the three-dimensional native conformation of a model protein from its amino acid sequence [16], provided the contact energies acting among the amino acids is known. The algorithm consists of three steps, namely 1) finding good candidates for the role of local elementary structures, 2) finding the folding nucleus and 3) finding the native conformation relaxing the residues not participating in the folding nucleus. This algorithm is based on the hierarchical sequence of events that allows the chain to fold fast and works because at each step only a limited portion of the configurational space has to be searched through.

In what follows we discuss in detail the 1D → 3D algorithm and apply it to representatives examples of notional proteins.

- *Step 1:* finding the local elementary structures (LES) which lead the process of protein folding. Elementary structures can be closed or open, depending on whether they contain interactions within themselves (outside for the peptidic bond), or not. Examples of closed elementary structures are provided by S_4^1, S_4^2 and S_4^3 (cf. fig. 2(a)) while examples of open elementary structures are shown in fig. 12(b). In keeping with this classification of LES, the present step is composed of two substeps.

  Substep 1a: Finding the open elementary structures. For each substring of the sequence, starting at monomer i and ending at monomer j ($0 < i < j < N$), we define the density of energy

  $$(9) \qquad \epsilon_s = \frac{1}{j-i} \sum_{i \leq l \leq j} \min_{k \notin (i,j)} U_{m(l)m(k)},$$

([8]) Consequently, this expression fails for longer loops, in which case one can use the standard analytical expressions available in the literature (cf., *e.g.*, ref. [53]).

where U is the matrix of contact energies used to design the notional protein. In other words, ϵ_s is the average energy with which each element of the substring (i,j) interacts with the rest of the chain. The substrings which are good candidates to be open elementary structures in the folding process have low values of ϵ_s. Among such substrings we select those with values of ϵ_s lower than a threshold ϵ_s^*.

Substep 1b: Finding the closed elementary substructures. For this purpose we evaluate, for each pair of monomers i and j, the function

$$p(i,j) = \frac{\exp\left[-U_{m(i)m(j)}/T_{\text{eff}}\right]}{(j-i)^\rho}, \tag{10}$$

where T_{eff} is an effective temperature which we set equal to the standard deviation of the interaction matrix U (e.g., $\sigma = 0.3$ for the case of the contact matrix displayed in fig. 1). The exponential factor $\rho = 1.7$ reflects the ratio between the number of conformations associated with the formation of a contact and the total number of conformations (cf. subsect. 3`2 and appendix B). If a substructure contains more than one interaction, the values of p associated with the different interactions are to be multiplied together. As possible (closed) local elementary structures, we select those composed of mononomers $i, i+1, \ldots, j-1, j$ and with $p(i,j) > p^*$, where p^* is a threshold value (see below).

- *Step 2:* Finding the folding nucleus. All the elementary structures (let S be the total number of such structures) found in steps 1a and 1b are moved in space and the conformational spectrum is found. This is done selecting all possible choices of $1, 2, \ldots, S$ local elementary structures, giving them all possible relative conformations and making a complete enumeration of their reciprocal positions in space. The conformations with lowest energy are selected as possible candidates for the (post-critical) folding nucleus of the protein.

- *Step 3:* Relaxing the remaining monomers around the folding core. This can be done through a complete enumeration of all the conformations displaying a given nucleus, as they are only few ($\sim 10^4$ for a 36 mer). Another way, which we found computationally attractive is to use a low-temperature Monte Carlo relaxation simulations, keeping fixed the monomers belonging to the folding core([9]). The (single) totally relaxed conformation with energy lower than E_c is the native conformation of the protein.

([9]) In some cases the system is non-ergodic, in the sense that from a given starting configuration it is not possible to reach all other configurations (with the folding core formed and fixed). In such cases several relaxation simulations are performed starting from different conformations (with the folding core formed and fixed). In keeping with this fact, the folding nucleus of a notional protein could be required not to be exceedingly stable, so as to avoid long-lived metastable states *en route* to folding.

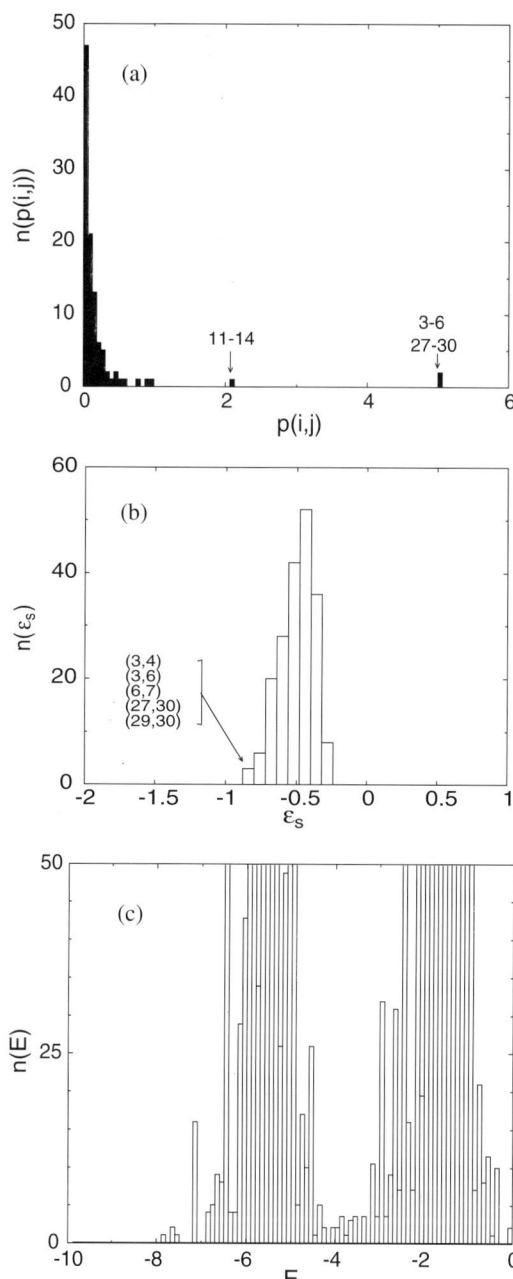

Fig. 11. – (a) The distribution of the parameter $p(i,j)$ (cf. eq. (10)), whose maximization allows to find the closed elementary structures. (b) The distribution of the energy density ϵ_s (cf. eq. (9)), employed to find open elementary structures. (c) The distribution of the energies associated with the possible folding nuclei of sequence S'_{36}, built out of the elementary structures 3–4–5–6, 11–12–13–14 and 27–28–29–30.

We have tested the algorithm described above on a number of sequences. Below we discuss results concerning the representative sequence S_{36} and on a sequence (cf. fig. 12(a)) optimized on a conformation poor of local contacts [36], which therefore provides an important test for the algorithm.

Starting with the sequence S_{36}, we display in fig. 11(a) the corresponding distribution of values of $p(i,j)$. Three bonds have a p-value which is remarkably larger than that associated with the rest of the possible bonds of the protein, and consequently are good candidates for stabilizing closed local elementary structures. The distribution of values of ϵ_s, displayed in fig. 11(b), shows a single peak, whose lowest points are associated with the same sites already involved in the closed elementary structures. It is thus likely that open elementary structures do not play any noticeable role in the folding process of S_{36}. We thus search for a folding nucleus composed of monomers $(3, 4, 5, 6)$, $(11, 12, 13, 14)$ and $(27, 28, 29, 30)$, and stabilized by the contacts 3–6, 11–14 and 27–30. A complete enumeration of all the conformations built out of these three elementary substructures gives the energy distribution displayed in fig. 11(c). The most stable of these conformations has energy -7.81 and is, in fact, the actual folding core. The relaxation of the other amino acids around it gives the right native conformation, with energy $E_n = -16.50$. The next low-energy conformations built out of the three elementary substructures have energy -7.75, -7.68 and -7.68. The relaxation of the other residues around these temptative folding nuclei lead to "native" energies -12.40, -12.58 and -14.05, respectively. The first two of them are larger than $E_c = -14.0$, so they correspond to states which belong to the set of structurally dissimilar conformations ($q \ll 1$) to the native conformation we are searching. The last of them has an energy just below E_c. Although it can hardly be confused with the native conformation, it corresponds to a metastable state which can slow down the folding process (cf. ref. [55]).

We now apply the 1D $\rightarrow$ 3D algorithm to the 36 mer designed sequence displayed in fig. 12(a), a sequence which has been optimized on a conformation characterized by very few local contacts [36]. Since we posit that local contacts are involved in the first level of the hierarchy of events that leads to the folded state, this example is important to assess the validity of our prediction scheme([10]). Following steps 1 and 2 above, the local elementary structures $(1, 2, 3, 4, 5, 6)$, $(20, 21, 22)$, and $(30, 31)$ are selected (cf. fig. 12(b)), the first of them being closed and the other two open. The lowest-energy folding nucleus built with these elementary structures (with energy -6.58) is shown in fig. 12(c). The relaxation of the remaining amino acids around this nucleus leads to the conformation shown in fig. 12(d). Having an energy $E_n = -15.99$, smaller than $E_c = -14.0$, it is the native conformation of the system.

4'1. *Caveats and limitations.* – To determine E_c in the calculations discussed above use was made of the Random Energy Model [56]. In this model the critical energy $E_c =$

([10]) Note that a strategy of complete non-locality in the design of notional proteins is doomed to be frustrated, in a similar way in which if the contact among amino acids is heterogeneous, no native conformation can have only favourable bonds.

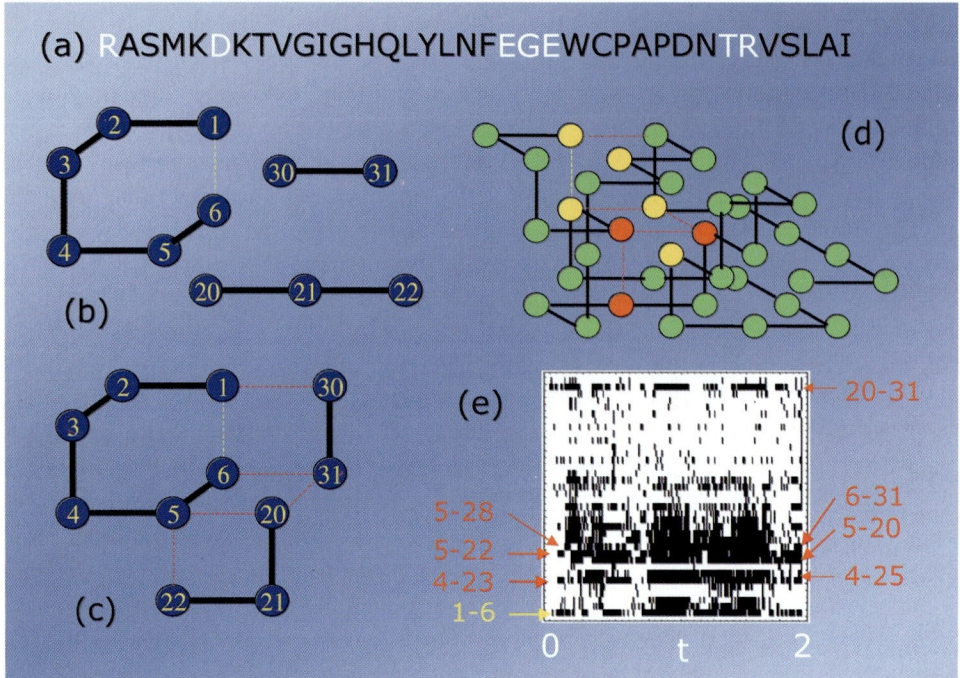

Fig. 12. – Prediction scheme for the sequence (a). In (b) the resulting local elementary structures are shown, two of which are open. All the amino acids participating in them are displayed in (a) in terms of white simbols. The same colour is used for the amino acid participating in the (internal) contacts of the closed structure. In (c) we show the disposition of these structure to form the (unique) folding core with which is associated the (single) completely compact conformation (d) with energy lower than E_c. This conformation coincides with the one used in the literature [36] to design the sequence shown in (a), the associated (native) energy being $E_n = -15.99$. Furthermore, Monte Carlo simulations testify the correctness of the predicted native conformation (d). This can be seen in (e), where the time evolution of the native contacts for a particular run is displayed as a function of the MC steps of the folding simulations. The sequence in (a) folds into the structure shown in (d) in approximately $2 \cdot 10^6$ MC steps. The normalized gap associated with sequence (a) is $\xi = 6.3$. The hot, warm and cold sites of the protein in its native conformation calculated according to [11] are displayed in terms of red, yellow and green beads, respectively.

$N_c\sigma(2\log\gamma)^{1/2}$ depends on the number N_c of contacts of fully compact conformations (40 and 56 for chains with $N = 36$ and $N = 48$, respectively), on the number of such conformations per monomer ($\gamma = 1.8$ [52], $\gamma = 2.2$ [57]) and on the variance σ of the contact energies (equal to 0.3 for the parameters suggested in table VI of ref. [18]). Using the average value $\gamma = 2$, one obtains $E_c = -14.1$ and $E_c = -19.8$ for $N = 36$ and $N = 48$, respectively, to be compared with the values of -14.0 and -21.5 calculated *a posteriori* making use of low-temperature Monte Carlo simulations. This uncertainty on E_c is of no consequence for the workings of the method in dealing with sequences with $\xi \gg 1$, but can limit its predictive power for sequences whose native energy is close to E_c.

4'2. *Hindsight*. – The final proof of the correctness of the hierarchy of events employed above, sequences of frames of a film obtained by analyzing the results of the inverse folding model, lies ultimately on the fact that by running it backwards we are able to solve the folding problem, that is, to read the three-dimensional structure of a protein from its amino acid sequence.

5. – Conclusions and perspectives

The studies discussed above allow for a number of conclusions which we shall state below and open new perspectives upon which we shall also try to elaborate. Although the most exciting new possibilities are connected with the study of real proteins making use of the arsenal of techniques and of the insight obtained from lattice models, we start by discussing some of the open questions lattice model simulations may help answer.

5'1. *Lattice model studies*. – Local elementary structures play an important role both in the folding as well as in the aggregation of proteins. Making use of these results, simple lattice model simulations may help to clarify the importance of the different mechanisms which are the basis of the folding of homo- and hetero-polymers [58, 59].

In the study of the folding of designed protein with native energies close to E_c we have found a number of sequences which fold both to the native and to another (unique) conformation. An interesting question, aside from that concerning the mechanism at the basis of this prion-like behaviour, is the effect of the presence of this type of sequences have in the folding of sequences associated with large values of the energy gap.

5'2. *Generalization of the* 1D → 3D *algorithm to real proteins*. – Designed lattice model sequences which in the native conformation display a large energy gap with respect to the lowest energy E_c of the structural dissimilar conformations measured in terms of the variance of the contact energies are known to be good folders. Designed sequences which preserve (to any extent) the energy gap share a common set of (conserved) contacts, the (post-critical) folding nucleus, in keeping with the fact that this is the minimum set of native contacts which ensure folding. Mutations on lattice model designed proteins displaying in the native conformation a large energy gap reveal the presence, in the native conformation, of "hot", "warm" and "cold" sites. Mutations, also multiple mutations, on the many cold sites are neutral, while mutations on the (few) hot sites completely denaturize the protein, the impact of mutations in the (also few) warm sites being intermediate between these two scenarios. Hot and warm site amino acids coincide with the conserved amino acids found in the folding nucleus, being essentially common to all the designed sequences which conserve (to any extent) the energy gap.

These results, found making use of lattice model Monte Carlo simulations within the framework of the inverse folding model strategy in which the global minimum (native state) is known, and the protein is designed by an annealing process in sequence space, provide a simple picture of the folding of a protein: a) formation of local elementary structures at the very early stages of the folding process, built out of the conserved amino acids (and/or conserved local contacts) participating in the folding nucleus, b) formation of the (post-critical) folding nucleus out of the (non-local) contacts among the local elementary structures, c) relaxation of the remaining amino acids into the native conformation.

Mimicking these (dynamical) steps within the framework of a static (thermodynamic) calculation to determine possible candidates of local elementary structures and of the folding nucleus, and enumerating the relatively few conformations obtained by relaxing the remaining amino acids keeping the chosen putative folding nucleus fixed, it is possible to read the three-dimensional structure of the lattice designed protein from its amino acid sequence, provided the contact energies used in the original sequence design (inverse folding model) are known. In fact, the (single) conformation which displays an energy smaller than E_c is the native conformation, in keeping with the fact that all the sequences (of the order of 10^{10} or more) obtained from the designed sequence in the inverse folding model through neutral mutations which conserve (to any extent) the energy gap, fold through exactly the same steps as the designed sequence. The fact that there can be degenerate sequences, *i.e.* mutated sequences which in the native conformation have the same energy, does not invalidate the argument.

In keeping with the above results, and to the extent that lattice model sequences designed within the framework of the inverse folding model do capture some of the essential features of the folding of small, single-domain, real proteins, the protein folding problem is essentially reduced at knowing the potential acting among the (very few) strongly interacting (hot) amino acids of the protein, amino acids which can either be known from protein engineering or theoretically determined with the help of empirical potentials.

Because the folding simulations and sequence design which are at the basis of the 1D → 3D lattice model strategy discussed above (sect. **4**) were carried out using the same set of potentials, the results were essentially independent of the choice of potential. While we expect that some predictions on real proteins may not be very sensitive to the extent with which the calculated potential agrees with the potential used by nature in the evolutionary design of real proteins, the question of how accurately the interaction among the amino acids participating in the local elementary structures has to be known is an open question.

We can envisage a number of scenarios, the most optimistic being one in which a description like that provided by CHARMM [60] will suffice to successfully generalize steps 1–3 (cf. sect. **4**) to the case of real proteins. Such a generalization will thus be simple to accomplish. The worst scenario, short of that in which lattice model predictions cannot be generalized to real proteins, is one in which an accurate description of the potential between the strongly interacting amino acids surrounded by thousands of water molecules is needed([11]). Even in this case the generalization of the 1D → 3D algorithm of sect. **4** to real proteins may be, if not immediately forthcoming, within reach by using

---

([11]) Because we are interested primarily in predicting the native structure of the protein from its amino acid sequence, and not in describing the dynamical evolution of the system, it is not obvious whether for that purpose the role of the solvent has to be explicitly taken into account or not.

a hybrid model [17] which combines a quantum-mechanical description of the solute with a classical molecular dynamical approach for the solvent (QM/MM). The quantum-mechanical fragment of a simulation system will be treated *ab initio* (LDA [61, 62] and TDLDA (see [63, 64] and references therein)). Long-range Coulomb interactions within the molecular mechanics fragment and between the quantum-mechanical and molecular mechanic fragments will be treated by a computationally efficient fast multipole method (like, *e.g.*, FMUSAMM) [17]. For the description of covalent bonds between the two fragments, use will be made of the scaled position link atom method (SPLAM). The Van der Waals interaction can be dealt with either making use of the recently developed generalizations of *ab initio* LDA [65], or through first-principle calculation of the linear response function of the interacting system making use of TDLDA (cf., *e.g.*, [63, 64, 66] and references therein).

Be as it may, the solution of the protein folding problem within lattice models reported in ref. [16] is likely to prove momentous in the quest of the 3D structure of a real protein from its 1D structure, allowing to combine the deep empirical insight developed during the last years in the study of protein folding (cf. in particular the lectures of W. Eaton, M. Oliveberg and E. Shakhnovich in this volume as well as [4] and refs. therein), with the powerfull theoretical techniques worked out in the study of infinite [67-71] and finite [72, 73] many-body systems and of their interaction with external fields [63, 64], theoretical tecniques which to a large extent find their origin in Feynman's formulation of quantum mechanics in general [74] and quantum electrodynamics in particular [75, 76].

Appendix A.

In this appendix we discuss the solution of eq. (5), which after some algebra one can find that it is given by the function

$$P_n(t) = \alpha(1 - \exp[-\lambda t]) + \beta(1 - \exp[-\mu t]), \tag{11}$$

where we have set

$$\lambda = \frac{1}{2}(w_{0 \to n} + w_{n \to 0} + w_{0 \to w} + w_{w \to 0} + \\
+ ((w_{0 \to n} + w_{n \to 0} + w_{0 \to w} + w_{w \to 0})^2 - \\
- 4(w_{0 \to n} w_{w \to 0} + w_{n \to 0} w_{w \to 0} + w_{n \to 0} w_{0 \to w}))^{1/2}), \tag{12}$$

$$\mu = \frac{1}{2}(w_{0 \to n} + w_{n \to 0} + w_{0 \to w} + w_{w \to 0} - \\
- ((w_{0 \to n} + w_{n \to 0} + w_{0 \to w} + w_{w \to 0})^2 - \\
- 4(w_{0 \to n} w_{w \to 0} + w_{n \to 0} w_{w \to 0} + w_{n \to 0} w_{0 \to w}))^{1/2}), \tag{13}$$

(14) $$\alpha = \frac{1}{\lambda - \mu} \left(\frac{-\mu w_{0 \to n} w_{w \to 0}}{w_{0 \to n} w_{w \to 0} + w_{n \to 0} w_{w \to 0} + w_{n \to 0} w_{0 \to w}} + w_{0 \to n} \right),$$

(15) $$\beta = \frac{1}{\lambda - \mu} \left(\frac{\lambda w_{0 \to n} w_{w \to 0}}{w_{0 \to n} w_{w \to 0} + w_{n \to 0} w_{w \to 0} + w_{n \to 0} w_{0 \to w}} - w_{0 \to n} \right),$$

for the value of the different parameters.

APPENDIX B.

In this appendix we deal with the number of conformations of a polymer chain. An ideal chain of N monomers, without the self-avoiding constraint, can assume γ^{N-1} conformations, $\gamma = 6$ being the coordination number of the cubic lattice. This because one can build all chains of length N starting from a point of the lattice, moving to one of the γ neighbouring sites, from here moving to one of the γ neighbouring sites of the second monomer, and so on $N-1$ times.

For a self-avoiding chain, the number of conformations can be approximated for long chains to γ^{N-1} with $\gamma = 4.68$ [53]. The total number of fully compact conformations can be obtained using a value of γ ranging from 1.8 [52] to 2.2 [57].

The number of conformations that a designed sequence has to search lies between the two extremes of self-avoiding chain and the fully compact chain. To find the value of γ for the unfolded chain, we have exploited the fact that the probability p to find the chain in a state with energy E and order parameter q is $p(E,q) = g(E,q) \cdot \exp[-E/T]/Z$, where $g(E,q)$ is the number of conformations with a given value of E and q. One can find the partition function Z from the condition of unicity of the native state, and consequently find the number of unfolded conformations $\int g(E,q) \, dE \, dq$ (the integral has been performed for $E > E_c$ and $0 \leq q \leq 1$). To find γ one has to pay attention that some local contacts are already formed and stable in the unfolded state, so that the length of the chain has to be renormalized (to 27, in this case, cf. subsect. **3**'2). The number of unfolded conformations is consequently associated to $\gamma = 3.45$.

It is also interesting to calculate the number of conformations with a bond between the i-th and j-th monomer ($i < j$). Such a system is composed of three parts coming out of the bond, namely two branches of length i and $N - j$, and a loop of length $j - i$. The total number of conformations is then the product of the numbers of conformations associated to the three parts. In keeping with the fact that the number of loops of length l can be extrapolated, for very small l, to $\gamma^l/l^{1.68}$, then the number of conformations associated with the bond is

(16) $$\frac{\gamma^{N-2}}{(j-i)^{1.68}}.$$

This expression is the one used in sect. **4** (eq. (10)).

REFERENCES

[1] ANFINSEN C. B., *Science*, **181** (1973) 223.
[2] CREIGHTON T. E., *Proteins* (W. H. Freeman and Co., New York) 1993.
[3] BRANDEN C. and TOOZE J., *Introduction to Protein Structure* (Garland Publishing Inc., New York) 1999.
[4] FERSHT A., *Structure and Mechanism in Protein Science* (W. H. Freeman and Co., New York) 1999.
[5] SHAKHNOVICH E. I., *Proteins with selected sequences fold to their unique native conformation*, Phys. Rev. Lett., **72** (1994) 3907.
[6] GOLDSTEIN R., LUTHEY-SCHULTEN Z. and WOLYNES P., *Optimal folding codes from spin-glass theory*, Proc. Natl. Acad. Sci. USA, **89** (1992) 4918.
[7] SALI A., SHAKHNOVICH E. I. and KARPLUS M., *Kinetics of protein folding: a lattice model study for the requirements for folding to the native state*, J. Mol. Biol., **235** (1994) 1614.
[8] SHAKHNOVICH E. I. and GUTIN A., *A novel approach to the design of stable proteins*, Protein Eng., **6** (1993) 793.
[9] BRYNGELSON J., ONUCHIC J. N., SOCCI N. D. and WOLYNES P., *Proteins: Struct. Funct. Gen.*, **21** (1995) 167.
[10] SHAKHNOVICH E. I., Curr. Opin. Struct. Biol., **7** (1997) 29.
[11] TIANA G., BROGLIA R. A., ROMAN H. E., VIGEZZI E. and SHAKHNOVICH E. I., *Folding and misfolding of designed proteinlike chains with mutations*, J. Chem. Phys., **108** (1998) 757.
[12] BROGLIA R. A., TIANA G., PASQUALI S., ROMAN H. E. and VIGEZZI E., *Folding and aggregation of designed proteins*, Proc. Natl. Acad. Sci. USA, **95** (1998) 12930.
[13] BROGLIA R. A., TIANA G., ROMAN H. E., VIGEZZI E. and SHAKHNOVICH E. I., *Stability of designed proteins against mutations*, Phys. Rev. Lett., **82** (1999) 4727.
[14] TIANA G., BROGLIA R. A. and SHAKHNOVICH E. I., *Hiking in the energy landscape in sequence space: a bumpy road to good folders*, Proteins: Struct. Funct. Gen., **39** (2000) 244.
[15] BROGLIA R. A. and TIANA G., *Hierarchy of events in the folding of model protein*, J. Chem. Phys., **114** (2001) 7267.
[16] BROGLIA R. A. and TIANA G., *Reading the three-dimensional structure of a protein from its amino acid sequence*, Proteins: Struct. Funct. Gen. (in press).
[17] EICHINGER M., TAVAN P., HUTTER J. and PARRINELLO M., *A hybrid method for solutes in complex solvents: density functional theory combined with empirical force fields*, J. Chem. Phys., **10** (1999) 10452.
[18] MIYAZAWA S. and JERNIGAN R., *Estimation of effective inter-residue contact energies from protein crystal structures: quasi-chemical approximation*, Macromolecules, **18** (1985) 534.
[19] SHAKHNOVICH E. I., Folding & Design, **1** (1996) R50.
[20] BOWIE J., LUTHEY-SHULTEN R. and EISENBERG D., *A method to identify protein sequences that fold into a known three-dimensional structure*, Science, **253** (1991) 164.
[21] VERDIER P. H., J. Chem. Phys., **59** (1973) 6119.
[22] HILHORST H. J. and DEUSTCH J. M., J. Chem. Phys., **63** (1975) 5153.
[23] SHAKHNOVICH E. I., FARZTDIMOV G., GUTIN A. M. and KARPLUS M., Phys. Rev. Lett., **67** (1991) 1665.
[24] SHAKHNOVICH E. I., ABKEVICH V. and PTITSYN O., Nature, **379** (1996) 96.
[25] ABKEVICH V. I., GUTIN A. M. and SHAKHNOVICH E. I., Biochemistry, **33** (1994) 10026.

[26] MIRNY L. A., ABKEVICH V. I. and SHAKHNOVICH E. I., *Proc. Natl. Acad. Sci. USA*, **95** (1998) 4976.
[27] KLIMOV D. K. and THIRUMALAI D., *J. Mol. Biol.*, **282** (1998) 471.
[28] SOCCI N. D. and ONUCHIC J. N., *J. Chem. Phys.*, **101** (1994) 1519.
[29] LI H., HELLING R., TANG C. and WINGREEN N., *Science*, **273** (1996) 666.
[30] SALI A., SHAKHNOVICH E. and KARPLUS M., *Nature*, **369** (1994) 248.
[31] SOCCI N. D., ONUCHIC J. N. and WOLYNES P. G., *Proteins: Struct. Funct. Gen.*, **32** (1998) 136.
[32] ABKEVICH V. I., GUTIN A. M. and SHAKHNOVICH E. I., *J. Chem. Phys.*, **101** (1994) 6052.
[33] KLIMOV D. and THIRUMALAI D., *Phys. Rev. Lett.*, **76** (1996) 4070.
[34] SOCCI N., BIALECH W. and ONUCHIC J., *Phys. Rev. E*, **49** (1994) 3440.
[35] SHAKHNOVICH E. I., *Phys. Rev. Lett.*, **72** (1994) 3907.
[36] ABKEVICH V. I., GUTIN A. M. and SHAKHNOVICH E. I., *J. Mol. Biol.*, **252** (1998) 460.
[37] BALDWIN R. L. and ROSE G. D., *Trends Biol. Sci.*, **24** (1999) 77.
[38] FERSHT A., *Phil. Trans. R. Soc. London Ser. B*, **348** (1995) 11.
[39] PRIVALOV P. L., *Adv. Prot. Chem.*, **33** (1979) 167.
[40] BRYNGELSON J. D. and WOLYNES P. G., *Proc. Natl. Acad. USA*, **84** (1987) 7524.
[41] ABKEVICH V. I., GUTIN A. M. and SHAKHNOVICH E. I., *Proteins: Struct. Funct. Gen.*, **31** (1998) 335.
[42] PTITSYM O. B., in *Protein Folding* (Freeman, New York) 1992, pp. 243-300.
[43] SHAKHNOVICH E. I. and FINKELSTEIN A. V., *Biopolymers*, **28** (1989) 1667.
[44] GO N. and ABE H., *Biopolymers*, **20** (1981) 991.
[45] GO N., *Annu. Rev. Biophys. Bioeng.*, **12** (1983) 183.
[46] PANDE V., GROSBERG A., ROKSHAR D. and TANAKA T., *Curr. Opin. Struct. Biol.*, **8** (1998) 68.
[47] PANDE V. S., GROSBERG A. YU. and TANAKA T., *Heteropolymer freezing and design: Towards physical models of protein folding*, *Rev. Mod. Phys.*, **72** (2000) 259.
[48] MCCUTEHEN S. L., COLON W. and KELLEY J. W., *Biochemistry*, **32** (1993) 12119.
[49] MCCUTEHEN S. L., LAI Z., MIROY G., KELLEY J. W. and COLON W., *Biochemistry*, **34** (1995) 13527.
[50] HAMILTON J. A., STEINRAUT L. K., BRADEN B. C., LIEPNIEKS J., BENSON M. D., HOLMGREN G., SANDGREN O. and STEEN L., *J. Biol. Chem.*, **268** (1993) 2425.
[51] TERRY C. J., DAMAS A. M., OLIVEIRA P., SARAIVIA M. J., ALVES I. L., COSTA P. P., MATIAS P. M., SAKAKI Y. and BLAKE C. C. F., *EMBO J.*, **12** (1993) 735.
[52] FLORY P. J., *J. Chem. Phys.*, **17** (1949) 303.
[53] DE GENNES P.-J., *Scaling Concepte in Polymer Physics* (Cornell University Press, New York) 1979.
[54] ODERMANN A., private communication.
[55] MIRNY L. A., ABKEVICH V. I. and SHAKHNOVICH E. I., *Folding Design*, **1** (1996) 103.
[56] DERRIDA B., *Phys. Rev. B*, **24** (1981) 2613.
[57] ORLAND H., ITZYKSON C. and DE DOMINICIS C., *J. Phys. (Paris) Lett.*, **46** (1985) L353.
[58] MATEN M. G., SANCHEZ DEL PINO M. M. and FERSHT A., *Mechanism of folding and assembly of a small tetrameric protein domain from the tumor suppressor p53*, *Nature Struct. Biol.*, **6** (1999) 191.
[59] SHAKHNOVICH E. I., *Nature Struct. Biol.*, **6** (1999) 99.
[60] BROOKS B. R., BRUCCOLERI R. E., OLAFSON B. D., STATES D. J., SWAMINATHAN S. and KARPLUS M., *J. Comp. Chem.*, **4** (1983) 187.
[61] KOHN W., *Rev. Mod. Phys.*, **71** (1999) 1253.
[62] CAR R. and PARRINELLO M., *Phys. Rev. Lett.*, **55** (1985) 2471.

[63] MAHAN G. D. and SUBBASWAMY K. R., *Local Density of Polarizability* (Plenum Press, New York) 1990.
[64] BERTSCH G. F. and BROGLIA R. A., *Oscillations in Finite Quantum Systems* (Cambridge University Press, Cambridge) 1994.
[65] KOHN W., MEIR Y. and MAKAROV D. E., *Phys. Rev. Lett.*, **80** (1998) 4153.
[66] ALASIA F., BROGLIA R. A., ROMAN H. E., SERRA L., COL'Ø G. and PACHECO J. M., *Single-particle collective degrees of freedom in C_{60}*, J. Phys. B, **27** (1994) L643.
[67] PINES D., *The Many Body Problem* (Benjamin, New York) 1962.
[68] ABRIKOSOV A. A., GORKOV L. P. and OZYALOSHINSKI I. E., *Methods of Quantum Field Theory in Statistical Physics* (Prentice-Hall, New Jersey) 1962.
[69] MARCH N. H., YOUNG W. H. and SAMPANTHAR S., *The Many-Body Problem in Quantum Mechanics* (Cambridge University Press, Cambridge) 1967.
[70] MAHAN G. D., *Many-Particle Physics* (Plenum Press, New York) 1981.
[71] ANDERSON P. W., *Basic Notions in Condensed Matter Physics* (Benjamin, Menlo Park, California) 1984.
[72] BOHR A. and MOTTELSON B. R., *Nuclear Structure*, Vol. **II** (Pergamon, New York) 1975.
[73] BORTIGNON P. F., BROGLIA R. A., BES D. R. and LIOTTA R., *Nuclear field theory*, Phys. Rep. C, **30** (1977) 305.
[74] FEYNMAN R. P. and HIBBS A. R., *Quantum Mechanics and Path Integrals* (McGraw-Hill, New York) 1965.
[75] FEYNMAN R. P., *Statistical Mechanics* (Addison-Wesley, Reading, Mass.) 1990.
[76] FEYNMAN R. P., *Theory of Fundamental Processes* (Benjamin, New York) 1975.

Complementary pictures for proteins: Designability and foldability, are they equivalent?

R. A. BROGLIA

Dipartimento di Fisica, Università di Milano - Via Celoria 16, I-20133 Milano, Italy
INFN, Sezione di Milano - Via Celoria 16, I-20133 Milano, Italy
The Niels Bohr Institute, University of Copenhagen - 2100 Copenhagen, Denmark

G. TIANA

Dipartimento di Fisica, Università di Milano - Via Celoria 16, I-20133 Milano, Italy
INFN, Sezione di Milano - Via Celoria 16, I-20133 Milano, Italy

1. – Introduction

The recent theoretical interest in understanding the folding of proteins arises from both developments in the study of polymers and the hope that important emergent features of a complex system can be captured by simple models. Thus, rather than focusing on the chemistry of the amino acid side chains, simple models of proteins attempt to unravel organizing principles starting from simple interactions on a lattice.

One principle which has inspired a large fraction of the work carried out following this approach is that of "foldability" (cf., *e.g.*, ref. [1-8] and references therein). The great majority of amino acid sequences have multiple ground states and hence may fold into different structures. Such sequences are unlikely candidates for coding functional proteins. Potentially good sequences are those with a unique ground state, separated by a large energy gap [9] from the lowest dissimilar configuration. This large gap implies that there are many excited states lying within the gap, which correspond to neutral mutations [10,11], leading to sequences (homologue families) which still fold to the native conformation although in somewhat longer times than that associated with the designed sequence.

Another principle, that of "designability" (cf., e.g., refs. [12,13], also ref. [14]), concentrates instead on the structure of the resulting protein. Polymer conformations which are highly designable, in the sense that they correspond to the non-degenerate ground state of a large number of amino acid sequences, are found to be thermodynamically more stable than other structures, and thus the best candidates for real protein configurations.

2. – The HP Model

An extensive study of model proteins designability has been carried out by Li, Helling, Tang and Wingreen [13] making use of 27 monomers chains and two kinds of residues (hydrophobic and polar), interacting through a matrix W of elements $W_{HH} = -2.3$, $W_{HP} = W_{PH} = -1.0$ and $W_{PP} = 0$. They enumerated all the 2^{27} sequences, counting how many of them have a non-degenerate ground state on each fully compact conformation. From this analysis, they observe that some structures are more designable than others (the best conformations can be designed by 3794 sequences, while 4256 conformations cannot be designed, cf. e.g., fig. 1(c)), that the highly designable conformations display geometrical features resembling β-strands, and that they have, as a rule, a bigger energy gap to the conformationally dissimilar state than non-designable folds.

This analysis is very interesting but has a basic problem, namely that sequences composed of only two kinds of residues are not very suitable to mimic real proteins. This is due to the inability these heteropolymers show to have a ground state which is unique and kinetically accessible at the same time, to the small value of the normalized gap, and to the instability of optimized sequences with respect to point mutations.

The sequences characterized by a non-degenerate ground state found by Li and coworkers have been obtained enumerating only fully compact conformations. Monte Carlo simulations show that, using the matrix W as it is, designed HP sequences get trapped in non-compact conformations whose energies are lower than the energy of the compact native conformation (see also [15]). For example, the optimal sequence of ref. [13], displayed in fig. 1(a), has an energy $E_N = -40.8$ (in arbitrary units, $\sigma_W = 0.9$ being the standard deviation of the interaction matrix) in the compact conformation on which it has been optimized. However, starting from an elongated chain the system can never reach this conformation (in 500 simulations of 10^8 MC steps each), being trapped in other conformations of energy $-41.2 \leq E \leq -40.8$ (cf. e.g., fig. 1(b)). To have fully compact conformations as actual ground states, it is then necessary to shift the matrix elements of the matrix W to more negative values, therefore making non-compact conformations unfavourable.

Considering the interaction matrix with elements $(W_{ij} + s)$, s being a negative constant, the conformational spectrum of each sequence is modified in such a way that the levels associated with non-compact conformations are raised above those associated with compact ones, restoring a fully compact conformation as non-degenerate ground state. The least negative value of s that accomplishes this task for the case discussed above is $s = -0.5$. A side effect of shifting the interaction matrix towards more negative values is to slow down the time the chain needs to find the ground state, starting from a random

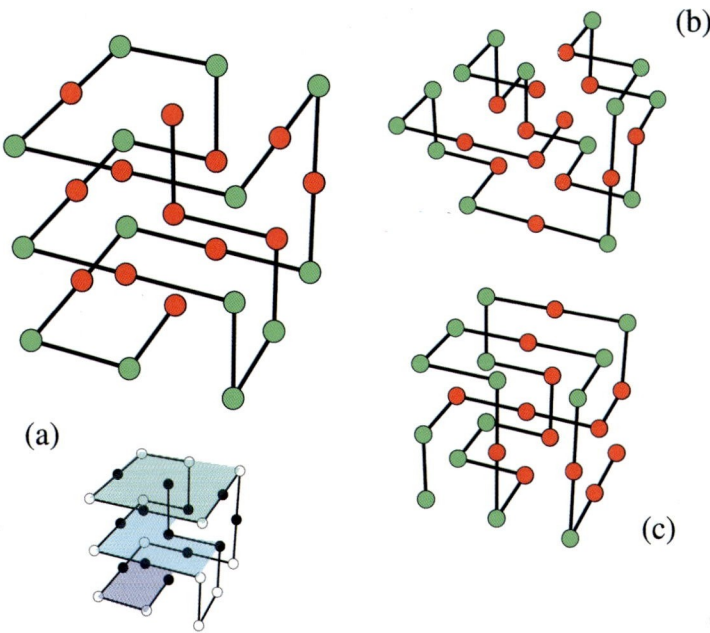

Fig. 1. – (a) Most designable conformation of a 27-monomer chain composed of two types of amino acids, namely hydrophobic (H) and polar (P) to which are associated the interaction energies $E_{HH} = -2.3$, $E_{HP} = -1.0$ and $E_{PP} = 0$ [13]. The corresponding sequence which minimizes this configuration with a fixed number of H and P residues (13 and 14, respectively) is $S_1 \equiv$ HPPHPHPHPHPHPHPPHPHPHPPPHHH, and displays an energy gap $\delta = 2.6$ to the lowest fully compact structurally different configuration. Note however that non-compact conformations with energy lower than the energy of the native state have been found. Changes in the monomer sites, plotted as red dots, correspond to the hot mutations discussed in the text (see also caption to fig. 3 and ref. [10]), in connection with the 36- 48- and the 80-monomer chains composed of 20 kinds of amino acids (cf. figs. 3–5). Mutations in these sites increase the energy of the target structure by an amount of the order of or greater than the so-called energy gap δ which, in this case, is of the order of the standard deviation of the interaction energies ($\sigma = 0.9$). In light blue the β-sheet associated with the conformation is shown. (b) Starting from a random, elongated configuration of the sequence denoted S_1, the chain compacts but does not fold, in the sense that it reaches different conformations which, within the statistic accumulated (10^2 coils followed through 10^6 MC steps), it never target into the most designable ("native") configuration shown in (a). Of all the random configurations targeted in the folding process by the sequence S_1, the one shown here is one, among many, which displays an energy which is lower than the most designable configuration shown in (a). (c) Example of the so-called "less-designable" configurations of Li et al. [13], whose minimum energy sequence is found for the sequence $S_2 \equiv$ PPHHHHPHPPHHPPHPHPPPHPPHPHH. Starting from random configurations of this sequence, the chain compacted, but again never folded, in the sense explained in (b).

conformation (cf. fig. 2). The reason for this behaviour is connected to the fact that for values of the mean interaction energy $\overline{(W+s)}$ between residues much more attractive

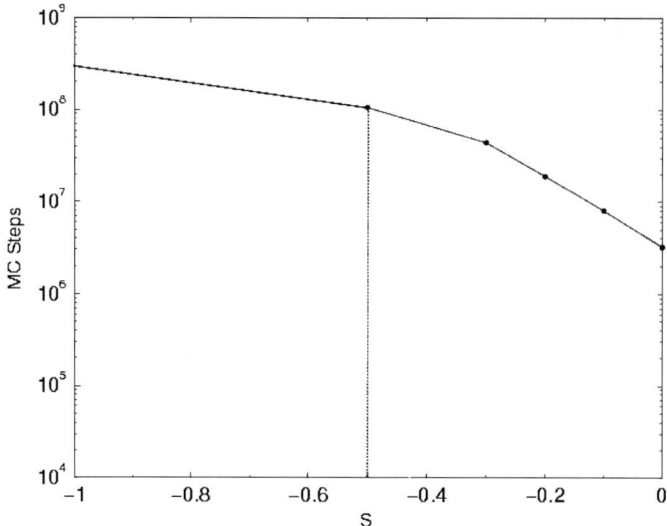

Fig. 2. – Relaxation times (that is, the time needed to reach the lowest-energy state in the HP model of the 27 mer folding to the conformation given in fig. 1(a)) as a function of the shift s of the interaction matrix. The points correspond to the time obtained at the temperature at which relaxation is fastest. The average is performed over twenty runs, and has the purpose of indicating only the order of magnitude of relaxation times. The native state is reached only for $S < -0.5$.

than the typical energy difference between monomers $-\sigma_{(W+s)}$ the chain loses its feature of designed heteropolymer. More precisely, there are two important temperature in the system, $T_1 \sim |\overline{(W + s)}|$ at which the chain builds and breaks its bonds in order to search for the native state in the conformational space essentially randomly, and $T_2 \sim \sigma_{W+s}$ at which the search for the ground state of a designed sequence is optimal because the specificity of the contacts becomes important. Consequently, if $T_1 \ll T_2$ the folding is optimal at T_1. Rising T_1, the search for the ground state has to compete with the unspecific binding of monomers. If $T_1 \gg T_2$ one can choose to run the simulation at two different temperatures: a) $T \approx T_1$, in which case the chain can build and break bonds, but the difference among residues (and, consequently, the specificity of the sequence) is essentially neglected, or b) at $T \approx T_2$, in which case the dynamics is essentially frozen. Accordingly, folding is faster if $T_1 \ll T_2$, or equivalently if $|\overline{(W + s)}| \ll \sigma_{(W+s)}$.

The relaxation times displayed in fig. 2 range from $\approx 10^6$ to $\approx 10^8$ MC steps. These times have to be compared with the random search time, which is (regardless of symmetries) $\approx (\gamma/e)^{N-1} = 10^7$ [16], where $\gamma = 5$ is the coordination number of the cubic lattice, N the length of the chain, e being the base of natural logarithm ($\ln e = 1$, $e = 2.7\ldots$). Consequently, at one of the extremes of fig. 3 ($s = -1.0$) the relaxation time is longer than that of a random search, while at the opposite extreme ($s = 0$) it is somewhat shorter, although still quite long. This is because the relaxation process is slowed down

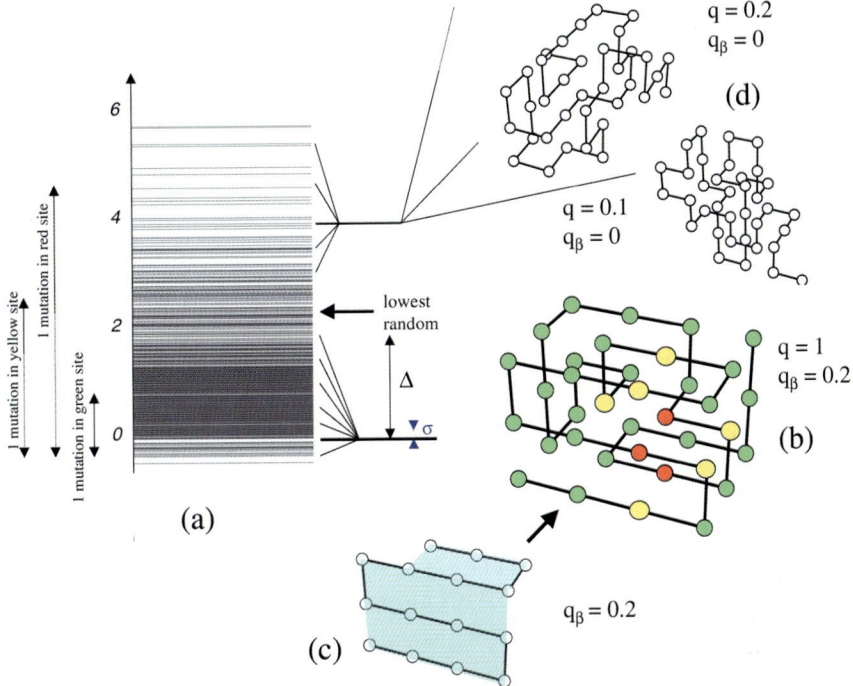

Fig. 3. – (a) Energy spectrum of the 36-monomer chain obtained mutating one or two amino acids in the sequence S_{36}. The sequence $S_{36} \equiv$ SQKWLERGATRIADGDLPVNGTYFSCKIMENVHPLA leads to the minimum energy, which we have set equal to zero, for the native conformation shown in (b). The sequence S_{36} folds, in the sense that starting from a random configuration, it always reaches the native configuration and this in a rather short time ($\sim 10^6$ MC steps). In (c) the β-sheet structure of the native conformation is shown. Introducing a *cold* mutation, that is changing amino acids in the green sites in the S_{36} sequence to produce the sequence $(S_{36})_g$, leads again to the folding of the sequence to the same "native" configuration (b), although the folding time increases by about 50% with respect to the folding time of the original S_{36} sequence, and the energy of the folded configuration is usually somewhat higher than that associated with sequence S_{36} (see (a) the energy range of average local energies corresponding to 1 cold mutation). There are few sequences which, when folded into (b), have an energy lower than that associated with the sequence S_{36}, even if this sequence was designed to have the lowest energy possible in the native configuration. This is because by changing an amino acid in S_{36}, one violates the constrain used in designing this sequence, namely that of having a given ratio of each kind of amino acid composition, as experimentally observed. A *warm* mutation (whose average local energy increase is shown in (a), that is a mutation obtained, typically, by replacing an amino acid in a yellow site of S_{36} to produce a sequence of type $(S_{36})_y$), leads again to protein folding into a configuration similar to the native one, and sometimes to the "native" conformation (b). The same behaviour observed for the sequences $(S_{36})_g$ is true for multiple mutations leading to a total energy change smaller than the gap δ. (d) Introducing a *hot* mutation (cf. (a) for the average value of the local energy increase), that is, changing amino acids in the red sites of the S_{36} sequence to produce the sequence $(S_{36})_r$ disrupts, as a rule, protein folding (cf. (d)). The corresponding sequence now targets, in the process of compacting itself, on a random (structurally dissimilar) configuration (dissimilarity quantified by a value of the similarity parameter $q \ll 1$). The similarity parameter is defined as the number of contacts which are the same as in the native conformation divided by the total number of contacts [19]. The parameter q_β, measuring the amount of secondary symmetry ((β-sheet)-like) displayed by a conformation, is defined as the number of contacts among residues belonging to parallel strands longer than 2, normalized with respect to the total number of contacts. Energetically, the sequence $(S_{36})_r$ in its compact configuration lies, as a rule, at an energy larger than 2.5 (in the units defined in the text) above the energy associated with the sequence S_{36} in its native configuration (cf. (a)). The high stability of S_{36} is associated with the presence of very few red sites (hot spots) occupied by highly hydrophobic amino acids, which lie protected inside the polymer chain.

by specific metastable trap conformations. In fact, a sequence composed of only two kinds of residues, lacking the specificity of the interactions within the nucleus observed for 20-letter sequences, can build a variety of rather different stable nuclei which compete with the folding nucleus.

For the smallest (negative) value of the shift of the interaction matrix which allows the sequence of ref. [13] to have a unique ground state (considering also non-compact conformations), that is $s = -0.5$, the folding time is still longer than the random search time ($\approx 10^8$, having run the simulation at $T = 1.0$). Consequently, two-letter sequences are not able to fold fast. This inability, which for other studies may not be of relevance, is in this case in which the kinetic behaviour of the system is important, a serious jeopardy.

Another feature typical of good folders, which is associated with both dynamical properties and stability, is the presence of a large energy gap between the native state and the lowest structurally dissimilar state [3]. The best sequence found (which is HPPHPHPH-PHPHPHPPHPHPHPPPHHH) displays an energy gap $\delta = 2.6$, equivalent to 2.9 times the standard deviation of the matrix elements σ_W. This is not much, considering for example the fact that good sequences made of 20 kinds of residues display gaps which are typically 10–15 times the standard deviation of the contact energies [18]. A consequence of this fact is the high sensitivity of HP sequences to point mutations. While in real proteins and in twenty-letter model proteins only a few percent of sites are sensitive to point mutations [10], within the HP model single mutations can disrupt the folded state in about half the sites of the sequence (in the case of the 27 mer, HP sequence mentioned above, for example, 13 sites are sensitive to mutations, *i.e.* "red" (hot) sites in fig. 1). These results suggest that sequences composed only of two kinds of residues hardly have the properties which are typical of real proteins.

3. – The twenty-letter model

In spite of these negative results, it is expected that the concept of designability, quantified by the number of sequences that uniquely fold into a particular conformation, will play a central role in the study of protein folding. This is because designability is, to a large extent, contained or better required by the foldability paradigm. In fact, thermodynamic stabilization of the native state through: a) sequence design, b) optimization of folding rate [19] (through mutations starting from a random sequence), constitute a necessary and sufficient condition for fast folding. While (a) embodies the strategy associated with the foldability paradigm, (b) implies that there is a large number of sequences [11], the same sequences which are obtained through "neutral" mutations of the designed sequence [10], which fold into the native conformation, being thus a realization of the designability paradigm.

Note that this large enhancement factor of the "cross-section" for a heteropolymer to target into the native conformation in the folding process is the one which upsets the odds against folding (Levinthal's paradox), and is at the basis of Levinthal's conjecture [20] that only evolutionary selected sequences are able to both be thermodynamically stable as well as to fold in a biologically suitable time.

In what follows we substantiate the picture presented above, making use of a model of heteropolymers where the monomers are constrained again to move on the vertices of a cubic lattice. We consider 20-letter chains of different lengths, and study their thermodynamic stability and the folding times following the "foldability" strategy. In these calculations the monomers are allowed to interact through different sets of inter-residue contact energies.

We start by discussing the results obtained for a twenty–amino-acid chain of 36 monomers with effective inter-residue contact energies obtained from the analysis of real protein data (disordered MJ contact energies, cf. table 6 of ref. [17]). A sequence with sufficiently low energy in the conformation chosen as native is denoted S_{36} (cf. caption to fig. 3). In the units we are considering ($RT_{\text{room}} = 0.6\,\text{kcal/mol}$), the energy of S_{36} in its native conformation is $E_{\text{nat}} = -16.5$. The value of the gap δ, that is, the energy difference between the native and the lowest dissimilar configuration (configuration with a similarity parameter q [19] much smaller than one (cf. fig. 3(d))) is large ($\delta = 2.5$), much larger than the variance ($\sigma = 0.3$) of the matrix elements. In other words, the normalized gap $\xi = \delta/\sigma$ (or the closely related z-score [1, 19, 21]) is much larger than one (in this case 8.3). Consequently, the designed protein is very stable with respect to mutations (change in amino acid sequence). In fact, introducing a mutation in any of the green sites (cf. fig. 3(b)), leads to an increase of the energy of the native conformation which is considerably smaller than δ (cf. fig. 3(a)). A mutation in a yellow site leads to an average increase of the energy of the configuration smaller than δ, although specific mutations are associated with an increase in energy of the native conformation of the order of δ. Mutations in red sites give, as a rule, rise to an increase of the energy of the native conformation larger than δ, leading to sequences which compact but always to different configurations (misfold) (cf. fig. 3(d)).

All single or double mutations associated with an energy increase of the native conformation smaller than the gap (levels within the energy gap in fig. 3(a)) lead to 159.854 sequences of this type, which fold to the native conformation, although the folding time may be somewhat longer than that associated with the folding of designed sequence S_{36} [11]. In mutating more than two sites of the chain, it is necessary to pay attention not to change in any important way the composition of the sequence. In fact, if too many hydrophobic residues are added to the "wild-type" composition, the chain cannot reach the native state any longer, getting trapped in non-compact conformations. Counting only the sequences with the same amino acid composition of the wild-type one (swapping of amino acids), which we shall call S'_{36}, we have found 10^{30} sequences folding to the same native structure as S_{36} (see appendix A). The above results indicate that there is a direct connection between the size of the normalized gap ξ and the number of sequences folding into the native conformation as their non-degenerate ground state, a fact which testifies the high degree of designability of the native conformation. Furthermore, it provides a concrete realization of Levinthal's hypothesis [20]. In fact, one knows that starting from a random sequence of 36 amino acids in which mutations are introduced biased by the condition that they speed up the folding rate, the sequence will evolve until it hits one of the S'_{36} sequences. It will then proceed by successive attempts which will lower both

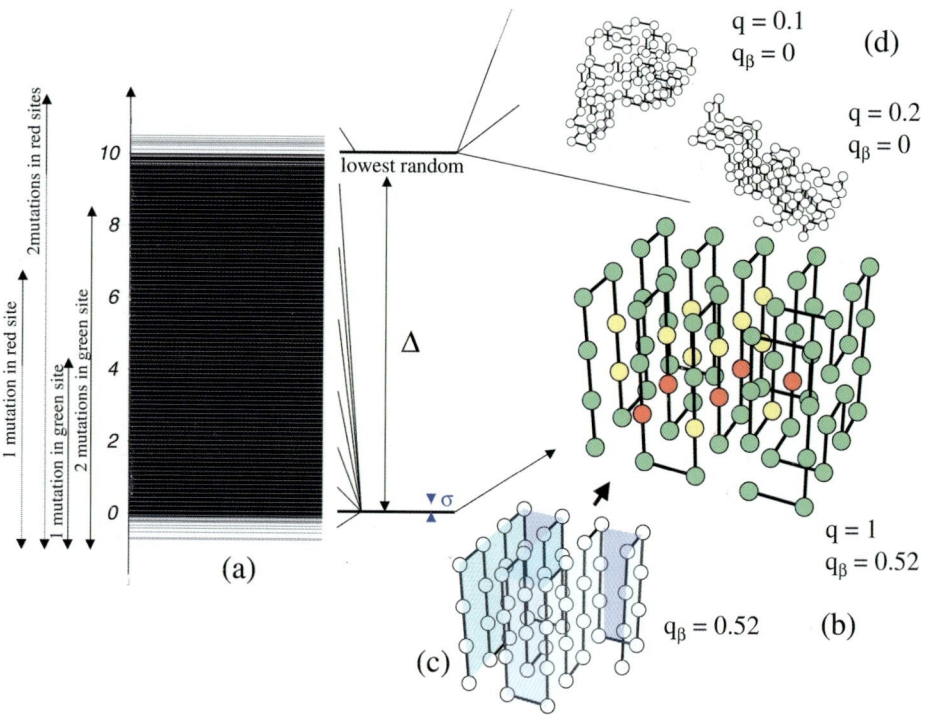

Fig. 4. – Same as in fig. 2, but for a 20-letter chain built out of 80 monomers.

the energy as well as shorten the folding time, until eventually S_{36} is reached [19]. In a way, the 10^{30} sequences of type S'_{36} lying within the gap (cf. fig 3(a)) act as a "funnel" which directs the selection of the sequences S'_{36} to eventually reach S_{36}.

Similar results were found in the study of a twenty-amino-acid chain composed of 80 monomers, the corresponding results being displayed in fig. 4. In this case the gap is so large ($\delta = 10$, $\xi = 33$) that to produce the lowest dissimilar (random) configuration two "hot" mutations have to be introduced in the designed sequence S_{80}. Consequently, the native structure allows for a very large number ($\sim 10^6$) of single and double mutations, the resulting sequence still folding to the original native conformation, a feature typical of a highly designable structure. The total number of composition-conserving mutated sequences, calculated as shown in Appendix A, has been found to be of the order of 10^{53}. Calculations carried out for a chain of length intermediate between those of the two chains studied above and containing 48 amino acids, leads to results in accordance with those displayed in figs. 3 and 4. In this case the gap $\delta = 6$ ($\xi = 20$) and there are 386.709 sequences with one or two mutations (cf. fig. 5), a total of 10^{34} composition-conserving sequences folding to the same native conformation.

It could be argued that the quoted number of sequences which fold to the same native conformation (10^{30}, 10^{34} and 10^{53} for 36, 48 and 80 monomers, respectively) provides, at being

$$(3) \quad n_m = n_m^{\text{tot}} \left(\pi m^2 \sigma_2^2/2\right)^{-1/2} \exp\left[-\frac{\overline{\Delta E_2}^2}{4(\sigma_2)^2}m\right]\frac{2(\sigma_2)^2}{\overline{\Delta E_2}}\left(\exp\left[\frac{\overline{E_2}}{2(\sigma_2)^2}\delta\right]-1\right),$$

where n_m^{tot} is the total number of sequences containing m mutations, $\overline{\Delta E_2}$ and σ_2 are the mean and the standard deviation of the energy distribution associated with mutations in two "cold" sites and δ is the energy gap associated with the lowest energy sequence found. Making use of this formula, we find the designability for the 36 mer, 48 mer and 80 mer discussed previously to be 10^{30}, 10^{34} and 10^{53}, respectively.

REFERENCES

[1] GOLDSTEIN R., LUTHEY-SCHULTEN Z. A. and WOLYNES P., *Proc. Natl. Acad. Sci. USA*, **89** (1992) 4918.
[2] SHAKHNOVICH E. and GUTIN A., *J. Chem. Phys.*, **93** (1990) 5967.
[3] SHAKHNOVICH E., *Phys. Rev. Lett.*, **72** (1994) 3907.
[4] MADDOX J., *Nature*, **370** (1994) 13.
[5] BRINGELSON J., ONUCHIC J. N., SOCCI N. D. and WOLYNES P., *Prot. Struct. Funct. Gen.*, **21** (1995) 167.
[6] SALI A., SHAKHNOVICH E. and KARPLUS M., *J. Mol. Biol.*, **235** (1994) 1614.
[7] GOVINDARAJAN S. and GOLDSTEIN R., *Biopolymers*, **36** (1995) 43.
[8] BRYNGELSON J., ONUCHIC J. N., SOCCI N. D., and WOLYNES P., *Prot. Struct. Funct. Gen.*, **21** (1995) 167.
[9] Large within the present context means larger than the standard deviation of the monomer contact energies (cf., e.g., ref. [17]).
[10] TIANA G., BROGLIA R. A., ROMAN H. E., VIGEZZI E. and SHAKHNOVICH E., *J. Chem. Phys.*, **108** (1998) 757.
[11] BROGLIA R. A., TIANA G., ROMAN H. E., VIGEZZI E. and SHAKHNOVICH E. I., *Phys. Rev. Lett.*, **82** (1999) 4727.
[12] FINKELSTEIN A. V., ABKEVICH V. and BATRETDINOV A., *Prot. Struct. Funct. Gen.*, **23** (1995) 142.
[13] LI H., HELLING R., TANG C. and WINGREEN N., *Science*, **273** (1996) 666.
[14] KARDAR M., *Science*, **273** (1996) 610.
[15] VENDRUSCOLO M., *Physica A*, **249** (1998) 576.
[16] FLORY P. J., *J. Chim. Phys.*, **17** (1949) 303.
[17] MIYAZAWA S. and JERNIGAN R., *Macromolecules*, **18** (1985) 534.
[18] Note that also in the case of 20 kinds of residues, there are some sequences displaying a small gap which still are able to fold fast [11]. The reason for this is that they share a set of native contacts and thus a set of conserved, strongly interacting amino acids which build the folding nucleus of the sequence which in the native structure displays a large gap. This sequence can be obtained from the small gap sequence by introducing swappings into the sequence.
[19] GUTIN A. M., ABKEVICH V. I. and SHAKHNOVICH E. I., *Proc. Natl. Acad. Sci. USA*, **92** (1995) 1282.

[20] LEVINTHAL C., *Mössbauer Spectroscopy of Biological Systems* (University of Illinois Press, Urbana) 1969.
[21] BOWIE J. U., LUTHY R. and EISENBERG D., *Science*, **253** (1991) 164.
[22] TIANA G., SHAKHNOVICH E. I. and BROGLIA R. A., *Prot. Struct. Funct. Gen.*, **39** (2000) 244.
[23] ABKEVICH V. I., GUTIN A. M. and SHAKHNOVICH E. I., *J. Mol. Biol.*, **252** (1995) 460.
[24] PLAXCO K. W., SIMMONS K. T. and BAKER D., *J. Mol. Biol.*, **277** (1998) 985.
[25] MICHELETTI A. M. C., BANAVAR J. R. and SENO F., *Phys. Rev. Lett.*, **82** (1999) 3372.
[26] CREIGHTON T. E., *Proteins* (W. Freeman and Co., New York) 1993.
[27] BORG J., JENSEN M. H., SNEPPEN K. and TIANA G., *Phys. Rev. Lett.*, **86** (2001) 1031.
[28] TIANA G., BROGLIA R. A. and PROVASI D., *Phys. Rev. E*, **64** (2001) 11904.

Dynamics of polymers: How fast can a protein fold?

H. ORLAND

Service de Physique Théorique, CEA-Saclay - 91191 Gif-sur-Yvette Cedex, France

The aim of this lecture is to give a brief introduction to the dynamics of a single polymer chain, and apply it to the problems of protein folding or expansion. A lot of the material presented in this lecture is detailed in several publications [1-3].

1. – Introduction

Recently, several experiments on the early stages of protein or polymer collapse or expansion [4-6] have been performed. The collapse of proteins is studied through fast mixing experiments, and the expansion through T-jump experiments. These experiments, carried out on cytochrome-C, measure the fluorescence yield due to the quenching of a tryptophan residue. The dead time of these experiments is very short, so that it becomes possible to measure dynamical quantities, such as end-to-end radii, down or below the microsecond scale. These are the early stages of collapse or expansion of the protein, and it is quite natural to assume that it is dominated by the hydrophobic effect, rather than any specific type of interactions. In addition, it is reasonable to believe that the short-time behavior should be quite universal, in the sense that it should not depend critically on the primary sequence.

In any case, the study of homopolymer collapse or expansion should set the limits on how fast a protein can fold or unfold. Indeed, in a homopolymer, all contacts are essentially equivalent, and thus any path that leads to a collapsed or expanded state is acceptable. In protein language, there is no entropic barrier to the folding or unfolding of a homopolymer, and thus, this will provide lower bounds to the folding or unfolding times of proteins.

According to de Gennes' theory [7], the process of collapse of a flexible coil leads to the formation of crumples on a minimal scale along the linear chain, which thickens and shortens under diffusion of the monomers, then forms new crumples of growing scale, until the final state of a compact globule is reached. In a refined model, K. A. Dawson et al. [8] consider a different two-step mechanism: first a fast formation of "pearls" along the chain, followed by a slower stage of compaction. This model has been revisited by A. Buguin et al. [9] who estimate the total time of collapse as $\tau_c = (\eta a^3/k_B\theta)(|\Delta T|/\theta)N^2$, where η is the viscosity of the solvent, θ is the θ temperature, a is the monomer size and ΔT is the temperature quench or jump. This time τ_c has a very strong dependence on molecular weight. For the case of proteins (see ref. [5]), $N = 300$, which yields a collapse time of $\tau_c \sim 1\,\mu$s.

Many theories have been devoted to this problem in the context of polymer theory [10-15]. In the present lecture, we will introduce the Langevin dynamics for a homopolymer chain. After briefly reviewing the dynamics of a non-interacting chain (Rouse theory), we will show how self-interactions can be included, and calculate the corresponding collapse or expansion times.

2. – Rouse model

The Rouse model [16] is the Langevin dynamics of an ideal chain. The polymer is modelled as a chain of beads of coordinates $\{\vec{r}_n\}$ and mass m_n attached by springs. Defining by a_0 the monomer length, the Hamiltonian of a system of N monomers is thus

$$(1) \qquad H = \frac{3}{2a_0^2} \sum_{n=1}^{N} (\vec{r}_{n+1} - \vec{r}_n)^2.$$

The Langevin equation, which describes the dynamics of the chain in a thermal reservoir at temperature T, is

$$(2) \qquad m_n \frac{d^2\vec{r}_n}{dt^2} + \gamma \frac{d\vec{r}_n}{dt} + \frac{\partial H}{\partial \vec{r}_n} = \vec{\xi}_n(t),$$

where the friction coefficient is related to the diffusion constant D of a single monomer by $\gamma = k_B T/D$. The 3 terms in the l.h.s. represent, respectively, the inertial term, the friction term and the force. The term on the r.h.s. is a Gaussian random force, which, together with the friction term, represents the coupling of the system to the heat bath.

The inertial term is known to produce vibrational modes, which are damped by the friction term. The time scale τ_0 after which one can neglect the inertial term can be obtained by a simple scaling argument. Comparing the inertial term with the friction term, one has

$$(3) \qquad \frac{m_n}{\tau_0^2} = \frac{\gamma}{\tau_0}$$

which reads $\tau_0 = mD/k_B T$. For a protein, typical values for the mass m of an amino-acid and for the diffusion constant D yield $\tau_0 \sim 10^{-13}$ s, therefore it is legitimate to neglect the inertial term in protein dynamics, and instead of (2), use the overdamped Langevin equation, which is called the Rouse equation:

$$\frac{d\vec{r}_n}{dt} = -\Gamma_0 \frac{\partial H}{\partial \vec{r}_n} + \vec{\eta}_n, \tag{4}$$

where $\Gamma_0 = D/k_B T$.

It can be shown that in order for the coordinates $\{\vec{r}_n(t)\}$ to sample a Boltzmann distribution in the large-time limit, the system should satisfy detailed balance. A simple and standard way to impose detailed balance is to assume that the random force $\vec{\eta}_n(t)$ is Gaussian distributed with correlation function given by

$$\langle \eta_m^\alpha(t)\eta_n^\beta(t')\rangle = 2D\delta_{\alpha\beta}\delta_{mn}\delta(t-t'). \tag{5}$$

It can then be proven that

$$\lim_{t\to\infty} P(\{\vec{r}_n(t)\}) \sim e^{-H(\{\vec{r}_n\})/k_B T}. \tag{6}$$

The Rouse model can be solved exactly. Equation (4) reads

$$\frac{d\vec{r}_n}{dt} = \frac{3\Gamma_0}{a_0^2}(\vec{r}_{n+1} - 2\vec{r}_n + \vec{r}_{n-1}) + \vec{\eta}_n. \tag{7}$$

Taking the continuous limit

$$\begin{cases} N \to \infty, \\ a_0 \to 0, \\ Na_0^2 \text{ finite}, \end{cases} \tag{8}$$

the coordinates of the chain become a continuous path denoted $\vec{r}(s,t)$ and eq. (7) becomes

$$\frac{d\vec{r}(s,t)}{dt} = \frac{3\Gamma_0}{a_0^2}\frac{d^2\vec{r}}{ds^2} + \vec{\eta}(s,t), \tag{9}$$

where the noise has Gaussian correlations given by

$$\langle \eta^\alpha(s,t)\eta^\beta(s',t')\rangle = 2D\delta_{\alpha\beta}\delta(s-s')\delta(t-t'). \tag{10}$$

This equation is easily solved by going to Fourier space. For simplicity, we assume that the chain is closed ($\vec{r}(N,t) = \vec{r}(0,t)$) so that the Fourier expansion is given by

$$\vec{r}(s,t) = \sum_{n=-\infty}^{\infty} e^{i\omega_n s}\vec{r}_n(t) \tag{11}$$

with frequencies

$$\omega_n = \frac{2\pi}{N} n. \tag{12}$$

Introducing the Rouse times

$$\tau_n = \frac{N^2 a_0^2}{12\pi^2 \Gamma_0 n^2}, \tag{13}$$

the radius of gyration can be expanded as

$$R_G^2(t) = 2 \sum_{n=1}^{\infty} e^{-2(t/\tau_n)} \langle \vec{r}_n(0) \vec{r}_{-n}(0) \rangle + \\ + 6 \frac{\Gamma_0}{N} \sum_{n=1}^{\infty} \tau_n \left(1 - e^{-2(t/\tau_n)}\right). \tag{14}$$

We can see that the large-time behavior is dominated by the Rouse mode with longest time, that is τ_1 and the relaxation of the radius of gyration is given by

$$R_G^2(t) \sim \frac{N a_0^2}{12} + C e^{-2(t/\tau_1)}. \tag{15}$$

The first term of the r.h.s. is simply the equilibrium radius of a Brownian chain. The relaxation is thus exponential, and dominated by one mode for $t \gg \tau_1$. Note that the characteristic Rouse time is of order $\tau_1 \sim 10^{-11} N^2$ in seconds. For a protein of length $N \sim 300$, this gives times of order 10^{-6} s.

By contrast, for short times, all Rouse mode contribute to (14). The asymptotic behavior of the radius of gyration is obtained by using the Poisson summation formula

$$\sum_{n=-\infty}^{\infty} f(n) = \sum_{m=-\infty}^{\infty} \tilde{f}(m), \tag{16}$$

where $\tilde{f}(m)$ is the Fourier coefficient of $f(x)$.
The result is

$$R_G^2(t) \sim R_G^2(0) + a_0 \sqrt{\Gamma_0 t} \tag{17}$$

and the behavior is not exponential, but rather diffusive at short times.

3. – Including interactions

For a protein in water, there are direct intramolecular interactions of the various atoms among themselves, as well as interactions of the atoms through the solvent. The net effect is the existence of a global effective monomer-monomer interaction, which would typically have a hard-core and a short-range attraction or repulsion. This type of short-range interaction can be modelled by 2- and 3-body interactions of the form

$$(18) \quad H = \frac{3}{2a_0^2} \sum_{n=1}^{N} (\vec{r}_{n+1} - \vec{r}_n)^2 + \frac{v}{2} \sum_{m \neq n} \delta(\vec{r}_m - \vec{r}_n) + \frac{w}{6} \sum_{m \neq n \neq k} \delta(\vec{r}_m - \vec{r}_n)\delta(\vec{r}_n - \vec{r}_k).$$

The first term has been discussed in the previous section. It represents the chain connectivity. The second term is a 2-body contact interaction, with $v > 0$ for repulsive interaction and $v < 0$ for attractive interaction. This term is related to the solvent quality or to the temperature. In particular, for a polymer, it can be shown that

$$(19) \quad v = (T - T_\theta)v_0,$$

where T_θ is the θ-temperature, which separates a swollen phase for $T > T_\theta$ from a globular phase for $T < T_\theta$. The last term is a 3-body repulsion which prevents the system from collapsing to infinite densities.

The high-temperature regime ($T > T_\theta$) is entropy dominated: the chain is swollen, with a swelling exponent $\nu \sim 3/5$. In fact, Flory theory [17] predicts

$$(20) \quad R_G \sim (T - T_\theta)^{1/5} N^{3/5}.$$

The low-temperature regime ($T < T_\theta$) is dominated by the hydrophobic energy, and the chain is collapsed into a globular state

$$(21) \quad R_G \sim \frac{N^{1/3}}{(T - T_\theta)^{1/3}}.$$

Finally at the θ-point, entropy and energy exactly compensate, and one finds

$$(22) \quad R_G \sim N^{1/2},$$

i.e. the chain is ideal, up to logarithmic corrections.

Using again the overdamped equation in the continuous limit, we have

$$(23) \quad \frac{\mathrm{d}\vec{r}(s,t)}{\mathrm{d}t} = -\Gamma_0 \frac{\partial H}{\partial \vec{r}(s,t)} + \vec{\eta}(s,t).$$

The idea is to introduce a reference Hamiltonian H_0 for which the trajectories $\vec{r}_0(s,t)$ can be calculated exactly, and then expand the true trajectories $\vec{r}(s,t)$ around it perturbatively. For simplicity, we assume that the chains are in a θ-solvent at $t = 0$, but this is not essential.

The reference chain, defined by $\vec{r}_0(s,t)$, satisfies the Langevin equation

$$\frac{\partial \vec{r}_0}{\partial t} = -\Gamma_0 \frac{\partial H_0}{\partial \vec{r}_0} + \vec{\eta}(s,t), \tag{24}$$

$$H_0 = \frac{1}{a^2(t)} \int_0^N \left(\frac{\partial \vec{r}_0}{\partial s}\right)^2 ds, \tag{25}$$

with the same diffusion constant and noise as the original equation, but with a much simplified Hamiltonian H_0. Indeed this Hamiltonian H_0 represents a Gaussian chain, but with a time-dependent Kuhn length $a(t)$.

Our method is a generalization of Edwards' Uniform Expansion Model [18] to dynamics. We expand the trajectory $\vec{r}(s,t)$ around $\vec{r}_0(s,t)$ to first order in $H - H_0$:

$$\vec{r}(s,t) = \vec{r}_0(s,t) + \vec{\chi}(s,t). \tag{26}$$

To first order, the equation for χ is

$$\frac{\partial \vec{\chi}}{\partial t} = \frac{\Gamma_0}{a^2(t)} \frac{\partial^2 \vec{\chi}}{\partial s^2} + \Gamma_0 \left[\left(\frac{1}{a_0^2} - \frac{1}{a^2(t)}\right)\frac{\partial^2 \vec{r}_0}{\partial s^2} + \vec{F}(\vec{r}_0(s,t))\right], \tag{27}$$

where $\vec{F}$ is the non-elastic part of the force.

The equation for $\vec{r}_0(s,t)$ and for $\vec{\chi}(s,t)$ can be easily solved. Expanding the radius of gyration in powers of χ, we obtain

$$R_G^2(t) = \int_0^N \frac{ds}{N} \langle \vec{r}^2(s,t)\rangle = \tag{28}$$

$$= \int_0^N \frac{ds}{N} \langle \vec{r}_0^2(s,t)\rangle_0 + 2\int_0^N \frac{ds}{N}\langle \vec{r}_0(s,t)\vec{\chi}(s,t)\rangle_0 + O(\chi^2).$$

We then determine the effective Kuhn length $a(t)$ by requiring that the first-order correction to the radius of gyration with respect to the reference one vanishes:

$$\int_0^N \frac{ds}{N}\langle \vec{r}_0(s,t)\vec{\chi}(s,t)\rangle_0 = 0. \tag{29}$$

The result of all this is a complicated equation for $a(t)$. Fortunately, this equation can be analyzed asymptotically in the limit of short and long time.

3˙1. Short-time behavior. – As in the non-interacting case, it can be easily seen that the short-time behavior is governed by a power law. It is easy to see that at short times, the expansion ($v > 0$) or collapse ($v < 0$) are identical, and the radius of gyration behaves like

$$R_G^2(t) \sim \left(1 \pm \left(\frac{t}{\tau_S}\right)^{3/4}\right) R_G^2(0) \qquad (30)$$

with a time scale

$$\tau_S \sim \frac{a_0^6}{Dv^{4/3}} N^{4/3} \sim \frac{N^{4/3} a_0^6}{D\Delta T^{4/3}}. \qquad (31)$$

3˙2. Long-time behavior. – At larger time, the relaxation of the radius of gyration is exponential in both cases. However, the expansion and collapse are not symmetric.

1) *Collapse.* At large times, the polymer collapses to a compact globule, with radius of gyration given by

$$R_G(t) \sim \left(\frac{w}{\Delta T}\right)^{1/3} N^{1/3} \left(1 + e^{-t/\tau_L}\right) \qquad (32)$$

and characteristic time given by

$$\tau_L \sim \frac{1}{4\pi^2 D} \left(\frac{w}{\Delta T}\right)^{2/3} N^{5/3}. \qquad (33)$$

As can be seen in this equation, the collapse time decreases with larger ΔT, that is with larger quenching from the θ-point.

2) *Expansion.* At large times, the polymer expands to a swollen coil, with radius of gyration given by

$$R_G(t) \sim \Delta T^{1/5} N^{3/5} \left(1 - e^{-t/\tau_L}\right) \qquad (34)$$

and characteristic time given by

$$\tau_L \sim N^{11/5}. \qquad (35)$$

We see that at large time scales, the collapse is much faster than the expansion. This is due to the fact that during the expansion, the density of the chain decreases and eventually goes to very small values, where the hydrophobic forces become inoperant.

When one puts in realistic numbers for a protein of length $N \sim 300$, one finds collapse times of order 10^{-7} s. This is somewhat smaller than what is actually measured in experiments (where the times are rather around 10^{-6} s), but again, the time scales we calculate are the fastest that could be observed.

4. – Conclusion

Simple homopolymeric models can be useful since they can be analyzed analytically and provide scaling laws for some of the time scales governing protein folding. The dependence of these scaling laws on both the chain length and the solvent quality might be checked experimentally. The times scales obtained from homopolymer models are usually shorter than those measured experimentally, but this is due to the fact that the homopolymer models ignore frustration and energy barriers. These theories might however be interesting in analyzing the non-specific part of the folding transition.

* * *

The author wishes to thank W. EATON for many useful discussions.

REFERENCES

[1] PITARD E. and ORLAND H., *Europhys. Lett.*, **41** (1998) 467.
[2] PITARD E., *Eur. Phys. J. B*, **7** (1999) 665.
[3] ORLAND H. and STEPANOW S., to be published.
[4] CHU B., YING Q. and GROSBERG A. Y., *Macromolecules*, **28** (1995) 180.
[5] CHAN C-K., HU Y., TAKAHASHI S., ROUSSEAU D. L., EATON W. and HOFRICHTER J., *Proc. Natl. Acad. Sci. USA*, **90** (1997) 1779.
[6] HAGEN S. J. and EATON W. A., to be published in *J. Mol. Biol.*
[7] DE GENNES P. G., *J. Phys. (Paris) Lett.*, **46** (1985) L-639.
[8] DAWSON K. A., TIMOSHENKO E. G. and KIERNAN P., *Nuovo Cimento D*, **16** (1994) 675.
[9] BUGUIN A., BROCHARD-WYART F. and DE GENNES P. G., *C.R. Acad. Sci. Paris, Ser. II b*, **322** (1996) 741.
[10] GROSBERG A. YU., NECHAEV S. K. and SHAKHNOVITCH E. I., *J. Phys. (Paris)*, **49** (1988) 2095.
[11] OSTROVSKY B. and BAR-YAM Y., *Europhys. Lett.*, **25** (1994) 409.
[12] MILCHEV A. and BINDER K., *Europhys. Lett.*, **26** (1994) 671.
[13] BYRNE A., KIERNAN P., GREEN D. and DAWSON K. A., *J. Chem. Phys.*, **102** (1995) 573.
[14] TIMOSHENKO E. G. and DAWSON K. A., *Phys. Rev. E*, **51** (1995) 492; TIMOSHENKO E. G., KUZNETSOV YU. A. and DAWSON K. A., *J. Chem. Phys.*, **102** (1995) 1816; KUZNETSOV YU. A., TIMOSHENKO E. G. and DAWSON K. A., *J. Chem. Phys.*, **103** (1995) 4807; KUZNETSOV YU. A., TIMOSHENKO E. G. and DAWSON K. A., *J. Chem. Phys.*, **104** (1996) 3338; TIMOSHENKO E. G., KUZNETSOV YU. A. and DAWSON K. A., *Phys. Rev. E*, **53** (1995) 3886.
[15] GANAZZOLI F., LA FERLA R. and ALLEGRA G., *Macromolecules*, **28** (1995) 5285.
[16] DOI M. and EDWARDS S. F., *The Theory of Polymer Dynamics* (Clarendon Press) 1988.
[17] FLORY P., *Principles of Polymer Chemistry* (Cornell University Press, Ithaca) 1971.
[18] EDWARDS S. F. and SINGH P., *J. Chem. Soc. Faraday Trans. II*, **75** (1979) 1001.

Geometrical aspects of protein folding

J. R. BANAVAR

Department of Physics and Center for Materials Physics, 104 Davey Laboratory
The Pennsylvania State University - University Park, PA 16802, USA

A. MARITAN and C. MICHELETTI

International School for Advanced Studies (SISSA) - Via Beirut 2-4, 34014 Trieste, Italy
INFM and the Abdus Salam International Center for Theoretical Physics, Trieste, Italy

F. SENO

INFM and Dipartimento di Fisica, G. Galilei, Università di Padova
via Marzolo 8, 35131 Padova, Italy

1. – Scope of the lectures

These lectures will address two questions. Is there a simple variational principle underlying the existence of secondary motifs in the native-state of proteins? Is there a general approach which can qualitatively capture the salient features of the folding process and which may be useful for interpreting and guiding experiments? Here, we present three different approaches to the first question, which demonstrate the key role played by the topology of the native state of proteins. The second question pertaining to the folding dynamics of proteins remains a challenging problem—a detailed description capturing the interactions between amino acids among each other and with the solvent is a daunting task. We address this issue building on the lessons learned in tackling the first question and apply the resulting method to the folding of various proteins including HIV protease and membrane proteins. The results that will be presented open a fascinating perspective: the two questions appear to be intimately related. The variety of results reported here all provide evidence in favour of the special criteria adopted by nature in the selection of viable protein folds, ranging from optimal compactness to maximum dynamical and geometrical accessibility of the native states.

2. – Introduction

A fascinating and open question challenging biochemistry, physics and even geometry is the presence of highly regular motifs such as α-helices and β-sheets in the folded state of biopolymers and proteins. Stimulating explanations ranging from chemical propensity to simple geometrical reasoning have been invoked to rationalize the existence of such secondary structures.

The realization that proteins have secondary structures arose with early crystallographic studies and the brilliant deduction of Pauling et al. [1] of the ability of an α-helix of the correct pitch to accommodate hydrogen bonds, thus promoting its stability. Inspired by the findings of Pauling, helix-coil transition models have been used to study the thermodynamics of helix formation [2]. It is interesting to note, however, that the number of hydrogen bonds is nearly the same when a sequence is in an unfolded structure in the presence of a polar solvent or in its native state rich in secondary-structure content [3]. It has also been suggested that the α-helix is an energetically favorable conformation for main-chain atoms but the side-chain suffers from a loss of entropy [3,4]. Nelson et al. [5] have shown both numerically and experimentally that non-biological oligomers fold reversibly like proteins into a specific three-dimensional structure with high helical content driven only by solvophobic interactions.

Recent studies have attempted to explain the emergence of secondary structure in proteins from geometrical principles rather than invoking detailed chemistry. Despite the concerted efforts of several groups, a simple general explanation remains elusive. A very natural line of investigation was undertaken by Yee et al. [6], Hunt et al. [3], and Socci et al. [7] who focused on the spontaneous emergence of secondary content from the mere requirement of overall compactness of homopolymeric chains. Their findings ruled out any significant relationship between the two, a fact also corroborated by the recent study of the kinetics of homopolymer collapse, where no evidence was found for the formation of local regular structures [8]. Despite the failure, this approach is particularly interesting due to the fact that optimal packing is a fundamental and fascinating problem in contexts ranging from everyday life to atomic physics. Perhaps, the best known packing problem is the one introduced by Kepler nearly four centuries ago, concerning the optimal packing of sphere. However, the packing of independent objects, like spheres, must be treated differently from the case of objects connected in a chain, such as beads in a string (an idealization of peptide chains). In sect. **3**, the packing problem is generalized to such chains and, remarkably, if one requires optimal packing uniformly along the chain, then a particular type of helix becomes the solution and, furthermore, it has the same geometrical characteristics as α-helices found in proteins.

In addition to packing considerations, dynamical effects also play a significant role when rapid packing/unpacking is entailed, as in the formation of amorphous glasses where crystallization is dynamically thwarted or in the more familiar problem of packing clothes in one's suitcase. The same question may be asked for protein-like structure. The fact that they contain motifs which are optimally compact does not imply that they can be easily reached from unfolded states. It is, however, widely believed that native states are, in general, highly accessible from the kinetic point of view. To investigate this issue, we formulate, in sect. **4**, a dynamical variational principle for selection in conformation space based on the requirement that the backbone of the native state of biologically viable polymers be rapidly accessible from the denatured state. The variational principle is shown to result in the emergence of helical order in compact structures, revealing a surprising accord with the compactness requirement discussed above.

Still concerning the folding dynamics, there are two key aspects distinguishing a protein from a generic heteropolymer: the specially selected sequence of amino acids and the three-dimensional structure which it folds reversibly into. Nature uses a rich repertory of twenty kinds of amino acids with sometimes major and other times subtle differences in their interactions with the solvent and with each other in order to design sequences that fit the putative native state with minimal frustration [9]. The chosen sequences are such that their target native states are reached through a funnel-like landscape [10-13] which facilitates the harmonious fitting together of pieces to form the whole. The three-dimensional structure impacts on the functionality of the protein and a fascinating issue is the elucidation of the selection mechanism in conformation space that picks out certain viable structures from the innumerable ones with a given compactness. Earlier studies have shown that there is a direct link between viable native conformations and high designability [14, 15].

A fruitful and general strategy for the study of protein folding would be to extract information on the folding process directly from the topology of the native state. This problem will be elucidated in sect. **5** and applied to HIV protease and in sect. **6** to membrane proteins. It will be shown that the natural folds of proteins have a much larger density of nearby structures than generic (artificial) conformations of the same character and that the exceedingly large geometrical accessibility of natural proteins may be related to the presence of secondary motifs [16]. It will be shown that a study of the influence of native state topology on the folding process can reveal information about the sites that are crucial to the folding process itself. As an application, we shall identify such sites for three proteins: 2ci2, barnase and HIV-1 Protease and show that they correlate very well with the key folding sites identified in experiments.

In sect. **6**, a general model based on topological properties of the native state is introduced to decipher the folding of membrane proteins. Nearly a quarter of genomic sequences and almost half of all receptors that are likely to be targets for drug design [17] are integral membrane proteins. Understanding the detailed mechanisms of their folding mechanism is a largely unsolved key problem in structural biology. By using our geometrical approach, we can investigate the equilibrium properties and the folding kinetics of a two-helix bundle fragment (comprising 66 amino acids) of bacteriorhodopsin. Once again, the approach seems to be extremely powerful and it appears to provide an efficient framework for understanding the variety of folding pathways of transmembrane proteins.

3. – Optimal shape of a compact polymeric chain

A fundamental problem in everyday life is that of packing with examples ranging from fruits in a grocery, clothes and personal belongings in a suitcase, atoms and colloidal particles in crystals and glasses, and amino acids in the folded state of proteins [18-22]. The simplest problem in packing consists of determining the spatial arrangement that accommodates the highest packing density of its constituent entities with the result being a crystalline structure.

A classic problem is the determination of the optimal arrangement of spheres in three dimensions in order to achieve the highest packing fraction. This problem first posed by Kepler has attracted much interest culminating in its recent rigorous mathematical solution [18, 19] that the answer for infinite systems is a face-centred-cubic lattice. This simply stated problem has had a profound impact in many areas [20-22], ranging from the crystallization and melting of atomic systems, to optimal packing of objects and subdivision of space.

The close-packed hard-sphere problem is simply stated: given N hard spheres of radius R, what is the arrangement which can be enclosed in the minimum volume, *e.g.* a cube of side L? This is solved by reformulating the problem, more convenient for numerical implementation, as the determination of the arrangement of a set of N points in a cube of linear size, L_0, that results in the minimum of half the distance between any pair of points or between the points and walls of the container, denoted by r_{min}, being as large as possible [23]. The linear size associated with the region enclosing the hard spheres follows from dimensional analysis and is given by $L = RL_0/r_{min}$, from which it follows that maximizing r_{min} is equivalent to minimizing L. It is notable that the resulting "bulk" optimal arrangement in the large N limit exhibits translational invariance in that, far from the boundaries, the local environment is the same for all points. In dimension $d = 2$ and $d = 3$ this corresponds to triangular and face-centred-cubic lattices, respectively.

Biopolymers like proteins, DNA and RNA have three-dimensional structures which are rather compact. Furthermore, they are the result of evolution and one may think that their shape may satisfy some optimality criterion. This naturally leads one to consider a generalization of the packing problem of hard spheres to the case of flexible tubes with a uniform cross-section. The packing problem then consists in finding the tube configuration which can be enclosed in the minimum volume without violating any steric constraints.

The problem can alternatively be formulated in a very simple and elegant way in terms of the curve which is the axis of the tube (the analog of the sphere centers in the hard-sphere packing problem) [24]. Consider a string (an open curve) in three dimensions. We will utilize a geometric measure [25] of the curve, the "rope-length", defined as the arc-length measured in units of the thickness, which has proved to be valuable in applications of knot theory [25-30]. The thickness Δ denotes the maximum radius of a uniform tube with the string passing through its axis, beyond which the tube either ceases to be smooth, owing to tight local bends, or it self-intersects. Our focus is on finding the optimal shape of a curve of fixed arc length, subject to constraints of compactness, which would maximize its thickness, or equivalently minimize its rope length.

Following the approach of Gonzalez and Maddocks [28], who studied knotted strings, we define a global radius of curvature as follows. The global radius of curvature of the curve at a given point is computed as the minimum radius of the circles going through that point and all other pairs of points of the string. It generalizes the concept of the local radius of curvature (the radius of the circle which locally best approximates the curve) by taking into account both local (bending of the string) and non-local (proximity to another part of the string) effects. For discretized curves, the local radius of curvature at a point is simply the radius of the circle going through the point and its two adjoining points. The minimum of all the global radii then defines the thickness, *i.e.* the minimum radius of the circles going through any triplet of discrete points. This coincides with the previous definition in the continuum limit, obtained on increasing the number of discretized points (assumed to be equally spaced) on the curve keeping the string length fixed [28]. Given a string configuration, the thickness is just the maximum allowed radius for the cross-section of a uniform tube that has the given curve as its axis [28]. We used several different boundary conditions to enforce the confinement of the string. The simplest ones discussed here are the confinement of a curve of length l within a cube of side L or constraining it to have a radius of gyration (which is the root-mean-square distance of the discretized points from their centre of mass) that is less than a pre-assigned value R. Even though different boundary conditions influence the optimal string shape, the overall features are found to be robust. Examples of optimal shapes, obtained from numerical simulations, for different ratios of l/L and l/R are shown in fig. 1. In both cases, two distinct families of curves, helices and saddles, appear. The two families are close competitors for optimality and different boundary conditions may stabilize one over the other. For example, if optimal strings of fixed length are constrained to have a radius of gyration less than R, then upon decreasing R, the curve goes from a regime where the trivial linear string is curled into an arc, then into a portion of helix and finally into a saddle. When the string is constrained to lie within a cube of size L, as L decreases first saddles are observed and then helices.

We have also been able to find bulk-like solutions which are not influenced by boundary effects. Such solutions can be obtained by imposing uniform local constraints along the curve. On imposing a minimum local density on successive segments of the string (for example, constraining each set of six consecutive beads to have a radius of gyration that is less than a preassigned value R), we obtained perfectly helical strings, corresponding to discretized approximations to the continuous helix represented in fig. 2, confirming that this is the optimal arrangement. Note that, in close analogy with the sphere-packing problem, a helix has translational invariance along the chain.

In all cases, the geometry of the chosen helix is such that there is an equality of the local radius of curvature (determined by the local bending of the curve) and the radius associated with a suitable triplet of non-consecutive points lying in two successive turns of the helix. This is a feature that is observed only for a special ratio c^* of the pitch, p, and the radius, r, of the circle projected by the helix on a plane perpendicular to its axis. When $p/r > c^* = 2.512$ the local radius of curvature, given by $\rho = r(1 + p^2/(2\pi r)^2)$,

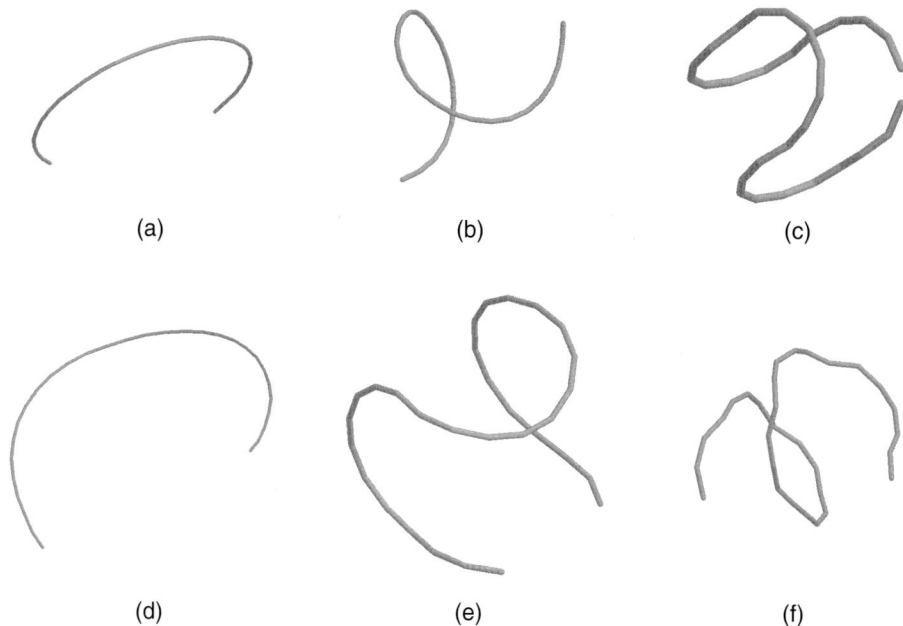

Fig. 1. – Examples of optimal strings. The strings in the figure were obtained starting from a random conformation of a chain made up of N equally spaced points (the spacing between neighboring points is defined to be 1 unit) and successively distorting the chain with pivot, crankshaft and slithering moves commonly used in stochastic chain dynamics [31]. A Metropolis Monte Carlo procedure is employed with a thermal weight, $e^{+\Delta/T}$, where Δ is the thickness and T is a fictitious temperature set initially to a high value such that the acceptance rate is close to 1 and then decreased gradually to zero in several thousand steps. Self-avoidance of the optimal string is a natural consequence of the maximization of the thickness. The introduction of a hard-core repulsion between beads was found to significantly speed up convergence to the optimal solution and avoid trapping in self-intersecting structures. We have verified that the same values (within 1 percent) of the final thickness of the optimal strings are obtained starting from unrelated initial conformations. Top row: optimal shapes obtained by constraining strings of 30 points with a radius of gyration less than R. (a) $R = 6.0$, $\Delta = 6.42$, (b) $R = 4.5$, $\Delta = 3.82$, (c) $R = 3.0$, $\Delta = 1.93$. Bottom row: optimal shapes obtained by confining a string of 30 points within a cube of side L. (d) $L = 22.0$, $\Delta = 6.11$, (e) $L = 9.5$, $\Delta = 2.3$, (f) $L = 8.1$, $\Delta = 1.75$.

is lower than the half of the distance of closest approach of points on successive turns of the helix. The latter is given by the first minimum of $1/2\sqrt{2 - 2\cos(2\pi t) + p^2 t^2}$ for $t > 0$. Thus $\Delta = \rho$ in this case.

If $p/r < c^*$, the global radius of curvature is strictly lower than the local radius, and the helix thickness is determined basically by the distance between two consecutive helix turns: $\Delta \simeq p/2$ if $p/r \ll 1$. Optimal packing selects the very special helices corresponding

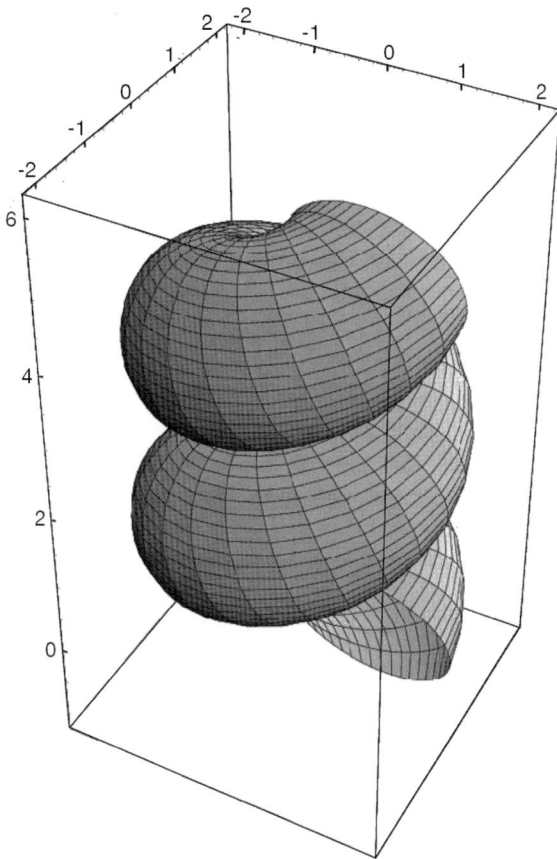

Fig. 2. – Shape of the optimal helix. The ratio of the pitch to radius of the centerline is 2.512.

to the transition between the two regimes described above. A visual example is provided by the optimal helix of fig. 2.

For discrete curves, the critical ratio p/r depends on the discretization level. A more robust quantity is the ratio f of the minimum radius of the circles going through each point and any two non-adjacent points and the local radius. For discretized strings, $f = 1$ just at the transition described above, whereas $f > 1$ in the "local" regime and $f < 1$ in the "non-local" regime. In our computer-generated optimal strings, f differed from unity by less than a part in a thousand.

It is interesting to note that, in nature, there are many instances of the appearance of helices. For example, many biopolymers, such as proteins and enzymes, have backbones which frequently form helical motifs. (Rose and Seltzer [32] have used the local radii of curvature of the backbone as input in an algorithm for finding the peptide chain turns in a globular protein.) It has been shown [16] that the emergence of such motifs in pro-

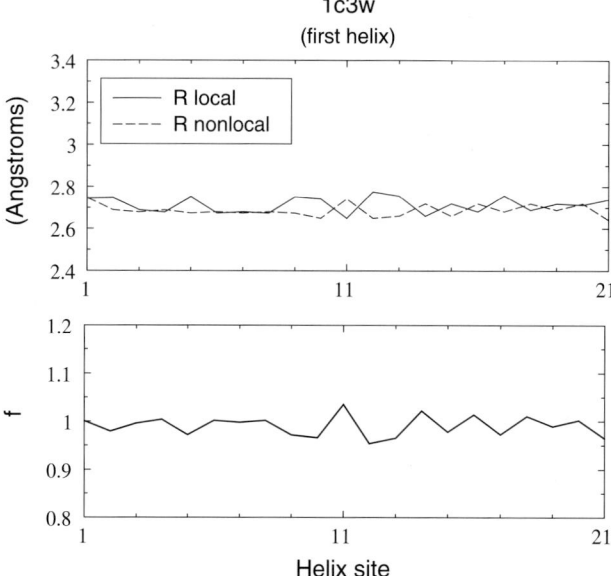

Fig. 3. – Top: local and non-local radii of curvature for sites in the first helix of bacteriorhodopsin (pdb code 1c3w). Bottom: plot of f values for the same sites.

teins (unlike in random heteropolymers which, in the melt, have structures conforming to Gaussian statistics) is the result of the evolutionary pressure exerted by nature in the selection of native state structures that are able to house sequences of amino acids which fold reproducibly and rapidly [33] and are characterized by a high degree of thermodynamic stability [34]. Furthermore, because of the interaction of the amino acids with the solvent, globular proteins attain compact shapes in their folded states.

It is then natural to measure the shape of these helices and assess if they are optimal in the sense described here. The measure of f in α-helices found in naturally occurring proteins yields an average value for f of 1.03 ± 0.01, hinting that, despite the complex atomic chemistry associated with the hydrogen bond and the covalent bonds along the backbone, helices in proteins satisfy optimal packing constraints. An example is provided in fig. 3 where we report the value of f for a particularly long α-helix encountered in a heavily investigated membrane protein, bacteriorhodopsin.

This result implies that the backbone sites in protein helices have an associated surrounding volume distributed more uniformly than in any other conformation with the same density. This is consistent with the observation [16] that secondary structures in natural proteins have a much larger configurational entropy than other compact conformations. This uniformity in the surrounding volume distribution seems to be an essential feature because the requirement of a maximum packing of backbone sites by itself does not lead to secondary-structure formation [6,7]. Furthermore, the same result also holds for the helices appearing in the collagen native-state structure, which have a rather differ-

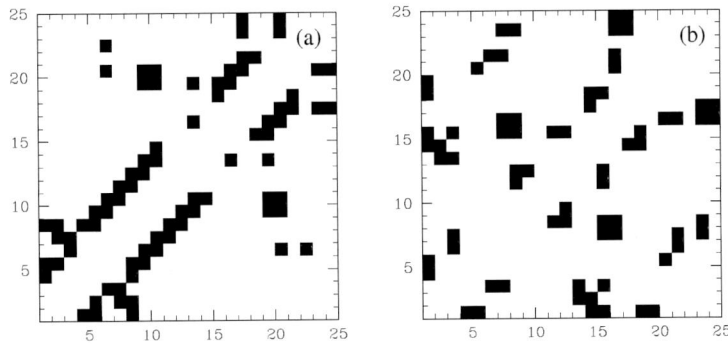

Fig. 7. – The panel on the left (right) shows the contact map of a structure with a very low (average) median folding time. The signature of helices in map (a) is shown by the thick bands parallel to the diagonal, while no such patterns are observed in the matrix (b).

in the number of residues in secondary motifs (secondary content). The bottom panel shows a milder decrease of the contact order (*i.e.* a larger number of short-range contacts) as the generations evolved, in agreement with the experimental findings of Plaxco *et al.* [42].

One of the optimal structures of length 25 is shown in fig. 6a). Due to the absence of any chirality bias in our structure space exploration, the helix does not have a constant handedness. The signature of the secondary motifs in the optimal structures is clearly visible in the contact maps of fig. 7, which are not sensitive to structure chirality. Strikingly, the variational principle selects conformations with significant secondary content as those facilitating the fastest folding. It is also noteworthy that the average value of f defined in sect. **3** is 0.9, which is very close to the ideal value, $f = 1$, despite the fact that the underlying FCC lattice prevents the structures from attaining a regular helical shape.

The correlation of the emergence of secondary structures with decrease of folding times is shown in the plot of fig. 8. We verified that the hybridization procedure is not biased towards low contact order by iterating it for various generations and pairing the structures at random. Even after dozens of generations, the generated structures had secondary contents of about 1/3–1/4 of the true extremal structures.

The very high secondary content in optimal conformations was found to be robust against changes in chain length or compactness of the target structure. On requiring that the structure be more compact, bundles of helices emerge (see fig. 6b)) along with an increase in contact order, signalling the presence of some longer-range contacts, which are necessitated in order to accommodate the shorter radius of gyration. It is noteworthy that our calculations lead predominantly to α-helices and not β-sheets, a fact accounted for by the demonstration that steric overlaps and the associated loss of entropy lead to the destabilization of helices in favor of sheets [4], the appearance of such sheets only in sufficiently long proteins [43] and the much slower folding rate of β-sheets compared to

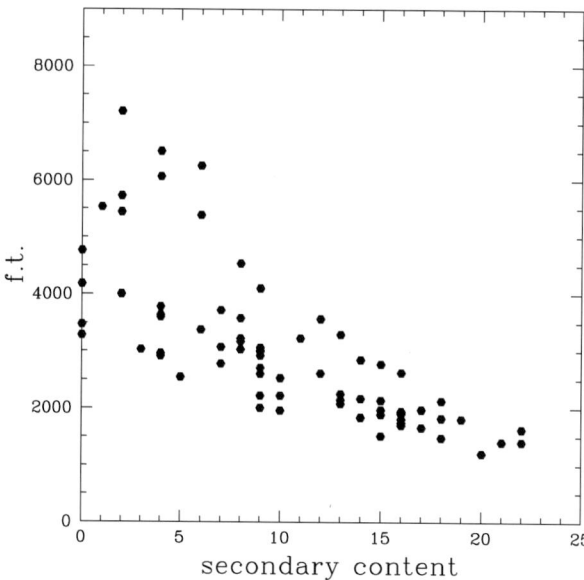

Fig. 8. – Scatter plot of folding time *vs.* secondary content for structures of length 25 collected over several generation of the optimization algorithm.

α-helices [44]. It is remarkable that the same requirement of rapid folding is sufficient to lead to a selection in both sequence and structure space underscoring the harmony in the evolutionary design of proteins. The results and strategies presented here ought to be applicable in protein-engineering contexts, for example by ensuring optimal dynamical accessibility of the backbone of proteins. A systematic collection of the rapidly accessible structures of various length should also lead to the creation of unbiased libraries of protein folds.

5. – Density of overlapping conformations for protein structures and role of native-state topology in the folding process

The rapid and reversible folding of protein-like heteropolymers into their thermodynamically stable native state [45] is accompanied by a huge reduction in conformational entropy [46, 47]. Evidence has been accumulating for an achievement of the entropy reduction through a folding funnel favouring the kinetic accessibility of the native state [9, 11, 48-50]. In sects. **3** and **4**, we have seen that secondary motifs of proteins may arise from the requirements of both being compact and easily kinetically accessible. Here we focus on a further characterization of the special role played by the native structure of proteins; again we will make no use of detailed information regarding amino acid sequences [16]. The study is carried out through a theoretical probe of the conformation space of proteins: a measure of the density of overlapping conformations (DOC) having a given overlap or percentage of contacts in common with a fixed native structure. We show with studies on chymotrypsin inhibitor (reference 2ci2 of the Protein Data Bank) and barnase (1a2p) that the DOC provides key information on the folding pathway. An analysis of the DOC for real protein structures and for artificially generated decoy ones suggests that an extremal principle is operational in nature, which maximizes the DOC at intermediate overlap, providing a large basin of attraction [9, 11, 13, 49, 51] for the native state and promoting the emergence of secondary structures.

Our study consists of determining the number of structures with a given structural similarity to a putative native state. The structural similarity between the native structure and another one is defined as the percentage of native contacts in the alternative conformation. It is well known that such a measure is a good coordinate characterizing the folding process [34, 37, 52]. To this purpose we adopt the Gō scheme introduced in the previous section (see eq. (1)). We also make again use of the FCC coarse-graining to avoid considering as distinct conformations that differ slightly.

The generation of conformations was carried out using a standard Monte Carlo procedure (see, e.g., refs. [34, 53]) which allows one to move simultaneously up to 7 randomly chosen C_α's to unoccupied FCC sites.

In order to minimize the effects of correlation between successively generated structures, we typically discarded 50 elementary moves before accepting each new conformation. A newly generated conformation was accepted with the usual Metropolis rule according to the change in the Boltzmann weight: $e^{\Delta/K_B T}$, where Δ is the change in contact overlap and T is a fictitious temperature. By choosing T appropriately, one can readily generate conformations with a desired average contact overlap, $\bar{q}$. At a given temperature, the true number of structures with overlap q is proportional to the number of conformations with overlap q obtained in the simulation multiplied by the Boltzmann weight. On undoing the Boltzmann bias, it is possible to recover the true density of conformations in a region around $\bar{q}$. In order to obtain the density of conformations for all values of overlap, we performed 2500 Monte Carlo samplings at different decreasing temperatures and then used standard deconvolution procedures [54]. Overall, for each distinct value of the overlap, more than 1000 structures were sampled. We have confirmed that the DOC curves are independent of the starting conformation and that the "folding" DOC obtained starting from a random conformation and cooling agrees to better than 3% with the "unfolding" DOC obtained starting from the target structure and increasing the temperature.

We begin with the backbones of the chymotrypsin inhibitor and barnase. We generated 2500 structures with a not too large overlap [55] ($\approx$ 40%) for each of them. It turned out that the most frequent contacts shared by the native conformation of 2ci2 with the others involved the helical residues 30-42 (see fig. 9, top). Contacts involving such residues were shared by 56% of the sampled structures. On the other hand, the rarest contacts pertained to interaction between the helix and β-strands and between the β-strands themselves. A different behaviour (see fig. 10, bottom) was found for barnase, where, again, for overlap of $\approx$ 40%, we find many contacts pertaining to the nearly complete formation of helix 1 (residues 8-18), a partial formation of helix 2, and bonds between residues 26-29 and 29-32 as well as several non-local contacts bridging the β-strands, especially residues 51-55 and 72-75.

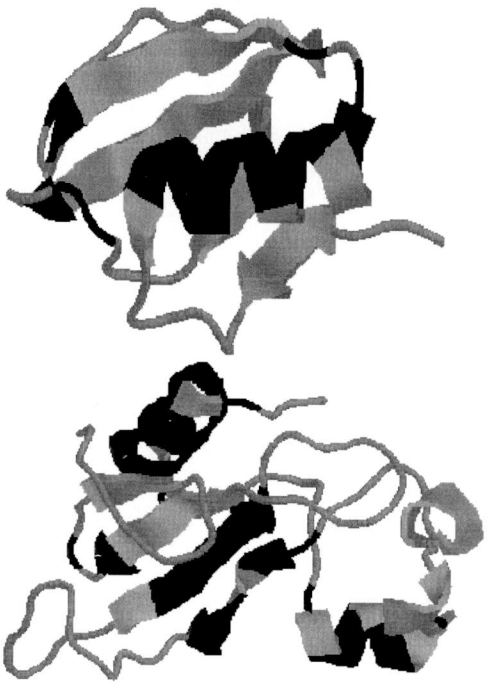

Fig. 9. – Ribbon plot (obtained with RASMOL) of 2ci2 (top) and barnase (bottom). The residues involved in the 12 [16] most frequent contacts of alternative structures with overlap ≈ 40% with the native conformations are highlighted in black. The majority of these coincide with contacts that are formed at the early stages of folding.

Both this picture and the one described for CI2 are fully consistent with the experimental results obtained by Fersht and co-workers in mutagenesis experiments [57, 58]. In such experiments, the key role of an amino acid at a given site is probed by mutating it and measuring the changes in the folding and equilibrium characteristics. By measuring the change of the folding/unfolding equilibrium constant one can introduce a parameter, termed ϕ-value, which is zero if the mutation is irrelevant to the folding kinetics and 1, if the change in folding propensity mirrors the change in the relative stability of the folded and unfolded states (intermediate values are, of course, possible). Ideally, the measure of the sensitivity to a given site should be measured as a suitable susceptibility to a small perturbation of the same site (or its environment). Unfortunately, this is not easily accomplished experimentally, since substitution by mutation can be rarely regarded as a perturbation. Notwithstanding this difficulty, from the analysis of the ϕ-values obtained by Fersht, a clear picture for the folding stages of CI2 and barnase emerges. In both cases, the crucial regions for both proteins are the same as those identified through the analysis of the DOC reported above. This provides a sound *a posteriori* justification that the main features of the folding of a protein can be followed from a study of the DOC. Remarkably, the method discussed above relies entirely on structure-related properties

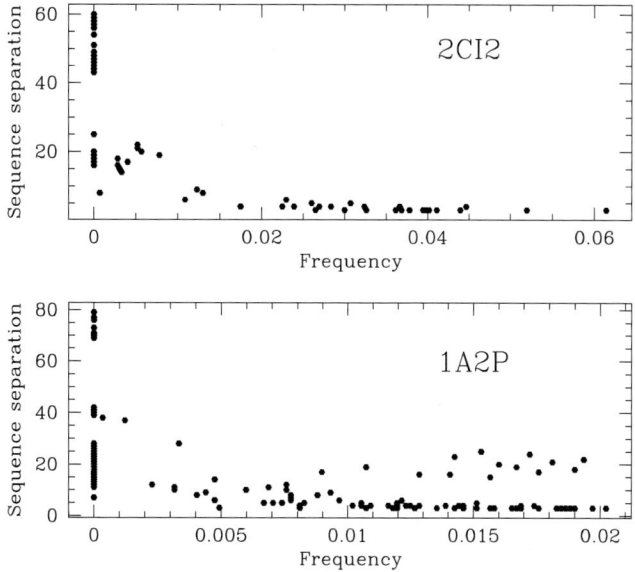

Fig. 10. – Distribution of sequence separation of contacts commonly found in the conformations that overlap with the native state structures of 2ci2 and 1a2p. The most frequent contacts for 2ci2 have a small sequence separation (3–4) and pertain to helix formation. The 1a2p case shows a very different behaviour with several contacts with very large sequence separation.

and suggests that the main features of the folding funnel are determined by the geometry of the "bare" backbone, while the finer details, of course, depend on the specific well-designed sequence. Since our own work in [16], other groups have used similar or alternative techniques to elucidate the role of the native state topology in the folding process [44, 59-61], confirming the picture outlined here.

Let us consider one way in which proteins, in general, are special and different from arbitrary compact polymers. To do so, we turn to an analysis of three proteins of length 51 (1hcg, 1hja and 1sgp) which have nearly the same number of native contacts (≈ 83). For each structure, we calculated the DOC with the constraint that the total number of contacts in the alternative structures do not exceed the number of contacts in the native state by more than 10% to avoid excessive compactness. To assess whether the DOC associated with naturally occurring proteins have special features, we generated three decoy conformations of the same length and number of contacts, but with different degrees of short and long-range contacts (in sequence separation). These decoys (subject to the afore mentioned "physical constraints") were generated with a simulated annealing procedure to find the structure with the highest overlap with a target contact matrix. By tuning the number of short-range *vs.* long-range entries in the target random contact matrix, we generated three structures with different degrees of compactness and local geometrical regularity.

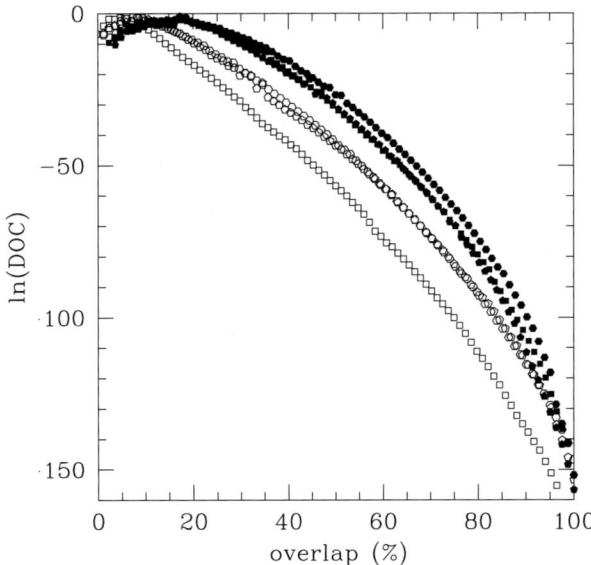

Fig. 11. – Density of overlapping conformations for proteins for 1sgp (filled squares), 1hja (filled pentagons) and 1hcg (filled hexagons). Curves for artificial decoy structures are denoted by the open symbols.

The logarithmic plots of the DOC are shown in fig. 11. A striking feature of the curves is that, for intermediate overlap, the DOC of the real proteins is enormously larger than that of the decoys and suggests that naturally occurring conformations have a much larger number of entryway structures than random compact conformations. Furthermore, for very high values of the overlap, the steepness of the protein curves is much larger than those of the decoys, showing that the reduction in the conformational entropy is also higher. This implies the existence of a funnel with a very large basin and steep walls. Another significant feature is the good collapse of the protein curves. We verified that this feature also obtains for 1bd0 and 2pk4 which each have 80 residues and 140 and 146 contacts, respectively. A simple explanation for the curve collapse could be that the DOC of real proteins is "extremal", in that it is close to the maximum possible value for intermediate values of the overlap.

The importance of the locality of contacts for folding kinetics was highlighted recently by Plaxco et al. [42] who found a correlation between folding rate and contact order, defined as the average sequence separation of contacts normalized to the total number of contacts and sequence length. With reference to fig. 11, the contact order values for proteins 1hcg, 1hja and 1sgp are 0.139, 0.214 and 0.204, respectively. For the decoy structures, they are 0.424, 0.222 and 0.179 for the curves denoted by open squares, pentagons and hexagons, respectively. The structure with an unusually high contact order has the lowest DOC curve and optimal sequences designed on it (or equivalently a Gō-like model) would be expected to exhibit slow-folding dynamics [62] in accord with the findings of Plaxco et al. [42].

Secondary-structure motifs [15,35] have characteristic signatures in the contact maps, such as bands parallel to the diagonal (α-helices and parallel β-sheets) or orthogonal to it (antiparallel β-sheets), as has been shown in the previous section. We have carried out some simple investigations to assess whether a correlation exists between the extremality of the DOC curve and the emergence of secondary-structure–like motifs. We considered a space of contact maps [63], within which each of the residues interacted with the same number of other residues, n_c (typically $n_c = 5$, as in the average case of a protein with about 100 residues and a cut-off distance of 6.5 Å). This space contains maps corresponding to both real structures and unphysical ones. Furthermore, to mimic the effects of the rigidity and geometry of the peptide bond, we disallowed contacts between residue i and the four neighboring residues along the sequence $i \pm 2$, $i \pm 1$.

In this context, the maximization of the density of states corresponds to finding the target matrix with the highest number of matrices sharing a given fraction of its contacts. Although it is difficult to solve this problem, for arbitrary values of the overlap, it is relatively easy to generate matrices with an overlap close to the maximum value, $\bar{q}_{\max}$ (for a $L \times L$ matrix, $\bar{q}_{\max} = L \cdot n_c$). To enumerate all matrices with overlap $\bar{q}_{\max} - 2$, one first identifies a pair of non-zero entries in the target matrix $\bar{m}$: $\bar{m}_{ij} = \bar{m}_{kl} = 1$. Then it is necessary to check whether entries $\bar{m}_{il}$, $\bar{m}_{kj}$ are both "free" (*i.e.* equal to zero) and do not correspond to forbidden contacts (*e.g.*, between i and $i+1$). If this is so, the old pair of entries (and their symmetric counterpart) are set to zero, and the new ones to 1. By considering, in turn, all possible pairs of non-zero entries one can generate all matrices

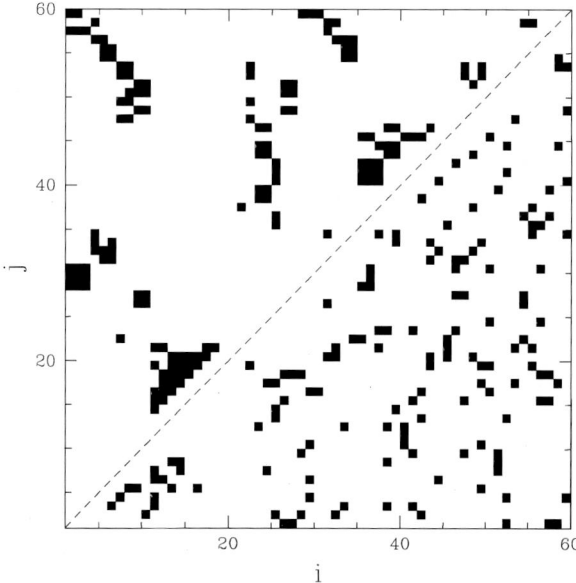

Fig. 12. – The upper (lower) triangle shows a target contact matrix with $L = 60$ that has a large (intermediate) number of contact maps with an overlap of $\bar{q}_{\max} - 2$ contacts.

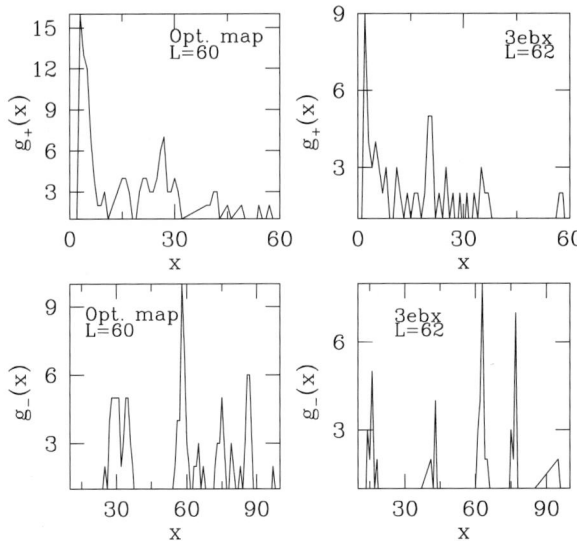

Fig. 13. – Correlation functions (see text) for an optimal target matrix of length 60 and for a protein of length 62 taken from the protein data bank.

of overlap $\bar{q}_{\max} - 2$. Then, by performing a simulated annealing in contact-map space one can isolate the map having the highest number of matrices with overlap $\bar{q}_{\max} - 2$.

We carried out our calculations for values of L around 60. The optimal matrices exhibit clustering reminiscent of α-helices and β-sheets, as shown in the upper triangle of fig. 12. A more quantitative measurement of the secondary-structure content of the optimal matrices can be obtained by considering the correlation functions, $g_{\pm}(x) = \sum_{i} m_{i,i\pm x}$, which show peaks in correspondence with the sequence separation of residues involved in α-helices and parallel β-sheets (g_+) or antiparallel β-sheets (g_-). Typical plots of the correlation functions for an optimal map of length 60 and for the protein 3 ebx (length 62) are shown in fig. 13. The similarity of the plots is striking, particularly because, in both cases, the height of the peaks in g_+ decreases with sequence separation, unlike the situation with g_-.

In summary, the geometry of protein backbones seems to have been optimized to provide a large basin of attraction to the native state. The results presented here are suggestive of an extremality principle underlying the selection of naturally occurring folds of proteins which, in turn, is shown to be possibly associated with the emergence of secondary structures. This observation complements the one made in the previous section, which highlighted how the presence of secondary motifs boosts the folding kinetics. Strikingly, by examining the DOC associated with a given native structure, it is possible to extract a wealth of information about the sites involved in crucial stages of the folding process. Despite the fact that such sites are determined from the analysis of their crucial topological role with respect to the native state with no input of the actual

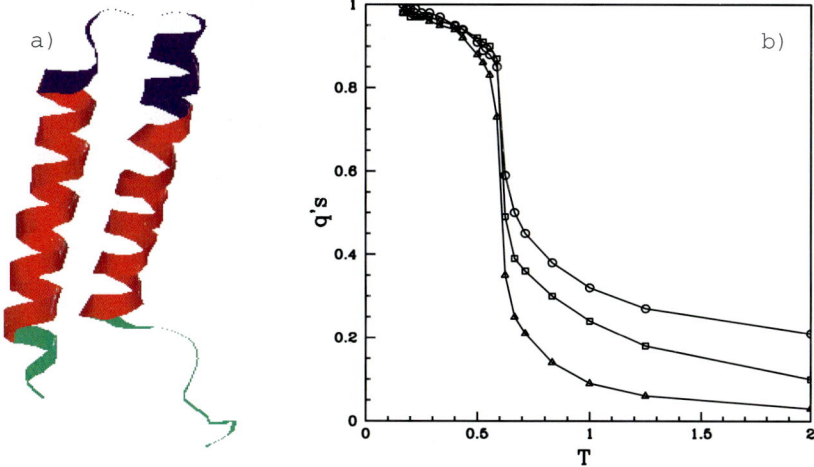

Fig. 16. – Structure and thermodynamics of the helical transmembrane protein. a) Ribbon representation of the two-helix fragment of bacteriorhodopsin formed by the first 66 amino-acids. The part inside the membrane is shown in red, the part above (below) the membrane in blue (green). b) Average equilibrium fraction of native contacts outside, q_b (○), inside, q_m (square), and across, q_s (△), the membrane as a function of the temperature T. All these quantities are expressed in energy unit of ϵ.

spond to the transmembrane segments which are inserted in the lipidic interior of the membrane [91]. These segments are predominantly made up of α-helices and β-sheets. The stability of α-helices and β-sheets inside the membrane follow from the formation of hydrogen bonds between the backbone atoms—other possibilities are excluded within the apolar environment [78, 82].

A detailed study of TMP has not yet been possible because little is known about the amino acids interactions among themselves, with the solvent and in particular with the lipidic interior of the membrane. Here [92], we present a simple strategy to decipher the folding kinetics of transmembrane proteins which is directly inspired by the geometrical approaches we have previously applied to globular proteins. This approach bypasses the details of the complex interactions of the protein in the lipid enviroment by introducing effective potentials, induced by the presence of the membrane and the associated interface region, that stabilize the native-state structure.

Due to the small number of degrees of freedom involved in our scheme, the dynamics of the system can be simulated for the full folding process. Moreover, the free energies of the most relevant intermediate states and free energy profiles along the reaction paths connecting them can be explicitly calculated by thermodynamic integration. Thus the model is able to quantitatively discriminate between the possible reaction paths envisaged for the insertion process of TMP across the membrane [78].

The TMP we considered is made up with the first 66 amino acids (each one represented by a fictitious residue located at the position of the C_α atom) of the first two

α-helices of bacteriorhodopsin (fig. 16a)). It has been shown that the first two helices of bacteriorhodopsin can be considered as independent folding domains [93] and that the side-by-side interactions between transmembrane helices play a key role in the stabilization of the protein structure [94]. The membrane is described simply by a slab of width $w = z_{\max} - z_{\min} = 26$ Å. Two non-bonded residues (i, j) are considered to form a contact if their distance is less than 6.5 Å. In the study of globular proteins, the topology of the native state is encoded in the contact map by considering all the pairs (i, j) of non-consecutive residues that are in contact. Here, in addition, the locations of such pairs with respect to the membrane have to be taken into account. The contacts are divided into three classes:

- *membrane contacts* where both i and j residues are inside the membrane;
- *interface contacts* with i and j in the interface region [78] outside the membrane;
- *surface contacts* with one residue inside the membrane and the other outside.

Thus a given protein conformation can have a native contact but improperly placed with respect to the membrane (*misplaced native contact*). The crucial interaction potential between non-bonded residues (i, j) is taken to be a modified Lennard-Jones (12-10) potential:

$$\text{(6)} \qquad \Gamma(i,j) \left[5 \left(\frac{r^N_{ij}}{r_{ij}} \right)^{12} - 6 \left(\frac{r^N_{ij}}{r_{ij}} \right)^{10} \right] + 5\, \Gamma_1(i,j) \left(\frac{r^N_{ij}}{r_{ij}} \right)^{12}.$$

r_{ij} and r^N_{ij} are the distance between the residues (i, j) and their distance in the native configuration, respectively. The matrices $\Gamma(i,j)$ and $\Gamma_1(i,j)$ encode the topology of the TMP in the following way: if (i,j) is not a contact in the native state $\Gamma(i,j) = 0, \Gamma_1(i,j) = 1$; if (i,j) is a contact in the native state but not at the proper location (*i.e.* a misplaced contact) $\Gamma(i,j) = \epsilon_1, \Gamma_1(i,j) = 0$; if (i,j) is a native state contact in the proper region $\Gamma(i,j) = \epsilon$, $\Gamma_1(i,j) = 0$. This model is intended to describe the folding process in the interface and in the membrane region. Our interaction potential (similar in spirit to the Gō model [37] introduced before) assigns two values to the energy associated with the formation of a native contact, ϵ and ϵ_1. The model captures the tendency to form native contacts. In addition, in order to account for the effective interactions between the membrane and the protein, the model assigns a lower energy, $-\epsilon$, to the contact which occurs in the same region as in the native state structure compared to $-\epsilon_1$ when the contact is formed but in the wrong enviroment. This mechanism proves to be crucial in driving the insertion of the protein across the membrane.

When $\epsilon = \epsilon_1$, the protein does not recognize the presence of the interface-membrane region and the full rotational symmetry is restored. The difference in the parameters $(\epsilon - \epsilon_1)$ determine the amount of tertiary structure formation outside the membrane. Our results are independent of the precise values of the energy parameters ϵ and ϵ_1 ($\epsilon > \epsilon_1$) as long as they are not too close to each other.

Our simulations have been performed with $\epsilon_1 = 0.1$ and $\epsilon = 1$. In order to account for the chirality of the TMP, a potential for the pseudodihedral angle α_i between the C_α atoms in a helix corresponding to four successive locations is added which biases the helices to be in their native-state structure.

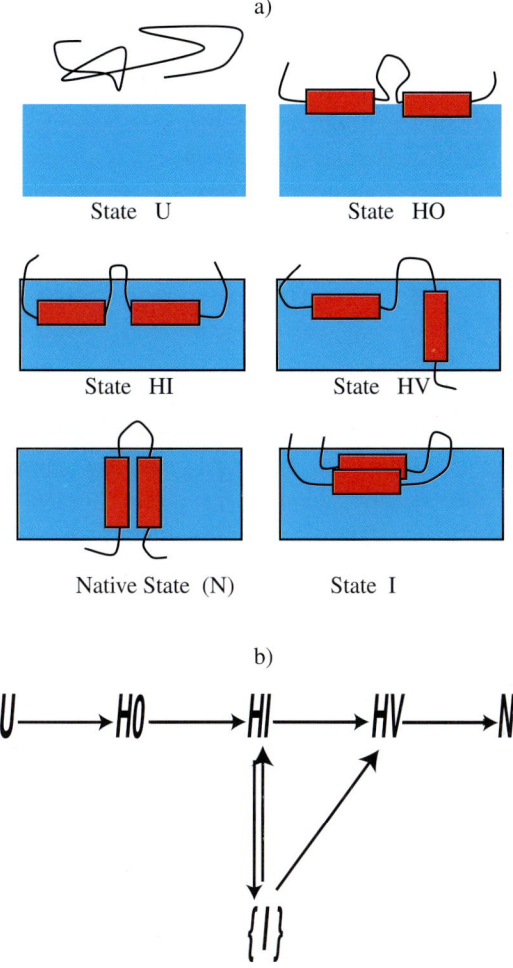

Fig. 17. – Schematic representation of states encountered by our model protein during the folding process. In a) the red cylinders denote α-helices that reside within the membrane in the native state. The region inside the membrane is in turquoise whereas the rest represents the interface region [78]. State U denotes the denatured state of the protein, HO is a state in which the helices have been formed but are not yet inside the membrane, whereas HI corresponds to a similar state but with the helices completely embedded in the membrane without any inter-helical contacts. HV denotes an obligatory intermediate and N depicts the native state. The state {I} represents an ensemble of long-lived conformations in which helices are formed inside the membrane with several inter-helical contacts, but with the two α-helices still incorrectly positioned. In b) the schematic pathways to the native state are shown.

The thermodynamics and the kinetics of the model were studied by a Monte Carlo method for polymer chains carried out in the continuum. The efficiency of the program (usually low for continuum calculations) has been increased by full use of the link cell technique [95] and by the multiple Markov chain method, a new sampling scheme, which has been proven to be particulary efficient in exploring the low-temperature phase diagram for polymers [96]. In our simulation 20 different temperatures (measured in dimensionless units) ranging from $T = 2$ to $T = 0.17$ have been studied.

The free energy difference $\mathcal{F}_B - \mathcal{F}_A$ between two states A and B has been estimated as the reversible work that has to be done in order to go from A to B [92]. The free energy differences obtained with this method are accurate to within $\sim 0.1T_C$ (T_C is defined below) for the various states, whereas the free energy barriers are accurate within $\sim 0.5T_C$. This error takes into account possible hysteresis effects due to the finite simulation time.

The structural similarity between the system equilibrated at temperature T and the native state is shown in fig. 16b) in terms of the average fraction of native-state contacts as a function of T and partitioned depending on their positions with respect to the membrane. The three curves correspond respectively to the average fraction of native contacts inside (q_m), outside (q_b) and across (q_s) the membrane. All these curves, well separated at high T, collapse for T below the transition temperature $T_C \sim 0.6$, indicating a cooperative effect in the folding. On monitoring the free energy as a function of the energy around T_C, one observes additional local minima (besides those corresponding to the unfolded and folded states) suggesting the presence of an intermediate.

The intermediate is characterized by having the two helices almost completely formed but not yet correctly inserted across the membrane. The presence of these extra minima suggests that non-constitutive membrane proteins would fold with multi-state kinetics corresponding to on-pathway intermediates. To establish the nature of the dominant folding pathway, we have performed a detailed analysis of the folding kinetics. Each independent kinetic folding simulation was started with the equilibrated denatured state at $T^* = 2.5$. The protein is placed initially outside the membrane in the interface region [78], at a distance comparable to the average size of the denatured protein and then suddenly quenched to a temperature ($T = 0.4$) well below the transition temperature. This case simulates the folding kinetics of non-constitutive membrane proteins, *i.e.* proteins that do not need a translocon providing a "tunnel" through which the protein is injected into the lipid bilayer. Folding to the native state occurs mainly through the states depicted in fig. 17a) with the dominant pathways shown in fig. 17b).

In all the pathways, the system goes from the unfolded state, U to state HI in which 80% of the secondary structure is formed (see q in fig. 18(c)) and disposed horizontally along the interface. The free energy of this state (measured with respect to the free energy of the fully folded state) is $\sim 2.4T_C$. This state corresponds to the formation of around 70% of the membrane contacts. The average time τ_{HI} to reach state HI is of the order of 500 Monte Carlo steps (see fig. 18; each Monte Carlo step corresponds to 50000 attempted local deformations). State HI turns out to be an obligatory on-pathway intermediate of the folding kinetics for non-constitutive MP in agreement with

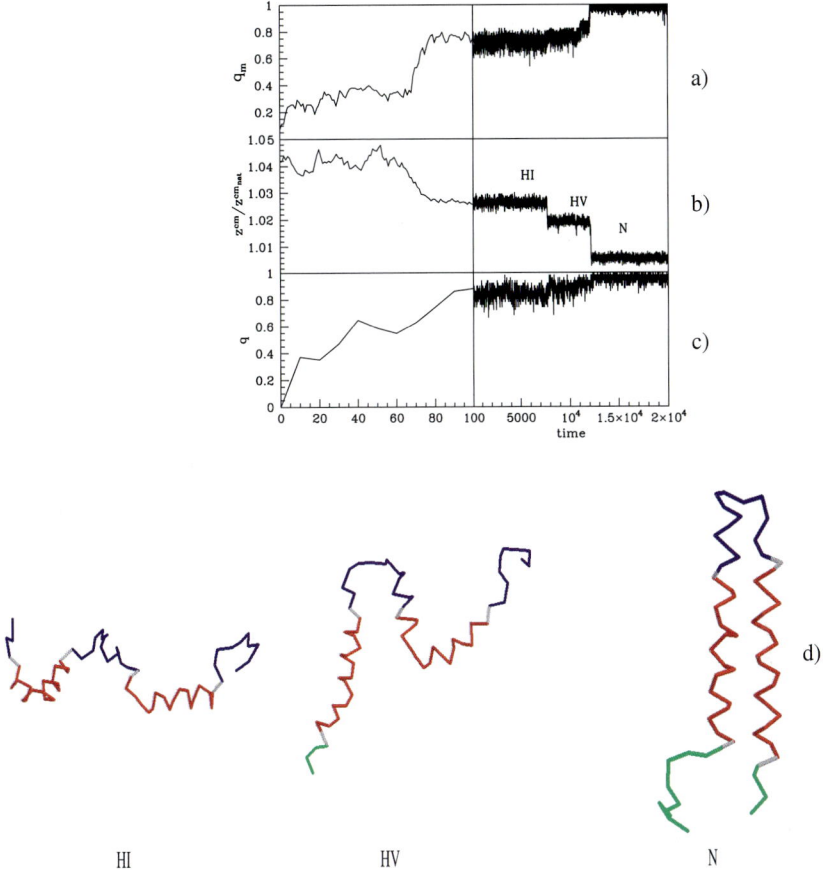

Fig. 18. – Typical time dependence of different parameters as a function of the Monte Carlo steps for the pathway U → HI → HV → N. Fraction of native contacts inside the membrane (a), normalized z-coordinate of the center-of-mass of the protein (with respect to that of the native state conformation) (b) and overall fraction of native helical contacts (c). (d) Protein conformations at different times during the folding. The colours red, green and blue have the same significance as in fig. 1a) with the grey bonds being the ones crossing the membrane.

the general argument mentioned above. Once the protein reaches state HI, it undergoes a relatively slow process of self-arrangement in order to insert and assemble the secondary structures across the membrane. This process is the rate-limiting step of the folding process, since it involves the translocation, through the lipidic layer, of a substantial number of hydrophilic residues. Among the possible pathways, starting from HI, the most frequent (60% of the cases) and the fastest turn out to be U → HI → HV → N. A quantitative characterization of this dominant pathway is presented in fig. 18 (for a single folding process). The intermediate HV is characterized by having one α-helix inserted across the membrane and is reached in an average period corresponding to a significant fraction of the total folding time (see fig. 16). The free energy in this state is $\sim 0.98 T_C$. The free energy barrier between HI and HV is at $\sim 4.31 T_C$ (hence, the rate constant of the transition HI $\to$ HV is proportional to $k_{\text{HI}\to\text{HV}} = \exp[-(4.31 - 2.4) T_C/T])$.

The last part of the folding process corresponds to the insertion of the second helix and the assembly of the two secondary structures into the native-state structure. This process lasts approximately one third of the folding time along the pathway U $\to$ HI $\to$ HV $\to$ N. The quasistatic free energy barrier between HV and the folded state is $\sim 1.66 T_C$. The rate constant of the transition HV $\to$ N is, therefore, proportional to $\exp[-(1.66-0.98) T_C/T]$. These results are consistent with the time scales observed in the unconstrained folding dynamics. At the end, the protein is completely packed (q_m saturates to 1 (fig. 18a) and the helices are correctly positioned across the membrane (note the second jump in the z coordinate of the center of mass in fig. 18b).

Much slower dynamics can occur when non-obligatory intermediates are visited by the system. These long-lived states ({I} in fig. 17a)) involve a distribution of misfolded regions that trap the system and are characterized by having most of the inter-helical contacts formed (assembly of the secondary structures) but with the two α-helices still incorrectly positioned. Note, for example, that in states {I}, only transmembrane contacts and some contacts outside the membrane are misplaced and they account for only a small fraction of the native-state energy. For this reason, in the states {I}, the free energy is $\sim 1.44 T_C$, only slightly higher than the free energy of HV. The folding can proceed from {I} either by disentangling the two helices and passing through the obligatory intermediate HV, or by the simultaneous translocation through the membrane of the two helices. These processes, however, entail the crossing of a big free energy barrier ($\sim 5.18 T_C$ for the first process and $6.1 T_C$ for the second) and happen with low probability. Indeed, at sufficiently low temperatures, the loss in energy of the interhelical contacts is not compensated by the gain in the configurational entropy due to the uncoupling of the α-helices. Thus below the folding temperature, I-states act as trapping regions for the system and when trapped, the protein spends most of the time during folding in this state.

In summary, we have shown that a topology based model can lead to a vivid picture of the folding process. Our approach predicts a folding process involving multiple pathways with a dominant folding channel. Further details not captured by the present approach may of course change the quantitative nature of the results. However, the model, which captures the bare essentials of a membrane protein, ought to provide a zeroth-order picture of the folding process. Also, as experimental data becomes available, the results could be benchmarked with models of this type to determine the nature of the other factors that matter.

* * *

We are indebted to F. CECCONI, A. LAIO, E. ORLANDINI, G. SETTANNI and A. TROVATO who have contributed to the results discussed in these lectures.

REFERENCES

[1] PAULING L., COREY R. B. and BRANSON H. R., *Proc. Natl. Acad. Sci USA*, **37** (1951) 205.
[2] ZIMM B. H. and BRAGG J., *J. Chem. Phys.*, **31** (1959) 526; PTITSYN O. B. and SKVORTSOV A. M., *Biophys.*, **10** (1965) 1007; LIFSHITZ I. M., GROSBERG A. Y. and KHOKHLOV A. R., *Rev. Mod. Phys.*, **50** (1978) 683.
[3] HUNT N. G., GREGORET L. M. and COHEN F. E., *J. Mol. Biol.*, **241** (1994) 214.
[4] AURORA R., CREAMER T. P., SRINIVASAN R. and ROSE G. D., *J. Mol. Biol.*, **272** (1997) 1413.
[5] NELSON J. C., SAVEN J. G., MOORE J. S. and WOLYNES P. G., *Science*, **277** (1997) 1793.
[6] YEE D. P., CHAN H. S., HAVEL T. F. and DILL K. A., *J. Mol. Biol.*, **241** (1994) 557.
[7] SOCCI N. D., BIALEK W. S. and ONUCHIC J. N., *Phis. Rev. E*, **49** (1994) 3440.
[8] HALPERIN A. and GOLDBART P. M., *Phys. Rev. E*, **61** (2000) 565.
[9] BRYNGELSON J. D. and WOLYNES P. G., *Proc. Natl. Acad. Sci. USA*, **84** (1987) 7524.
[10] ONUCHIC J. N., LUTHEY SCHULTEN Z. and WOLYNES P. G., *Annu. Rev. Phys. Chem.*, **48** (1997) 545.
[11] LEOPOLD P. E., MONTAL M. and ONUCHIC J. N., *Proc. Natl. Acad. Sci. USA*, **89** (1992) 8721.
[12] WOLYNES P. G., ONUCHIC J. N. and THIRUMALAI D., *Science*, **267** (1995) 1619.
[13] DILL K. A. and CHAN H. S., *Nature Struct. Biol.*, **4** (1997) 10.
[14] BUCHLER N. E. G. and GOLDSTEIN R. A., *Proteins: Struct. Funct. Genet.*, **34** (1999) 113; MICHELETTI C., MARITAN A., BANAVAR J. and SENO F., *Phys. Rev. Lett.*, **80** (1998) 5683; MICHELETTI C., MARITAN A. and BANAVAR J., *J. Chem. Phys.*, **110** (1999) 9730.
[15] LI H., HELLING R., TANG C. and WINGREEN N., *Science*, **273** (1996) 666.
[16] MICHELETTI C., BANAVAR J., MARITAN A. and SENO F., *Phys. Rev. Lett.*, **82** (1999) 3372.
[17] *Nature Structural Biology: Editorial*, **6** (1999) 1.
[18] SLOANE N. J. A., *Nature*, **395** (1998) 435.
[19] MACKENZIE D., *Science*, **285** (1999) 1339.
[20] WOODCOCK L. V., *Nature*, **385** (1997) 141.
[21] CAR R., *Nature*, **385** (1997) 115.
[22] CIPRA B., *Science*, **281** (1998) 1267.
[23] STEWART I., *Sci. Am.*, **278** (1998) 80.
[24] MARITAN A., MICHELETTI C., TROVATO A. and BANAVAR R., *Nature*, **406** (2000) 287.
[25] BUCK G. and ORLOFF J., *Topol. Appl.*, **61** (1995) 205.
[26] KATRITCH V., BEDNAR J., MICHOUD D., SCHAREIN R. G., DUBOCHET J. and STASIAK A., *Nature*, **384** (1996) 142.
[27] KATRITCH V., OLSON W. K., PIERANSKI P., DUBOCHET J. and STASIAK A., *Nature*, **388** (1997) 148.
[28] GONZALEZ O. and MADDOCKS J. H., *Proc. Natl. Acad. Sci. USA*, **96** (1999) 4769.
[29] BUCK G., *Nature*, **392** (1998) 238.
[30] CANTARELLA J., KUSNER R. B. and SULLIVAN J. M., *Nature*, **392** (1998) 237.
[31] SOKAL A. D., *Nucl. Phys. B*, **47** (1996) 172.
[32] ROSE G. D. and SELTZER J. P., *J. Mol. Biol.*, **113** (1977) 153.
[33] MARITAN A., MICHELETTI C. and BANAVAR J. R., *Phys. Rev. Lett.*, **84** (2000) 3009.
[34] SALI A., SHAKHNOVICH E. and KARPLUS M., *Nature*, **369** (1994) 248.
[35] CREIGHTON T. E., *Proteins - Structures and Molecular Properties* (W. H. Freeman and Company, New York) 1993, pp. 182-188.

[36] LEVINTHAL C., *J. Chem. Phys.*, **65** (1968) 44.
[37] GO N. and SCHERAGA H. A., *Macromolecules*, **9** (1976) 535.
[38] COVELL D. G. and JERNIGAN R., *Biochemistry*, **29** (1990) 3287.
[39] PARK B. H. and LEVITT M., *J. Mol. Biol.*, **249** (1995) 493.
[40] The native state structures of monomeric proteins of length between 50 and 200 show an excellent correlation of this form, when two non-consecutive amino acids along the sequence are defined to be in contact when they are within 6.5 Å of each other.
[41] HOLLAND J. H., *Adaptation in Natural and Artificial Systems* (MIT Press) 1992.
[42] PLAXCO K. M., SIMONS K. L. and BAKER D., *J. Mol. Biol.*, **277** (1998) 985.
[43] CAPALDI A. P. and RADFORD S. E., *Curr. Opin. Struct. Biol.*, **8** (1998) 86.
[44] MUÑOZ V., HENRY E. R., HOFRICHTER J. and EATON W. A., *Proc. Natl. Acad. Sci. USA*, **95** (1998) 5872.
[45] ANFINSEN C., *Science*, **181** (1973) 223.
[46] KARPLUS M. and WEAVER D. L., *Nature*, **260** (1976) 404; *Protein Sci.*, **3** (1994) 650.
[47] PTITSTYIN O. B., *FEBS Lett.*, **285** (1991) 176.
[48] CHAN H. S. and DILL K. A., *J. Chem. Phys.*, **99** (1994) 2116.
[49] ONUCHIC J. N., WOLYNES P. G., LUTHEY-SCHULTEN Z. and SOCCI N. D., *Proc. Natl. Acad. Sci. USA*, **92** (1995) 3626.
[50] NYMEYER H., GARCIA A. E. and ONUCHIC J. N., *Proc. Natl. Acad. Sci. USA*, **95** (1998) 5921.
[51] BRYNGELSON J. D., ONUCHIC J. N., SOCCI N. D. and WOLYNES P. G., *Proteins: Struct. Funct. Gen.*, **21** (1995) 167.
[52] CAMACHO C. J. and THIRUMALAI D., *Proc. Natl. Acad. Sci. USA*, **90** (1993) 6369.
[53] KOLINSKI A. and SKOLNICK J., *J. Chem. Phys.*, **97** (1992) 9412.
[54] FERRENBERG A. M. and SWENDSEN R. H., *Phys. Rev. Lett.*, **63** (1989) 1195.
[55] The degree of the overlap may be benchmarked against the typical overlap of any two compact conformations, which is about 10–20% (consistent with the unfolding simulations of ref. [56]). A value of $q \approx 20\%$ thus corresponds to a "molten globule" state.
[56] LAZARIDIS T. and KARPLUS M., *Science*, **278** (1997) 1928.
[57] FERSHT A. R., *Proc. Natl. Acad. Sci. USA*, **92** (1995) 10869.
[58] ITZHAKI L. S., OTZEN D. E. and FERSHT A. R., *J. Mol. Biol.*, **254** (1995) 260.
[59] GALZITSKAYA O. V. and FINKELSTEIN A. V., *Proc. Natl. Acad. Sci. USA*, **96** (1999) 11299.
[60] ALM E. and BAKER D., *Proc. Natl. Acad. Sci. USA*, **96** (1999) 11305.
[61] CLEMENTI C., NYMEYER H. and ONUCHIC J. N., *J. Mol. Biol.*, **298** (2000) 937.
[62] WOLYNES P. G., *Proc. Natl. Acad. Sci. USA*, **94** (1997) 6170; PLOTKIN S., WANG J. and WOLYNES P. G., *J. Chem. Phys.*, **106** (1997) 2932.
[63] LEVITT M., *J. Mol. Biol.*, **104** (1976) 59.
[64] CECCONI F., MICHELETTI C., CARLONI P. and MARITAN A., *Proteins: Struct. Funct. Genet.*, **43** (2001) 365.
[65] CLEMENTI C., CARLONI P. and MARITAN A., *Proc. Natl. Acad. Sci. USA*, **96** (1999) 9616.
[66] EVANS D. J., HOOVER W. G., FAILOR B. H., MORAN B. and LADD A. J. C., *Phys. Rev. A*, **28** (1983) 1016.
[67] BROWN A. J., KORBER B. T. and CONDRA J. H., *AIDS Res. Hum. Retroviruses*, **15** (1999) 247.
[68] ALA P. J. *et al.*, *Biochemistry*, **37** (1998) 15042.
[69] MOLLA A. *et al.*, *Nat. Med.*, **2** (1996) 760.
[70] MARKOWITZ M. *et al.*, *J. Virol.*, **69** (1995) 701.
[71] PATICK A. K. *et al.*, *Antimicrob. Agents Chemother.*, **40** (1996) 292.
[72] CONDRA J. H. *et al.*, *Nature*, **374** (1995) 569.

[73] TISDALE M. et al., *Antimicrob. Agents Chemother.*, **39** (1995) 1704.
[74] JACOBSEN H. et al., *J. Infect. Dis.*, **173** (1996) 1379.
[75] REDDY P. and ROSS J., *Formulary*, **34** (1999) 567.
[76] FERSHT A. R., *Structure and Mechanism in Protein Science* (W. H. Freeman, New York) 1999.
[77] KARPLUS M. and SALI A., *Curr. Opin. Struct. Biol.*, **5** (1995) 58.
[78] WHITE S. H. and WIMLEY W. C., *Ann. Rev. Biophys. Biomol. Struct.*, **28** (1999) 319.
[79] OSTERMEIER C. and MICHEL H., *Curr. Opin. Struct. Biol.*, **7** (1997) 697.
[80] VON HEIJNE G., *Prog. Biophys. Molec. Biol.*, **66** (1996) 113.
[81] BOOTH P. J., *Folding & Design*, **2** (1997) R85.
[82] POPOT J. L. and ENGELMAN D. M., *Biochemistry*, **29** (1990) 4031.
[83] PAPPU R. V., MARSHALL G. R. and PONDER J. W., *Nature Struct. Biol.*, **6** (1999) 50.
[84] MILIK M. and SKOLNICK J., *Proc. Natl. Acad. Sci. USA*, **89** (1992) 9391.
[85] MILIK M. and SKOLNICK J., *Proteins: Struct. Funct. Gen.*, **15** (1993) 10.
[86] JACOBS R. E. and WHITE S. H., *Biochemistry*, **26** (1987) 6127.
[87] ROSEMAN M. A., *J. Mol. Biol.*, **200** (1988) 513.
[88] WIMLEY S. C. and WHITE S. H., *Designing Transmebrane α-Helices That Insert Spontaneously*, preprint (University of California, Irvine) 2000.
[89] BONACCINI R. and SENO F., *Phys. Rev. E*, **60** (1999) 7290.
[90] BIGGIN P. C. and SANSOM M. S. P., *Biophysical Chemistry*, **76** (1999) 161.
[91] DEBER C. M. and GOTO N. K., *Nature Struct. Biol.*, **3** (1996) 815.
[92] ORLANDINI E., SENO F., BANAVAR J. R., LAIO A. and MARITAN A., *Deciphering the Folding Kinetics of Transmembrane Helical Proteins*, *Proc. Natl. Acad. Sci. USA*, **97** (2000) 14229.
[93] KAHN T. W., STURTEVANT J. M. and ENGELMAN D. M., *Biochemistry*, **31** (1992) 8829.
[94] KAHN T. W. and ENGELMAN D. M., *Biochemistry*, **31** (1992) 6144.
[95] GEROFF I., MILCHEV A., BINDER, K. and PAUL W., *J. Chem. Phys.*, **98** (1993) 6256.
[96] TESI M. C., VAN RENSBURG E. J., ORLANDINI E. and WHITTINGTON S. G., *J. Stat. Phys.*, **29** (1996) 2451.

The folding thermodynamics and kinetics of crambin using an all-atom Monte Carlo simulation

J. Shimada, E. L. Kussell and E. I. Shakhnovich

Department of Chemistry and Chemical Biology, Harvard University
12 Oxford Street, Cambridge, MA 02138, USA

1. – Introduction

Previous lattice and off-lattice folding simulations [1-10] featured coarse-grained protein representations where amino acids are often modeled as spheres. While such simplified representations have their advantages, such as the reduction in the number of degrees of freedom, there are several shortcomings. First, one could argue that such models do not capture the full complexity of conformational space, thereby not truly addressing the Levinthal problem. Second, coarse-grained models may lack some realistic features of secondary-structure elements. Finally, such models do not address the packing of sidechains in the protein interior, which is often viewed as an important aspect of the folding process, given the diversity of sidechain shapes and their dense packing in the native state [11].

Molecular dynamics approaches have attempted to bridge the gap between simulation and reality by using atomic representations of proteins with explicit solvent molecules. Unfortunately, the computational times for such minimally coarse-grained simulations still remains prohibitively high as evidenced by a recent work that yielded only a partial folding trajectory [12] even with the aid of a massively parallel supercomputer. An ensemble of complete folding trajectories is required in order to gain meaningful physical insights, especially as our theoretical understanding of protein folding has become increasingly grounded on statistical mechanical principles [13]. Other molecular dynamics research efforts have tried to address this issue by obtaining multiple unfolding runs [14,15] at extremely high temperatures.

We present a Monte Carlo (MC) simulation [16] which combines an all-atom description of the protein with coarse-grained motions and energetics. In this simulation, (1) all heavy atoms in the protein are represented as impenetrable spheres, (2) all backbone and sidechain torsions, which account for all degrees of freedom relevant to folding, are allowed to move, and (3) a square well, Gō potential is used for the interaction energy. There are several advantages to this method. First, a statistically significant number of complete folding trajectories can be collected using conventional computational resources. Second, the atomic-level resolution of the simulation yields microscopic descriptions of the folding process of actual protein structures including sidechain packing. Finally, this coarse-grained approach allows for a systematic investigation of the physical principles dictating protein folding. If one begins with a detailed energy function as in molecular dynamics, it may be difficult to deconvolute exactly which energy terms were essential (or inessential) for folding. On the other hand, as was demonstrated in theoretical investigations [17], by thoroughly investigating the successes and limitations of coarse-grained models, it is possible to test which features of a model are necessary and/or sufficient for describing the complex physics of heteropolymers.

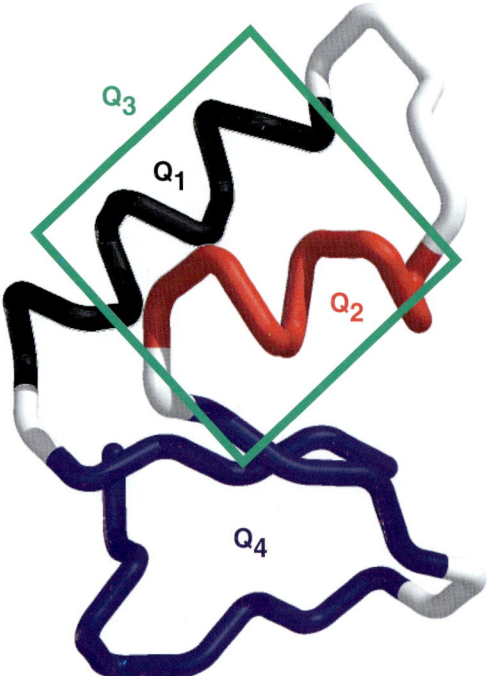

Fig. 1. – The 46-residue protein crambin. The important secondary/tertiary structural elements are indicated by different colors. Black: helix 1 (residues 6–18); red: helix 2 (23–30); green: interhelix contacts; blue: β-sheet (1–5, 32–35, 38–46). As shown by the matching colors, the Q_i parameters are defined as the fraction of native contacts in specific structural elements: Q_1-helix 1; Q_2-helix 2; Q_3-interhelix contacts; Q_4-β-sheet.

Using our simulation, we have repeatedly folded the 46-residue α-β protein crambin (fig. 1) to within 1 Å backbone dRMS (computed over C-α atoms). Furthermore, the thermodynamic and kinetic folding properties obtained from our simulation are consistent with experimental studies of other single domain proteins [18, 19], for which the folding transition generally exhibits two-state behavior with no accumulating intermediates between the denatured and native states [20].

One of the more successful methods reported in the literature generated folded structures for crambin to ≈ 3 Å C-α dRMS [21] using a sequence-based potential. These final structures were obtained from a two-step heuristic approach: simplified structures obtained from folding runs on a coarser lattice were then refined on a finer lattice. In contrast, our method uses a potential based on knowledge of the native structure, but generates the entire folding transition without altering key conditions (such as the move set, temperature, or protein representation) once the simulation is initiated. As such, it produces an ensemble of trajectories which can yield thermodynamic and kinetic information. Although the potential we employ cannot be used to fold arbitrary protein sequences, we demonstrate that the full conformational search problem can be solved using standard computer resources.

2. – Simulation method

Full atom representation. Each non-hydrogen atom present in the crambin crystal structure [22] (Brookhaven PDB accession code: 1AB1) was represented by a hard sphere, whose size was given by scaling the relevant VdW radius (r) from ref. [23] by a factor $\alpha (< 1)$.

Move set. A single MC step consisted of a backbone move followed by 10 sidechain moves. Each backbone and sidechain move was accepted according to the Metropolis criterion [24]. A backbone move consisted of rotating the ϕ-ψ angles of up to 3 non-proline residues from a randomly selected window of 6 consecutive residues. A sidechain move consisted of rotating all sidechain torsion angles (χ) of a randomly selected non-proline residue. The size of the backbone and sidechain rotations were obtained from a Gaussian distribution with zero mean and standard deviation 2 and 10 degrees, respectively.

Square-well Gō potential. We used an atomic square-well potential [25] with the well depths given by Gō energetics [26]. In particular, for two atoms A and B separated by a distance R, the energy $\epsilon(A, B)$ was calculated according to

$$\epsilon(A, B) = \begin{cases} \infty, & R < \sigma, \\ \Delta(A, B), & \sigma \leq R < \lambda\sigma, \\ 0, & R \geq \lambda\sigma, \end{cases}$$

where $\sigma = \alpha(r_A + r_B)$ is the hard-core distance, λ is a scaling factor > 1, and $\Delta(A, B) = -1$ if A and B are in contact in the native conformation and 1 otherwise. The total energy of a conformation was computed as the sum over all pairs:

$$E = \sum_{\text{all pairs}} \epsilon(A,B).$$

All atom pairs of $i - i+1$ residues were excluded to eliminate any biases towards local structure, and all backbone-backbone contacts were ignored to eliminate non-specific interactions. The energies of the disulfide bonds were treated no differently from any other contact. We chose $\alpha = 0.75$ because it was the largest value for which the native structure exhibited no steric clashes. Furthermore, with $\alpha = 0.75$, we could not fold crambin with the sidechain torsions held fixed at their native values, suggesting that α was sufficiently large to enforce excluded-volume constraints. The selection of small λ values (≤ 1.6) significantly increased the time of collapse, while large λ values (≥ 2.0) made sidechain packing more degenerate. We therefore selected $\lambda = 1.8$ in order to balance the two effects. This makes the contact distance for methyl carbons to be $5.08\,\text{Å}$.

Folding simulations. Random coils were generated by unfolding from the native state for 3×10^5 steps with only the excluded-volume interaction turned on. Both the average energy ($E = 77$, $Q = 0.02$) and average structure ($R_g = 19.5$, backbone dRMS = 17.0) of the random coils indicate that these conformations are completely unfolded and unstructured. Each random coil was then simulated with the square-well Gō potential turned on at a particular temperature until it folded or 10^8 MC steps elapsed. 2.5×10^6 MC steps approximately took 1 hour of computation time on a Pentium III 550 Mhz PC.

3. – Results

We considered the protein folded if the following three criteria were satisfied for at least 5×10^6 steps: 1) the fraction of native contacts (Q) exceeded 0.7; 2) the backbone dRMS was less than $1.25\,\text{Å}$; 3) the fraction of native contacts in the four secondary/tertiary structural features (Q_i, $1 \leq i \leq 4$) exceeded 0.5 (see fig. 1). Criterion 1), which is similar to the one used to signify a folding event in lattice simulations [27], could not by itself distinguish conformations that one might intuitively consider folded (*i.e.*, properly formed secondary structure and low overall dRMS) from obvious misfolds. Although our definition for folding did not measure when equilibrium was attained, it was useful for identifying when the major folding event—the transition from the random coil to a near-native state—occurred.

Thermodynamics. The folding thermodynamics is aptly described by a first-order–like transition [28] (fig. 2), with the MC folding transition temperature (T_f) estimated to be between 2.1 and 2.3. For high temperatures ($2.0 < T < 3.0$), the data is well fit by the exponential function typically used to describe two-state folding behavior [29]. This is in general agreement with the folding thermodynamics obtained with lattice models [2]. By matching the backbone dRMS value from NMR measurements ($1.48\,\text{Å}$) [30], we estimate

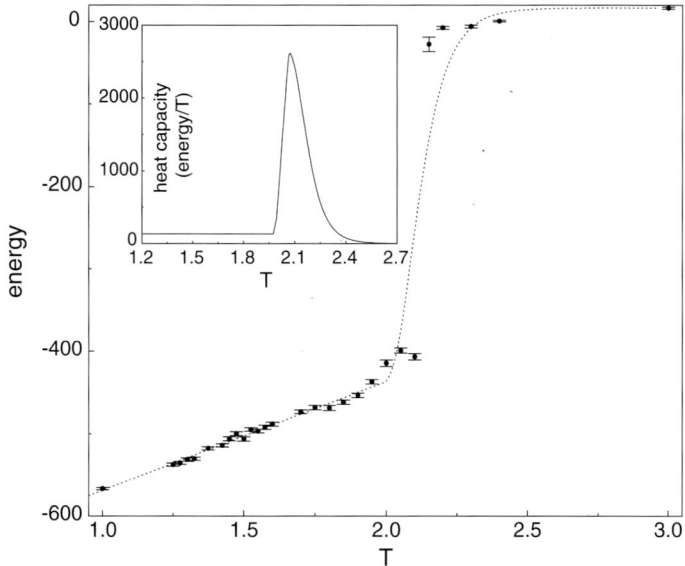

Fig. 2. – The energy and heat capacity (shown as insert) of crambin as a function of temperature (T). The energy data (•) were collected from uncorrelated structures sampled from simulations of 10^8 MC steps initiated from the native state. For $T < 2.0$, the energy data was fitted to a linear function ($R^2 = 0.987$), while for $T \geq 2.0$, the exponential function $f(E,T) = E_n + (E_u - E_n)\exp[-C/T+D]/(1+\exp[-C/T+D])$ ($C = 70.82$, $D = 34.16$, E_n = native energy = -658, E_u = unfolded energy = 16.67; estimated $\sigma \approx 90$) was used. The heat capacity (C_V) was obtained by evaluating $C_V = \mathrm{d}E/\mathrm{d}T$ on the fitted curves. We note that because our simulation does not explicitly model solvent-protein interactions, the computed heat capacity falls to zero for $T > T_f$, in contrast to experimental observations.

that room temperature corresponds to $2.1 < T < 2.2$, which is consistent with the general observation that most proteins are marginally stable at room temperature [31].

Kinetics. Representative folding runs are shown in fig. 3a)-e). Of the 250 runs completed for $T = 1.25$ to 1.875, 165 folded within our observation window of 10^8 steps. For $1.4 \leq T \leq 1.8$, 135 out of 160 (= 84 %) runs folded. Although different folding pathways were observed (fig. 4), a fast-folding pathway was found where kinetic traps were avoided. The ensemble of trajectories following this fast-folding pathway was characterized by a definite sequence of events: 1) formation of the interhelix contacts (event 2 in fig. 4); 2) formation of the two helices (event 3); and 3) formation of the β-sheet (events 4 and 5). This observation can be rationalized from simple topological considerations: the helices are most easily formed when the two ends of the polymer are not constrained by the β-sheet.

At high temperatures ($1.7 < T < 1.8$), at least half of the folding trajectories did not collapse within 10^8 steps, making clear that collapse is the rate-limiting step

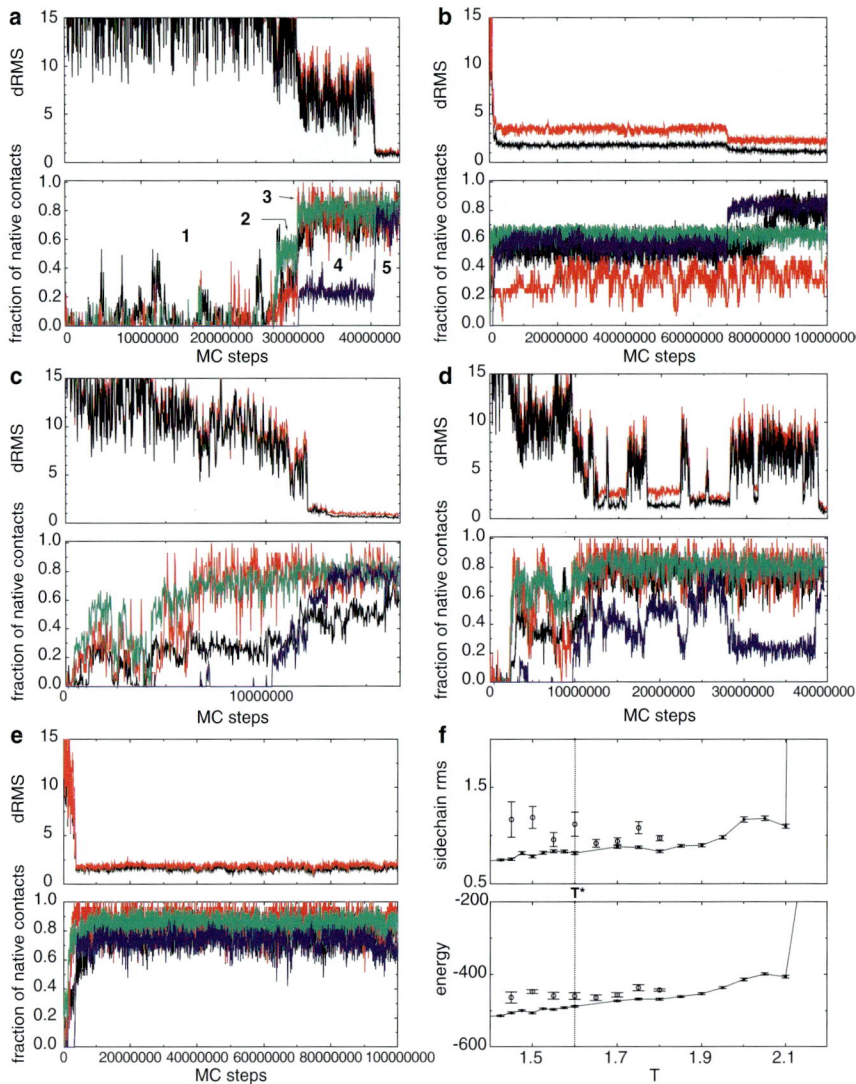

Fig. 3. – Typical folding runs at various temperatures. The upper panel for each run shows backbone (black) and sidechain (red) dRMS as the runs progresses. The lower panel tracks the four Q_i values, with the color coding shown in fig. 2a). Note that the lengths are different for each run. a) Collapse-rate limited cooperative folding at high temperature ($T = 1.875$). The events labeled 1–5 in the Q_i plot correspond to those noted on the fast-folding pathway in fig. 4. b) A trajectory ending with a helix 2 misfold at low temperature ($T = 1.25$). c) Successful folding after encountering a helix 1 misfold at high temperature ($T = 1.775$). d) Successful folding after encountering a metastable β-sheet misfold at high temperature ($T = 1.825$). e) A trajectory ending at a low temperature sidechain-packing trap ($T = 1.425$). f) Broken ergodicity for $T < T^* \approx 1.6$. For each temperature, 52 runs which folded to near-native conformations (dRMS < 1.25 and Q_i's > 0.6) were each extended for an additional 25×10^6 steps. The average of these runs is indicated by ∘. The average obtained from simulations started from the native state (see fig. 2) is indicated by •.

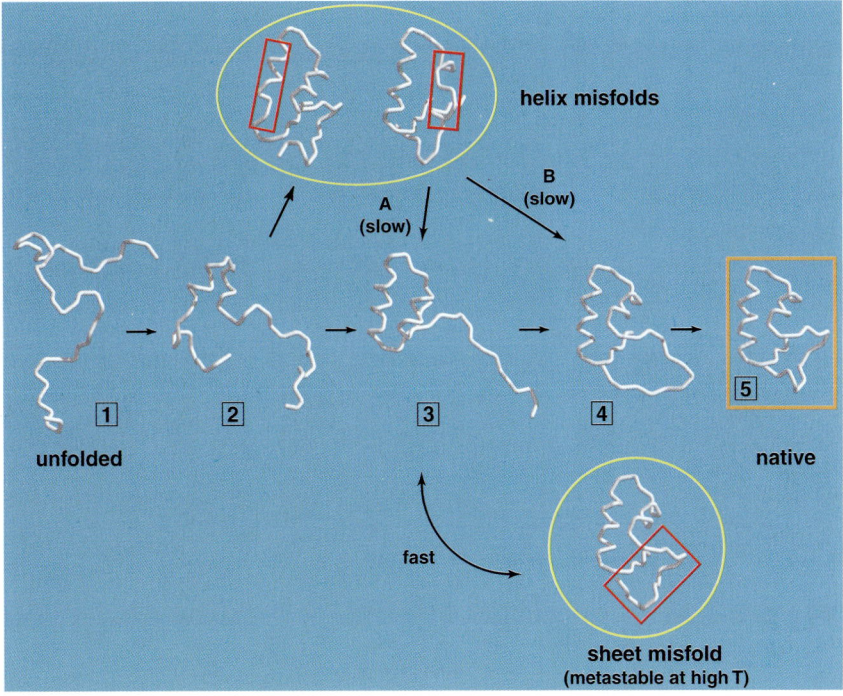

Fig. 4. – Summary of the folding kinetics. The successive events observed along the fast-folding pathway are marked by boxed numbers. Note that the sidechain packing trap is not indicated in this figure.

(figs. 3a), 5a)). The kinetics appear to be two state, as evidenced by the narrowly distributed near-native and unfolded populations, rapid collapse events, and no accumulating intermediates. The transiently populated state with energy -150 corresponds to event 2 of the fast-folding pathway (fig. 4) where a subset of the interhelix and helix contacts have been formed. At middle temperatures ($1.55 < T < 1.675$), this intermediate accumulates for longer times and the collapsed state ensemble broadens (fig. 5b)). At low temperatures ($T < 1.525$) this effect becomes particularly pronounced, with high energy, collapsed states accumulating for very long times (fig. 5c)). While collapse is very fast at these temperatures, the persistently broad distribution of low dRMS (< 2 Å) states indicates that the compact state is riddled with deep traps. Finally, the median folding time rapidly increased for $T < 1.4$ and $T > 1.8$, and a broad minimum existed at $1.5 < T < 1.65$. More runs are needed to narrow the fastest folding temperature to a smaller range.

In separate simulations of the isolated helix 1 (residues 6-18), we observed a sharp, cooperative coil-helix transition at $T_f^{\text{helix 1}} \approx 1.1$. This result is in complete agreement with the zero helical propensity predicted by AGADIR [32] (at $pH = 7$ and $T = 298$ K), since the temperatures relevant for our study are considerably higher than $T_f^{\text{helix 1}}$. This likely explains why the diffusion-collision scenario [33] was not observed as the major

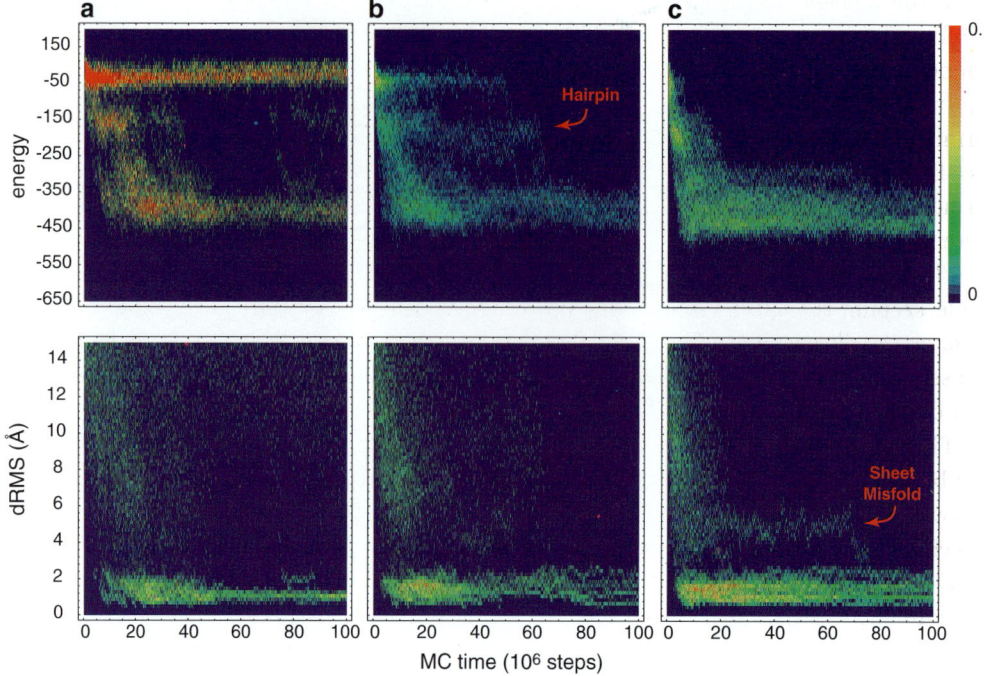

Fig. 5. – Energy and backbone dRMS kinetics histograms as a function of MC time of all runs at high ($1.7 < T < 1.8$; panel a), middle ($1.5 < T < 1.7$; panel b), low ($T < 1.5$; panel c) temperatures. The color indicates the fraction of runs at a particular energy or dRMS at a given MC time. The color scale goes from blue (0.0) to red (0.1). Because the runs were terminated as they folded, after a run has folded it no longer contributes to the histogram. For this reason, as time progresses and more runs fold, the color of the histograms moves towards blue. At all times, the histogram is normalized by the total number of runs at step 0. The data may be viewed as being obtained by a hypothetical experiment that records the energy and dRMS of all non-folded structures as the simulation progresses. The label "hairpin" refers to the formation of interhelix contacts.

fast-folding pathway; in fact, only 1 out of over 150 runs which folded had the helices forming prior to the interhelix contacts. This is in contrast to a recent study using a similar Gō potential [9]. Although there may be several reasons for this discrepancy (such as differences in helical stabilities), it is difficult to directly compare the two simulations because ref. [9] used a coarse-grained structural description while we did not. We note that the move set we used is not likely to contain any bias against diffusion-collision kinetics because unlike lattice simulations, in which helices can reorient only by partial unfolding, our model allows entire secondary structure elements to move as units.

Characterization of traps. – The kinetic traps observed were either backbone misfolds, where the secondary structure elements were not properly formed, or compact conformations with incorrectly packed sidechains. The first type of backbone misfold resulted when the β-sheet folded incorrectly after the helix-turn-helix moiety was formed. Correction of this misfold required partial or complete unfolding of the β-sheet. At high temperatures ($T > 1.7$), this misfold was metastable, as the β-sheet repeatedly folded and unfolded until the correct topology and packing was achieved (fig. 3d)). In general, forming the sheet was observed to be the rate-limiting step at low temperatures ($T < 1.5$) (see fig. 5c)). The other type of backbone misfold occurred when the two helices were not formed properly prior to the formation of the β-sheet. Two pathways (A and B in fig. 4) were available at all temperatures to correct these helix misfolds. Pathway A required the β-sheet to partially or completely unfold. In contrast, pathway B corrected the helix while keeping the sheet intact. This pathway was accelerated with increasing temperature, as the "breathing motion" resulting from greater backbone fluctuations facilitated reinsertion of the helix sidechains (compare, for example, the Q_4 fluctuations in fig. 3c) and d)).

The traps resulting from incorrectly packed sidechains (fig. 3e)) are characterized by low backbone dRMS and correct backbone topology (high Q_i's) and were observed only at low temperatures. At higher temperatures, after the major folding event, equilibrium sidechain packing is rapidly achieved. As shown in fig. 3f), for $T < T^* \approx 1.6$, near-native backbones (dRMS < 1.25 and Q_i's > 0.6) obtained from folding runs would not relax further to match the energies attained in runs initiated from the native state. This suggests that T^* may signify a kinetic transition temperature, where ergodicity is broken and a gap emerges between measurements taken from finite but long unfolding and folding runs. At low temperatures, it appears that the major backbone folding event traps sidechains in disordered non-native conformations, which cannot be readily relaxed because of insufficient backbone fluctuations.

4. – Discussion

Compared to molecular dynamics studies of solvated folding [12], the approach used in this study is still minimalist. Yet, we have demonstrated that important insights into the folding process, such as the role of sidechain packing, may be obtained by properly combining an all-atom description with simple, atomic resolution energetics. Importantly, our simulation can record a statistically significant number of folding events at atomic resolution for real protein sequences, thereby allowing a relatively direct comparision between simulation and experimental results.

Unfortunately, because of the lack of experimental folding studies on crambin, our results presently cannot be directly verified. They should be viewed as theoretical predictions which may be tested by future experiments. We selected crambin for this study because of its small size and its non-trivial α-β structure. We note that as the temperature approaches T_f, the folding kinetics of crambin approaches two-state behavior, in agreement with experimental studies of most small single domain proteins [20]. It is plausible that under a Gō potential, where all native interactions are treated identically, the folding properties of crambin should be consistent with those of similar size and topology. More specifically, the modern viewpoint that nucleation is a major event in the folding kinetics [34-36] is consistent with our results: at temperatures just below $T_{\rm f}$, collapse via formation of the β-sheet is the major kinetic barrier (fig. 3a)). A transiently populated, partially collapsed intermediate in which the interhelix contacts are formed is seen (fig. 5a)), and is probably necessary for proper sheet formation. Once the sheet has formed, the chain is compact, and folding proceeds rapidly with concomitant sidechain packing. Furthermore, structural elements that bring together residues that are closer in position along the chain (such as the helix-turn-helix) appear to form faster than those which bring together residues distant in sequence position. We believe this is similar to the observation made by Plaxco et al. that folding kinetic rates are correlated with the relative contact order of the native state structure [37].

Even with a potential strongly biased towards proper folding, the presence of diverse sidechain geometries and excluded-volume interactions can lead to the presence of severe kinetic traps, as we observe at very low temperatures. However, our results demonstrate that at reasonable temperatures sidechains can be successfully packed in a manner consistent with a low dRMS native-like backbone conformation. We believe that the move set we employed contributed to the success of our simulation. Efficient sampling in the compact state, as evidenced by the $> 10\%$ acceptance rate even for very low temperatures, resulted from both non-local and local conformational changes being permitted, with non-local changes becoming more available with higher temperatures. The qualitative folding behaviour is consistent with experimental observations of small single domain proteins, suggesting that the essential features of a polypeptide with all torsional degrees of freedom are captured by this move set.

We intend to carry out similar studies on larger single domain proteins for which there are extensive experimental data. Taking our results as a proof-of-concept, we believe that high-resolution investigations of folding pathways may finally be possible using ordinary computational resources.

* * *

We thank H. ANGERMAN, G. BERRIZ, M. MORRISSEY, E. PITARD, and J. WILDER for helpful discussions. This work was supported by NIH grant R0152126.

REFERENCES

[1] SALI A., SHAKHNOVICH E. I. and KARPLUS M., *Nature*, **369** (1994) 248.
[2] SHAKHNOVICH E. I., *Curr. Opin. Struct. Biol.*, **7** (1997) 29.
[3] PANDE V. S., GROSBERG A. Y., ROKSHAR D. and TANAKA T., *Curr. Opin. Struct. Biol.*, **8** (1998) 68.
[4] DILL K. A., BROMBERG S., YUE K., FIEBIG K. M., YEE D. P., THOMAS P. D. and CHAN H. S., *Proteins*, **4** (1995) 561.
[5] BRYNGELSON J., ONUCHIC J. N., SOCCI N. D. and WOLYNES P. G., *Proteins*, **21** (1995) 167.
[6] KLIMOV D. and THIRUMALAI D., *Phys. Rev. Lett.*, **76** (1996) 4070.
[7] HONEYCUTT J. D. and THIRUMALAI D., *Biopolymers*, **32** (1992) 695.
[8] BERRIZ G., GUTIN A. M. and SHAKHNOVICH E. I., *J. Chem. Phys.*, **106** (1996) 9276.

[9] ZHOU Y. and KARPLUS M., *Nature*, **401** (1999) 400.
[10] CLEMENTI C., NYMEYER H. and ONUCHIC J. N., *J. Mol. Biol.*, **298** (2000) 937.
[11] PONDER J. W. and RICHARDS F. M., *J. Mol. Biol.*, **193** (1987) 775.
[12] DUAN Y. and KOLLMAN P., *Science*, **282** (1998) 740.
[13] PANDE V. S., GROSBERG A. Y. and TANAKA T., *Rev. Mod. Phys.*, **72** (2000) 259.
[14] LI A. and DAGGETT V., *J. Mol. Biol.*, **257** (1996) 412.
[15] LAZARIDIS T. and KARPLUS M., *Science*, **278** (1997) 1928.
[16] BINDER K. and HEERMAN D. W., *Monte Carlo Simulation in Statistical Physics*, second edition (Springer-Verlag, Berlin) 1992.
[17] GROSBERG A. Y. and KHOKHLOV A. R., *Statistical Physics of Macromolecules* (AIP Press, New York) 1994.
[18] GRANTCHAROVA V. P. and BAKER D., *Biochemistry*, **36** (1997) 15685.
[19] JACKSON S. E. and FERSHT A. R., *Biochemistry*, **30** (1991) 10428.
[20] JACKSON S. E., *Folding & Design*, **3** (1998) R81.
[21] KOLINSKI A. and SKOLNICK J., *Proteins*, **18** (1994) 353.
[22] TEETER M. M., ROE S. M. and HEO N. H., *J. Mol. Biol.*, **230** (1993) 292.
[23] TSAI J., TAYLOR R., CHOTHIA C. and GERSTEIN M., *J. Mol. Biol.*, **290** (1999) 253.
[24] METROPOLIS N., ROSENBLUTH A. W., ROSENBLUTH M. N., TELLER A. H. and TELLER E., *J. Chem. Phys.*, **21** (1953) 1087.
[25] MCQUARRIE D. A., *Statistical Mechanics*, first edition (Harper Collins, New York) 1976.
[26] GO N. and ABE H., *Biopolymers*, **20** (1981) 991.
[27] ABKEVICH V. I., GUTIN A. M. and SHAKHNOVICH E. I., *J. Chem. Phys.*, **101** (1994) 6052.
[28] HAO M. and SCHERAGA H. A., *Acc. Chem. Res.*, **31** (1998) 433.
[29] FERSHT A., *Structure and Mechanism in Protein Science*, first edition (W. H. Freeman and Company, New York) 1999.
[30] XU Y., WU J., GORENSTEIN D. and BRAUN W., *J. Mag. Res.*, **136** (1999) 76.
[31] CREIGHTON T. E. (Editor), *Protein Folding* (W. H. Freeman and Company, New York) 1992.
[32] MUNOZ V. and SERRANO L., *Nature Struct. Biol.*, **1** (1994) 399.
[33] KARPLUS M. and WEAVER D. L., *Nature*, **260** (1976) 404.
[34] FERSHT A. R., *Proc. Natl. Acad. Sci. USA*, **92** (1995) 10869.
[35] SOSNICK T. R., MAYNE L. and ENGLANDER S. W., *Proteins*, **24** (1996) 413.
[36] ABKEVICH V. I., GUTIN A. M. and SHAKHNOVICH E. I., *Biochemistry*, **33** (1994) 10026.
[37] PLAXCO K. W., SIMONS K. T. and BAKER D., *J. Mol. Biol.*, **277** (1998) 985.

Protein sidechain packing in ubiquitin

E. KUSSELL

Department of Biophysics, Harvard University
240 Longwood Ave., Boston, MA 02115, USA

J. SHIMADA and E. I. SHAKHNOVICH

Department of Chemistry and Chemical Biology, Harvard University
12 Oxford Street, Cambridge, MA 02138, USA

1. – Introduction

One remarkable characteristic of proteins is that sidechains comprising the hydrophobic core are as closely packed as organic crystals [1]. In most proteins for which 3D structures have been obtained, the core residues are rarely disordered and adopt one of a small number of alternative conformations. This almost unique packing is effected by a combination of steric interactions (excluded-volume effects) and energetic stabilization (hydrophobic, polar, and charge interactions). The proportions by which these interactions contribute to the overall stability is unknown, but several studies suggest that steric and hydrophobic interactions are of primary importance [2,3].

Sidechain packing has been studied intensively for various reasons, notably: 1) it is thought to be a crucial piece of the protein folding puzzle [4]; 2) the selection of protein sequences through evolution may have been influenced by how well a sequence can be packed for a particular fold [3]; 3) accurate packing algorithms are necessary for completing the final stages of a protein structure prediction; and 4) existing threading and homology modeling algorithms may be significantly improved by a better understanding of how sidechains are stabilized in the core [5,6]. A plethora of computational methods for modeling protein sidechains and energetics exist [2,5,7,8], and yet which ones best capture the underlying physics is unclear. This is due, in part, to the use of energy functions containing many different types of interactions.

In this paper, we attempt to untangle a single type of sidechain packing interaction. We investigate the role of sterics in isolation of all other types of interactions. Our hope is that by taking a bottom-up approach to the problem, that is, by building up the complexity of our model in a step-wise fashion, we can begin to understand the relative importance of each interaction. The ability to make quantitative statements in this regard is critical to each of the areas of research outlined above.

To this end, we use an all-atom, rotamer-based model of the protein ubiquitin to obtain repacked conformations of its interior sidechains. By choosing the most realistic representation of sidechain and backbone geometries, we eliminate the errors encountered in coarse-grained models. Because there is no consensus on how excluded-volume interactions should be modeled in the context of proteins, we consider three different models. The simplest model treats all heavy atoms as hard spheres which are not allowed to overlap. The second model further restricts the sterically allowed space by adding a second shell around the hard spheres which tolerates only a limited number of overlaps throughout the entire protein. The third model features a continuous r^{-12} potential (which is the repulsive contribution from a Lennard-Jones potential), and therefore does not suffer from the somewhat arbitrary choices (such as the hard-sphere radii) inherent in the first two models. Since a potential which decays faster than our chosen r^{-12} potential would allow more sidechain conformations, and since a slower decaying function is generally not used to model sterics [9], the three we have chosen should cover most of the possibilities as far as models featuring spherically symmetric, atom-based potentials go.

For each model, we obtain the distribution of rms values for sidechain conformations satisfying excluded-volume constraints. Our results show that sterics alone cannot stabilize a unique native conformation; in fact, we find that there is a vast number of significantly different conformations with native-like packing density and meeting steric constraints. *En route*, we provide an estimate for the amount of additional stabilization that must be provided through interactions other than sterics. Finally, we find that the results obtained using three different models are consistent with each other, suggesting that our main conclusions are model-independent.

2. – Methods

All-atom protein representation. All heavy atoms of the protein ubiquitin present in the crystal structure (PDB accession code: 1ubq) were represented. Forty non-surface residue positions (see fig. 5a below) were identified by requiring that they have greater than the average number of contacts and/or have $C\alpha$-$C\beta$ bonds pointing away from the solvent. Each sidechain torsion of these residues was allowed to attain only one of three rotamer values [10] plus a random noise of $\pm 10°$ (Rotamer probabilities [10] were not used in our study). At all times, the atoms of the other 36 surface sidechains and the entire backbone were kept fixed in their original crystal structure positions.

TABLE I. – *The atomic groups, vdW radii (r), and the hard-core distances ($\alpha_h r$) used in our study. The atom groups and vdW radii were obtained from Table 2 of Tsai et al. (1999). The atomic group AnHm refers to an atom of element A with a chemical valence of n and m hydrogen atoms bonded to it. A methyl carbon ($-CH_3$), for example, falls under the C4H3 atomic group.*

Atomic group	r	$\alpha_h r$
C3H0	1.61	1.21
C3H1	1.76	1.32
C4H1	1.88	1.41
C4H2	1.88	1.41
C4H3	1.88	1.41
N3H0	1.64	1.23
N3H1	1.64	1.23
N3H2	1.64	1.23
N4H3	1.64	1.23
O1H0	1.42	1.07
O2H1	1.46	1.10
S2H0	1.77	1.33
S2H1	1.77	1.33

Models of sterics. The excluded-volume interactions of protein sidechains were modeled in three ways:

1) In the *hard-sphere model* atoms were treated as impenetrable spheres of given radii. These hard-sphere radii are necessarily smaller than the van der Waals (vdW) radii, and were defined by scaling the vdW radii by some factor α_h. All atom-atom interactions at distances smaller than the sum of the hard-sphere radii ("hard clashes") were strictly forbidden. A pair of atoms i and j separated by a distance r_{ij} is said to be a hard clash if $r_{ij} < \alpha_h(r_0(i) + r_0(j))$, where $r_0(i)$ is the vdW radius of atom i. The set of vdW radii determined in [11] was used (see table I). α_h was taken to be the largest value such that all hard clashes in the native conformation of ubiquitin can be eliminated within the $\pm 10°$ allowed at each torsion; we found that $\alpha_h = 0.75$. The steric energy, U_h, of a conformation in this model is the number of hard clashes. Only conformations with $U_h = 0$ are sterically allowed in this model.

2) Since results could depend strongly on the choice of α_h, we explored a *soft-sphere model* in which atoms consisted of two radii—the hard-sphere radii fixed by α_h, and somewhat larger, soft-sphere radii. Soft radii were defined by scaling vdW radii by a parameter α_s, where $\alpha_h < \alpha_s \leq 1$. A soft clash occurs when $\alpha_h(r_0(i) + r_0(j)) < r_{ij} < \alpha_s(r_0(i) + r_0(j))$. All hard clashes were forbidden as before, while only a limited number of soft clashes are allowed over the entire protein. The steric energy, U, of a conformation is given by $U = U_h + U_s$, where U_h is the number of hard atom-atom clashes, and U_s is the number of soft atom-atom clashes above

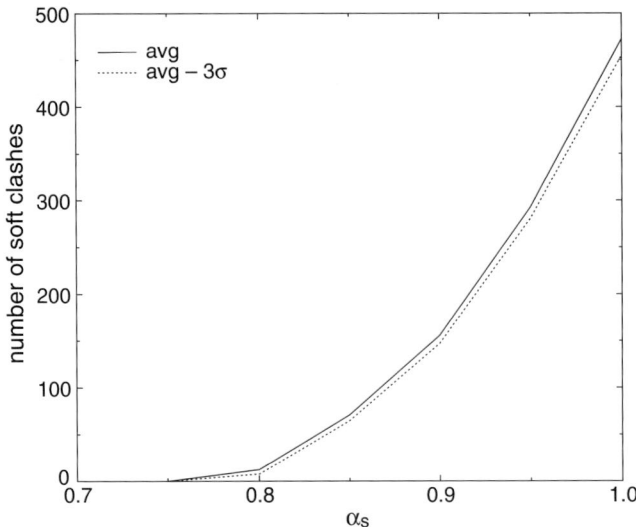

Fig. 1. – The number of soft clashes in near-native conformations as a function of α_s. The solid line denotes the average number observed in an ensemble of near-native conformations generated by adding random noise to the native sidechains. The dotted line is determined by taking three standard deviations below the average, and corresponds to the soft clash threshold used in this study.

threshold. The threshold number of soft clashes allowed is a function of α_s, as described in fig. 1. Only conformations with $U = 0$ are sterically allowed in this model.

3) The third model, *Lennard-Jones sterics*, used the repulsive term of a Lennard-Jones potential. For each pair of atoms we fit the LJ potential, $A/r^{12} - 1/r^6$, so that the minimum coincides with the sum of the vdW radii taken from [11]. With this model, the steric energy of a conformation is given by $U_{\rm LJ} = \sum_{ij} A_{ij}/r_{ij}^{12}$, where the sum is taken over all pairs of atoms, and $A_{ij} = (r_0(i) + r_0(j))^6/2$. Since there is no energy cut-off in this model, the sterically allowed conformations were defined by determining a range of allowable energies. By allowing only the 10° noise at each sidechain torsion about the native ubiquitin conformation, the median steric energy was found to be 0.36, with approximately 95% of the sampled energies lying between 0.3 and 0.4. Thus, a conformation was defined to be sterically allowed if its energy was between 0.3 and 0.4. The median was used because, due to the form of this potential, the energies of rare, badly clashed conformations can dominate the average.

Monte Carlo packing algorithm. Initial random protein conformations were packed to a given rms interval $[R_{\min}, R_{\max}]$ by a Metropolis Monte Carlo simulation [12,13]. At each step of the simulation, a random position was selected and changed to another rotamer. The move was accepted with probability p given by

$$p = \begin{cases} 1, & \delta H \leq 0, \\ \exp\left[-\delta \frac{H}{T}\right], & \text{otherwise}, \end{cases}$$

with

$$\delta H = \delta U + \begin{cases} C(R_{\min} - R), & R < R_{\min}, \\ 0, & R_{\min} \leq R \leq R_{\max}, \\ C(R - R_{\max}) & R > R_{\max}, \end{cases}$$

where U is the steric energy, R is the rms, and $C = 10$. All reported rms values were computed over the 40 non-surface sidechains atoms, including the $C\beta$ atoms as in [2].

Umbrella sampling. Umbrella sampling [14] was used to determine the number of states $\Omega(R)$ for each of the 3 models. First, different sterically allowed conformations were collected for rms intervals of size 0.05 Å over the entire rms range (0–4.5 Å). The rms values of conformations differing only by random noise were averaged. Next, the probability to observe a conformation with rms R, $p_i(R)$, was obtained for the i-th rms interval, $[R_i, R_{i+1}]$. Since the distribution of conformations $\Omega(R)$ must be a continuous function of R, $\Omega(R)$ can be determined by appropriately scaling the probabilities $p_i(R)$ to ensure continuity at interval boundaries. Specifically, the probabilities $p_i(R)$ are sequentially scaled by $p_{i-1}(R_i)/p_i(R_i)$, starting from $i = 2$. $\Omega(R)$ is then obtained by multiplying all rescaled p_i's by $\Omega(R_1)/p_1(R_1)$.

3. – Results

Figure 2 shows the distribution of *packed* (sterically allowed) and *random* conformations, obtained over a range of rms values using umbrella samping. The *random* curve is shown as a control, and corresponds to the space of all sidechain conformations, packed and not packed. With $\alpha_s = \alpha_h$ (the hard-sphere model), the number of packed conformations is approximately 10^{30}. If $\alpha_s > \alpha_h$ (the soft-sphere model) the number of packed conformations decreases. When the soft radii are equal to the van der Waals radii (at $\alpha_s = 1.0$), the number of repackings is approximately 10^{20}. Using the steric LJ model, the number of repackings is 10^{27}.

The packing density of a conformation is usually defined as the volume of the van der Waals envelope of a protein divided by its Voronoi volume. The density of a random sample of 1000 repacked conformations of ubiquitin was computed and found to vary by only ±1% of the native density. Given this minimal variation in packing density, we conclude that the requirement of having native-like density does not significantly reduce the number of packed sidechain conformations. In addition, we were able to build

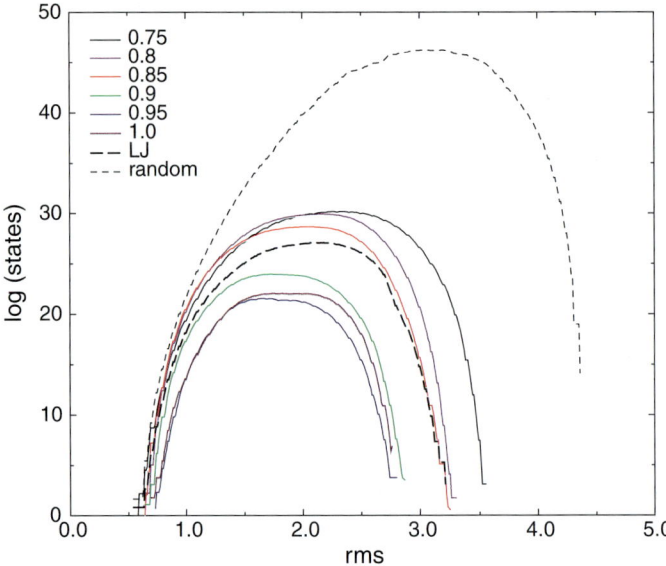

Fig. 2. – The density of packed and random conformations as a function of rms for various values of α_s. Since the distribution for $\alpha_s = 0.75$ should always be greater than those of all $\alpha_s > 0.75$, the error is estimated to be ± 2 log units. It should be noted that the van der Waals radii are strictly greater than the appropriate excluded-volume radii, as can be seen from numerous studies of gas phase organic molecules [9]. The appropriate value of α_s is, therefore, strictly less than 1.0, and we include $\alpha_s = 1.0$ in our plots only as an upper limit on the radii.

hydrogen atoms (using CHARMM [15]) on 1300 randomly selected, packed conformations (see fig. 3), confirming that the conformations were not too compact.

While the number of conformations is large, it is conceivable that most repackings are structuraly similar to one another, and thus that the number of truly different repackings may be much smaller than 10^{20}. To determine how different these repacked conformations are, we uniquely identified each conformation by the rotamers of the 40 non-surface positions. The difference between two conformations was measured by counting the number of single-bond torsions which had to be rotated (by $120°$) in order to obtain one conformation from the other. Since there are a total of 93 torsions in the non-surface positions of ubiquitin, we expect two random conformations to differ by 61 ($= 93 \times 2/3$) torsion moves.

The average pairwise difference between repacked conformations (fig. 4) shows that the conformations are truly different from each other. Even when the soft radii are maximally large ($\alpha_s = 1.0$), an average pair of conformations differs at approximately 25% of the torsions while in the steric LJ model, the conformations differ at 40% of the torsions. In general, smaller soft radii lead to more different repackings.

The variability (entropy) of each sidechain was computed (fig. 5) to verify that all positions are contributing significantly to the multitude of packed conformations. We

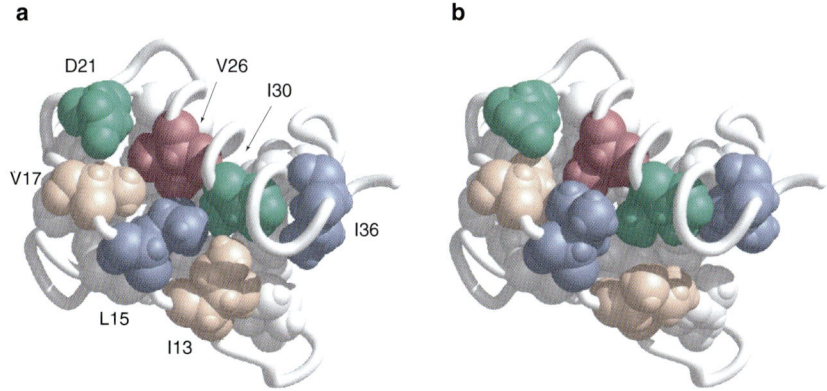

Fig. 3. – The native (a) and repacked (b) 20 core residues of ubiquitin. The repacked structure (2.5 Å rms) shown was minimized with CHARMM [15] after hydrogens were added with the HBUILD facility. All such minimizations, including those performed for other randomly selected packed structures, did not change the rotamer states of the sidechains, but did eliminate close contacts (as defined by CHARMM) resulting from the addition of hydrogens. Colors are used solely to aid visualization, with corresponding residues colored identically. All seven labelled residues have changed rotamers under repacking. Note that the cavities seen in both structures are filled by non-core residues which are not shown. A set of 1300 packed conformations for which hydrogens were added and minimized with CHARMM are available for public download at http://www-shakh.harvard.edu/~shimada/index.html.

found that almost all positions change upon repacking, and in particular that 7 buried residue positions (5, 7, 13, 15, 17, 21, and 26) exhibit large variability. Although the remaining 13 buried positions show less variation, they are nevertheless found in multiple conformational states even at rms values as low as 1.0 Å.

We also examined the correlation between the rotamer changes made at a pair of positions x and y by measuring the number of states lost, $\exp[\Delta S]$, due to interactions between x and y, where $\Delta S = S_{xy} - (S_x + S_y)$, S_{xy} is the joint entropy, and S_x and S_y are the individual entropies. For any given rms, we found no more than 4 residue pairs exhibiting even mild correlation, of which very few were in the core. We conclude that repacking at a single position can be accommodated by the surrounding residues in many different ways.

4. – Discussion

A clear picture of sidechain packing emerges from these data: geometry and packing density alone are not sufficient to stabilize the native sidechain conformation. Upon folding to the native ubiquitin backbone topology, the conformational space of the sidechains is cut by approximately 25 orders of magnitude (fig. 2) due to excluded volume, a dramatic but far from complete reduction. The native backbone can still accommodate at least 10^{20} different sidechain conformations, while maintaining its high degree of com-

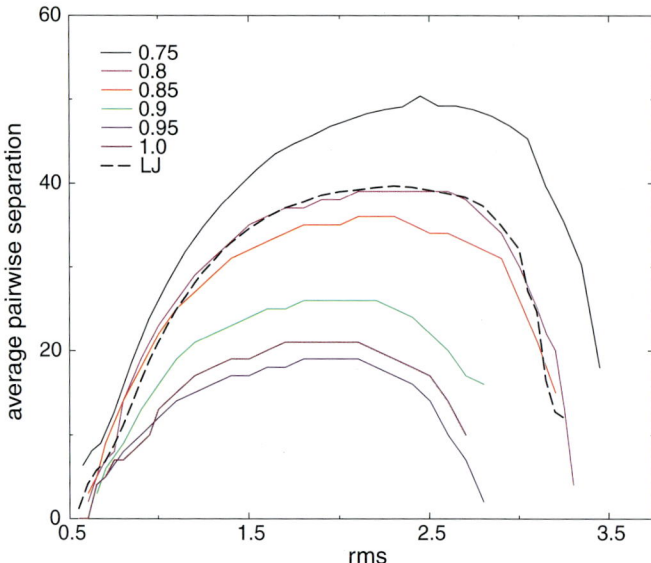

Fig. 4. – The average number of bond rotations needed to bring one conformation to another as a function of rms for various values of α_s. The average is taken over 1000 randomly selected pairs of conformations found in a given rms window.

pactness. Other sources of stabilization (such as attractive van der Waals interactions, hydrogen bonding, polar interactions) are necessary to overcome the high entropic cost associated with fixing the sidechains in their native conformations.

The three models employed in this study yield consistent results. The average rms in all models is located between 1.8 Å and 2.2 Å. We see a jump in total number of states using the soft-sphere model between $\alpha_s = 0.85$ and $\alpha_s = 0.9$, suggesting that perhaps 0.9 is a bit too large for modeling sterics (because the radii are approaching vdW sizes). The LJ curve lies between these two, and being a continuous model, it may be the most appropriate one. The consistency of models with $0.75 \leq \alpha_s \leq 0.85$ with the steric LJ model lends credence to these simpler models.

The packing of sidechains in a native backbone has often been described as a jigsaw-puzzle: the shapes of the sidechains (the pieces) are enough to uniquely determine how they will fit in the protein interior [16]. This argument has been used to explain the high density observed in proteins. The large number of repackings (fig. 2) and the diversity of these conformations (fig. 4), all with native-like density, suggest that a precise fitting of residues is not necessary for a high packing density. The rotamer variability calculations demonstrate that repacking is occurring at all positions. Furthermore, the correlation calculations show that a conformational change at one position does not necessitate a particular change at spatially proximate positions. We conclude that while sidechains must fit together to avoid steric clashes and maintain native-like density, these constraints alone do not give rise to precise and unique fitting of residues.

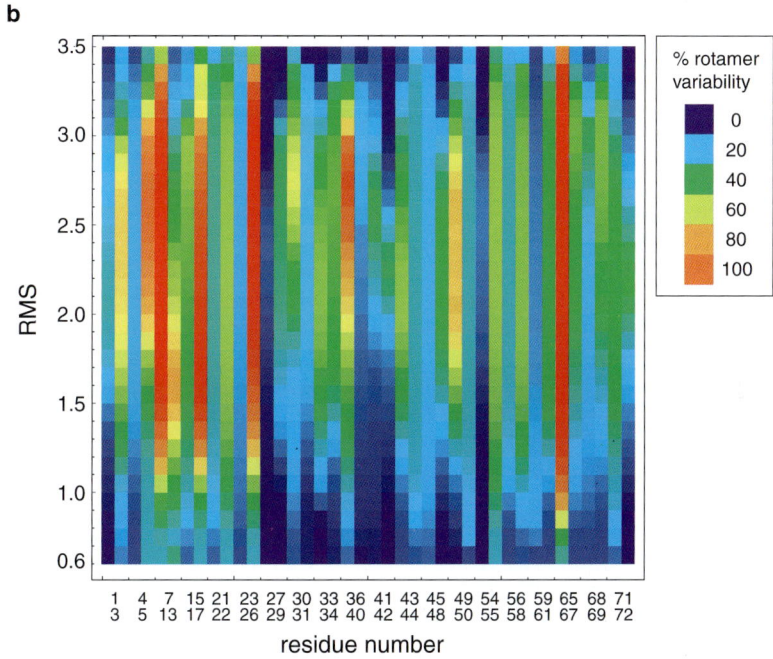

Fig. 5. – a). The native sequence of ubiquitin. The 40 non-surface residues are indicated in bold face, while the 20 core residues are starred. b). Dependence of rotamer variability on rms for $\alpha_s = 0.75$. Packed conformations collected in each rms window were used to calculate rotamer entropy, S, at each position ($S = -\sum p_i \log p_i$, where p_i is the frequency of occurrence of the i-th rotamer at the given position). The effective number of observed rotamers, e^S, is plotted as a percentage of the total number of possible rotamers at each position.

A natural extension of our work is to add an attractive interaction to the potential which would stabilize the native packing. From the distribution of packed conformations, we estimate that the native conformation must be about $(kT \ln 10^{30})/40 = 1.7 kT$ per sidechain lower in energy than the average packed conformation with native backbone. This estimate can be used to guide efforts towards designing better potentials for sidechain packing. Furthermore, the techniques presented here, combined with a stepwise increase in the complexity of potentials, can determine how much stabilization each additional interaction provides. With the systematic design of a successful sidechain packing potential, we may finally gain insight into how important is sidechain packing to the folding process. It will be interesting to see whether a potential designed with only packing in mind can be used to fold a protein using all-atom folding simulations.

∗ ∗ ∗

We are grateful for stimulating discussions with G. BERRIZ, A. FINKELSTEIN, L. MIRNY, and M. MORRISSEY, and especially thank K. DORAN for his assistance. Financial support from NIH (grant 52126) and NSF (graduate fellowship, E.K.) is acknowledged.

REFERENCES

[1] FERSHT A., *Structure and Mechanism in Protein Science*, 1st edition (W. H. Freeman and Company, New York) 1999.
[2] LEE C. and SUBBIAH S., *J. Mol. Biol.*, **217** (1991) 373.
[3] PONDER J. W. and RICHARDS F. M., *J. Mol. Biol.*, **193** (1987) 775.
[4] SHAKHNOVICH E. I. and FINKELSTEIN A. V., *Biopolymers*, **28** (1989) 1667.
[5] BOWER M. J. and COHEN F. E. and DUNBRACK R. L., *J. Mol. Biol.*, **257** (1997) 1268.
[6] MIRNY L. A. and SHAKHNOVICH E. I., *J. Mol. Biol.*, **283** (1998) 507.
[7] PETRELLA R. J. and LAZARIDIS T. and KARPLUS M., *Folding & Design*, **3** (1998) 353.
[8] MENDES J., BAPTISTA A. M., CARRONDO M. A. and SOARES C. M., *Proteins*, **37** (1999) 530.
[9] MCQUARRIE D. A., *Statistical Mechanics*, 1st edition (Harper Collins, New York) 1976.
[10] DUNBRACK R. L. jr. and COHEN F. E., *Protein Sci.*, **6** (1997) 1661.
[11] TSAI J., TAYLOR R., CHOTHIA C. and GERSTEIN M., *J. Mol. Biol.*, **290** (1999) 253.
[12] METROPOLIS N., ROSENBLUTH A. W., ROSENBLUTH M. N., TELLER A. H. and TELLER E., *J. Chem. Phys.*, **21** (1953) 1087.
[13] BINDER K. and HEERMAN D. W., *Monte Carlo Simulation in Statistical Physics*, 2nd editon (Springer-Verlag, Berlin) 1992.
[14] CHANDLER D., *Introduction to Modern Statistical Mechanics*, 1st edition (Oxford University Press, New York) 1987.
[15] BROOKS B. R., BRUCCOLERI R. E., OLAFSON B. D., STATES D. J., SWAMINATHAN S. and KARPLUS M., *J. Comput. Chem.*, **4** (1983) 187.
[16] LEVITT M., GERSTEIN M., HUANG E., SUBBIAH S. and TSAI J., *Annu. Rev. Biochem.*, **66** (1997) 549.

PROTEIN EVOLUTION

Evolutionary perspectives on protein structure, stability, and functionality

RICHARD A. GOLDSTEIN

Biophysics Research Division and Department of Chemistry, University of Michigan
Ann Arbor, MI 48109-1055, USA

1. – Introduction

A number of different perspectives have recently changed the way that we understand evolution and its effect on biological macromolecules. The first perspective is an increased appreciation for the relationship between evolution and observed molecular properties. This has resulted from a greater availability of sequence data encouraging evolutionary studies at the molecular level, combined with increased sensitivity to the evolutionary heritage encoded in these molecules.

A second perspective was the understanding of the role of chance in evolution. Specifically, with the rise of the neutral theory we can see how adaptation is only one aspect of evolutionary change and that this change, especially at the molecular level, may represent random movement among approximately equally fit genotypes. Biological systems are characterized by a many-to-few mapping from genotype to phenotype, with selection occurring primarily at the phenotype level. This results in large "neutral networks" of equivalent genotypes. Although mutations between these sequences may not cause appreciable changes in fitness, they can have a dramatic impact on the large-scale evolutionary process including at the phenotype level. This perspective also highlighted the need to understand the genotype-to-phenotype mapping and the nature of the fitness landscape in order to model the evolutionary process.

A third perspective was the introduction of the theory of reaction dynamics into evolutionary biology. Eigen's incorporation of the chemical-engineering theory of flow reactors allowed quantitative analysis of the behavior of populations of self-replicating molecules. In addition to allowing us to approach evolution in terms of creation of information, and demonstrating how life could have evolved from initial concentrations of RNA, these theories showed that we cannot think of a species as a homogeneous set of identical genotypes but rather as a collection of similar but different genotypes. The intrinsic variability of the genotype in the population can have a dramatic effect on the evolutionary dynamics.

In this paper, we will first briefly review these new perspectives, first discussing classical gene dynamics in infinite systems, then the neutral theory of evolution, and finally some of the consequences of Eigen's theories as developed by him and others. We will then discuss how these perspectives can help to explain some of the observed properties of proteins including the distribution of observed structures and their thermal properties. While a number of investigators have looked at these phenomena, I will necessarily concentrate on theoretical modeling performed by my group over the past few years.

2. – Principles of molecular evolution

2˙1. *Classical theory of gene dynamics.* – In order to understand how the process of evolutionary change can influence the observed properties of proteins, we first need to consider the process of evolutionary change itself. The dominant philosophy is of course Darwin's theory of random variation followed by natural selection.

First a few definitions. (This material is covered in depth in a number of books [1-3].) A *locus* refers to a location of a given gene in the genome. Different variations of the genes are called *alleles*. Let us consider that there is a wild-type allele, that is, one version of the gene that describes every member of the population. A different version of the gene, a different allele, can result from random variation due to errors in conservation or replication. There are three possible fates of this gene. The individual or the descendents of the individual can fail to reproduce, in which case this different allele is removed from the population. Alternatively, this mutant allele can spread throughout the population so that it now becomes the new "wild-type". This process is called *fixation*. Or these gene can achieve a certain representation in the population, neither dying out nor replacing the wild type. We then say that this locus is *polymorphic*. Let us see how these alternatives are possible.

Consider a diploid organism, that is, with two copies of each gene, one of which is transferred to the next generation. We will call the wild-type allele A and the mutant B. There are then three diploid genotypes possible, AA, AB, and BB. An organism is called a *homozygote* if both alleles are the same (AA or BB in this example); otherwise it is a *heterozygote*. If the fraction of the gene A in the population is p and the fraction of gene B in the population is q (so that $q = 1 - p$), then at equilibrium where these ratios are maintained and where mating is random the population will have genotype AA with frequency p^2, genotype AB with frequency $2pq$, and genotype BB with frequency q^2—the so-called *Hardy-Weinberg equilibrium*. Often these ratios will not be maintained and the Hardy-Weinberg equilibrium will not hold. This is the case when these different genotypes correspond to different genotypes with different fitnesses, where the fitness

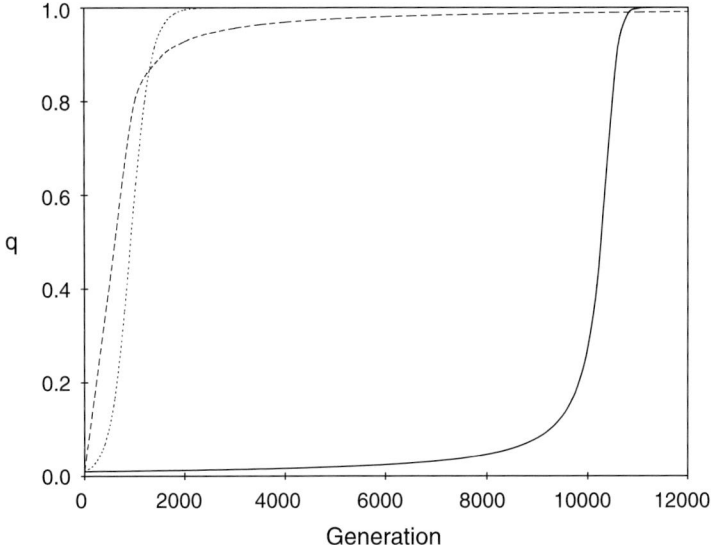

Fig. 1. – Gradual fixation of an advantageous mutation starting with an initial frequency of 0.01. The relative fitnesses of the various genotypes are $s_{BB} = 0.01$ and either $s_{AB} = 0.00$ (solid line, corresponding to a recessive mutation), $s_{AB} = 0.005$ (dotted line), or $s_{AB} = 0.01$ (dashed line, corresponding to a dominant mutation). The recessive mutation requires longer for fixation due to the slow buildup of BB homozygotes with a competitive advantage. Conversely, it is difficult to achieve total fixation of a dominant advantageous mutation because of the relative fitness of the heterozygotes.

measures the relative probability of contributing to the next generation. Let us consider the relative fitnesses of these three genotypes as ω_{AA}, ω_{AB}, and ω_{BB}, respectively. Often we consider fitnesses relative to the wild type AA, so $\omega_{AA} = 1$, $\omega_{AB} = 1 + s_{AB}$, and $\omega_{BB} = 1 + s_{BB}$. For an infinite population and current values of p and q we can calculate how much of each allele will be present in the next generation. The fraction of allele B in the next generation, $\acute{q}$, is given by

$$\acute{q} = q + \frac{pq[p(\omega_{AB} - \omega_{AA}) + q(\omega_{BB} - \omega_{AB})]}{p^2 \omega_{AA} + 2pq\omega_{AB} + q^2 \omega_{BB}}. \tag{1}$$

Figures 1 and 2 show two different examples. In fig. 1, we consider the case where the new mutant gene is advantageous with $s_{BB} = 0.01$. The new allele B is fixed in the population with probability 1 with dynamics that depend upon the heterozygote fitness s_{AB}. In fig. 2 we have what is called *overdominant selection* where the heterozygote AB has the highest fitness: in this case $s_{AB} = 0.02$ and $s_{BB} = 0.01$. (The classical example of overdominance is the mutation for sickle-cell anemia, where the mutant homozygote (BB) is lethal, yet the heterozygote (AB) has increased resistance to malaria.) The result is relaxation to a constant steady-state population of some B for any initial population.

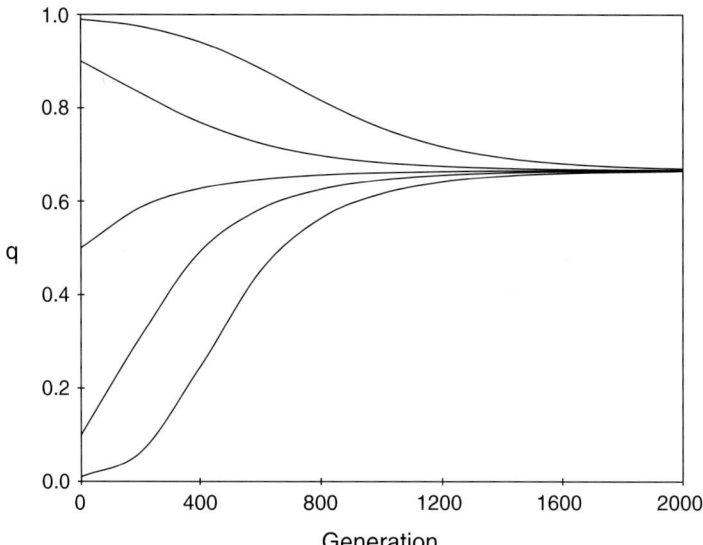

Fig. 2. – Population dynamics with overdominant selection where the heterozygote has increased fitness relative to either homozygote, for a range of initial gene frequencies. The relative fitnesses of the various genotypes are $s_{AB} = 0.02$ and $s_{BB} = 0.01$.

2˙2. *Finite populations and the neutral theory*. – The eventual fate of a mutant gene is deterministic if the population is infinite: there are differential equations that can be solved to give the behavior described above. In reality, populations are quantized and finite. This can have a strong impact on the fate of various alleles. For instance, imagine that neither allele has a selective advantage so that the fitnesses of all three possible genotypes are equal. In this case, the population dynamics of infinite populations would result in the proportion of the two alleles remaining constant in a Hardy-Weinberg equilibrium. In a finite population eventually either the mutant allele would be eliminated or achieve fixation, at least if additional copies of the mutant allele are not produced by further mutations. This is because random fluctuations in the allele fractions would occur; with a finite probability that any allele frequency would decrease to zero from where it cannot recover. The discreteness of the population is key to this process, as there have to be an integral number of copies of each allele. Consider a single copy of a new allele in a population of size N (so that $q = 1/2N$, with the $2N$ coming from the fact that the individuals are diploid). In the simplest case the heterozygote has the average fitness of the two homozygotes so we can write $\omega_{AA} = 1$, $\omega_{AB} = 1 + s$, and $\omega_{BB} = 1 + 2s$. Kimura derived the probability of eventual fixation of B [4]:

$$P_{\text{fixation}} = \frac{1 - e^{-(2N_e s)/N}}{1 - e^{-4N_e s}}, \qquad (2)$$

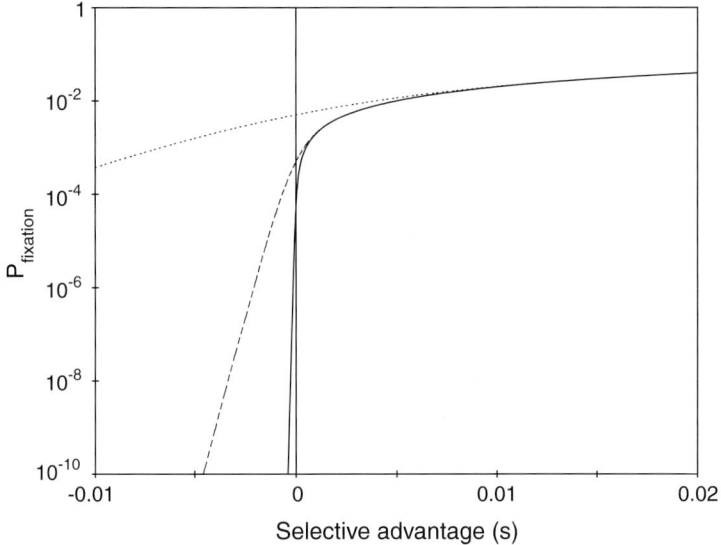

Fig. 3. – Probability of fixation of a mutation with initial frequency $1/2N$ as a function of s for various population sizes: $N = 100$ (dotted line), $N = 1000$ (dashed line), and $N = 10000$ (solid line). The difference between population size and effective population size is ignored.

where N_e is the effective population size, that is, the population that is actually reproducing at any one time. (For human populations, $N_e \sim N/3$ [5].) Ignoring the difference between N and N_e results in curves of P_{fixation} as a function of s for different population sizes as shown in fig. 3. As would be expected, for a neutral mutation ($s = 0$) the probability of eventual fixation represents the initial fraction of the population, $P_{\text{fixation}} = 1/2N$. In fact, all mutations with values of $|s| < 1/2N$ have approximately probability $1/2N$ of fixation; these mutations are essentially neutral. For larger values of s, the probability of fixation becomes $P_{\text{fixation}} = 2s$. For finite populations there is a chance of a negative mutation becoming fixed in the population, just as there is a chance of a positive mutation being removed.

In classical evolutionary theory, the process of evolutionary change involved chance advantageous mutations that became fixed, what is called *adaptive* evolution. Kimura and Jukes and King proposed the *neutral theory*, which postulated that the vast majority of possible mutations are either deleterious or neutral ($|s| < 1/2N$) [4,6]. As the deleterious mutations will generally be removed from the population by purifying selection, most observed substitutions would be neutral or slightly deleterious. This was used to explain four observations. One observation had to do with the large variation in genotypes in a typical population. It was observed that many genes were *polymorphic*, that is, multiple alleles exist in the population. In the classical theory, polymorphism could result from overdominant selection (as shown in fig. 2) or from frequency-dependent selection—where there was an advantage to being different from others in the population. In these cases there was selective pressure towards polymorphism. According to the neutral theory, polymorphism was generally a temporary result of a nearly neutral mutation that had not yet been either eliminated nor fixed. Fixation and elimination times for neutral mutations are quite long, about $4N$ generations. Under these conditions it would be natural to have a large amount of polymorphism in the population. Kimura claimed that the amount of polymorphism in observed populations was too large to be explained by positive adaptation.

A second observation was that any particular gene often tends to evolve at a roughly constant rate in different organisms, that there is a *molecular clock*. (This is not to say that different genes evolve at the same rate—there are quite large variations in the rate of substitution of different genes.) Figure 4, for instance, shows the relationship between the time of divergence of various species from humans according to the fossil record compared with the dissimilarity in the sequences of alpha-hemoglobin; the near straight-line behavior across such different organisms (with the exception of chicken) is striking. This molecular clock is a natural result of the neutral theory. Neutral mutations in a population should arise at a rate proportional to the number of genes in the population, $2\mu N$. The probability of fixation is $1/2N$. Multiplying these two factors together, the total rate of introduction and fixation of neutral mutations should be μ, independent of the population size. This would explain the constant rate of genetic evolutionary change. Conversely, for adaptive change the fixation probability is $2s$, resulting in a total rate of introduction and fixation of $4\mu Ns$, proportional to the population size. This would be incompatible with the molecular clock hypothesis.

A third observation was that important genes evolve slower than less important genes, which evolve slower than regions of the genome that do not seem to serve any purpose. If most mutations were either neutral or disadvantageous, the importance of the gene will correlate with the likelihood that changing the gene will be deleterious. As a result, mutations in less significant regions of the genome will have a smaller probability of being removed by purifying selection, and thus more chance at fixation.

The last observation has come with the rise of genetic manipulation. Many substitutions at the DNA level do not change the amino acids of expressed proteins—these are called *synonymous substitutions*. It is likely that the vast majority of these substitutions are essentially neutral. Additionally, is now clear how plastic the resulting amino acid sequences are, that it is not difficult to find many amino acid substitutions that results in proteins with seemingly identical properties [7]. Many of the changes that we can make in the laboratory seem to be neutral, at least within the accuracy and the context of the experiments.

The neutral theory does not downplay the role of adaptive change. Obviously the characteristics of living creatures show that we are highly adapted to our surroundings. The argument is that adaptive changes, though important, are extremely rare and that most substitutions are either neutral or deleterious. As will be discussed below, one reason for the presence of neutral change at the genetic level is the many-to-one mapping of genotype to phenotype.

One additional concept introduced by Gould is the idea of the *spandrel* [8]. According

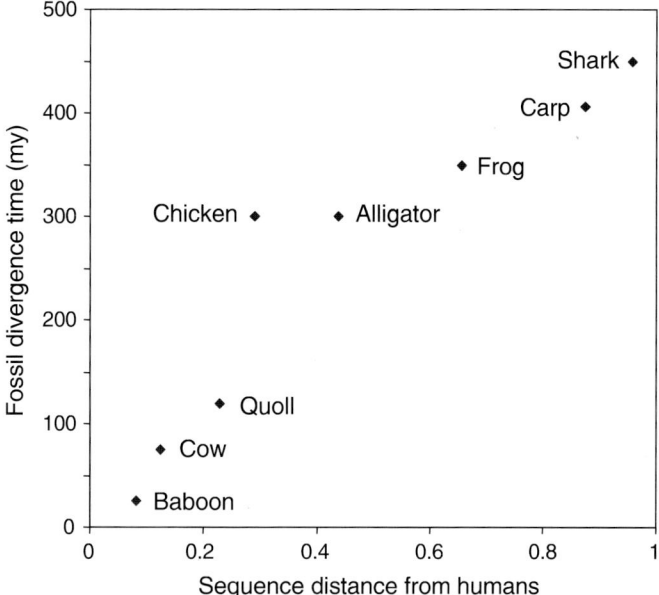

Fig. 4. – Evidence for a molecular clock. Plot shows the relative time of divergence from man according to the fossil record compared with the sequence divergence. Adapted from [2].

to Gould, certain features of an organism might arise from reasons having nothing to do with selection, either through random neutral drift or as an unavoidable consequence of some other modification. The organism might then be able to adapt this feature for adaptive purpose. Accordingly, it is dangerous to explain the existence of this feature as arising as a positive adaptation towards its eventual purpose, what he called the "Panglossian paradigm" after the character in Voltaire's *Candide* who sees everything as optimal in this "best of all possible worlds".

Is the neutral hypothesis correct? This is still a topic of much debate. The degree of polymorphism seems to be somewhat between the rate predicted by adaptionists and neutralists. While there seems to be some degree of constancy to the rate of evolution of each gene across different evolutionary lines, the molecular clock seems to run somewhat erratically. Neutralists claim that these irregularities can be explained by accounting for different rates of mutation, different generational times, and some adaptive bursts. Adaptionists can also explain why important regions of the genome evolve slower than less important regions: genetic changes in the less important regions of the genome result in smaller changes in fitness, and smaller changes in fitness are more to be advantageous than a larger changes [9]. While it is true that it is not difficult to make genetic changes that have no observable effect on the phenotype or on the organisms chance for survival and reproduction, effects too small to be seen in the laboratory may still have a large impact on the evolutionary process. Even synonymous substitutions in the DNA code

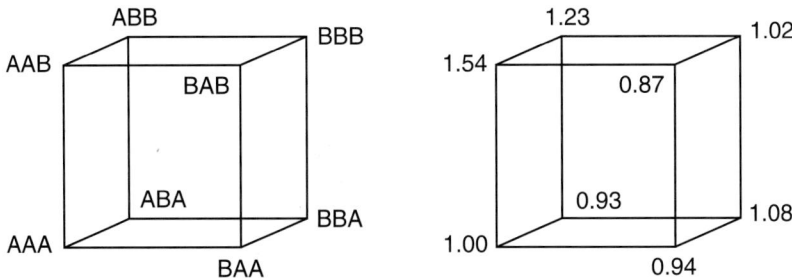

Fig. 5. – An example of a fitness landscape of a trimer written in a two-letter alphabet. The left of the diagram shows the available sequences and their connectivities, while the right of the diagram shows the corresponding fitnesses.

that have no effect on the amino acid sequence of the expressed proteins can affect the probability of survival by affecting the rate of protein transcription.

2˙3. Eigen's theory of quasi-species. – Kimura's neutral theory includes aspects of random change due to the finite size of populations, resulting in stochastic effects that are not included in the infinite-population differential equations. But this model still involves a homogeneous population in which mutations occurs. The polymorphism is a result of the evolutionary dynamics, rather than a critical component. These aspects were to change with the introduction of the theory of quasi-species.

It is easiest to consider the situation by considering a fitness landscape, a term first introduced by Wright [10]. The fitness landscape consists of a sequence space, that is, the space of all possible sequences, combined with a fitness value for each point. Each dimension in the fitness landscape corresponds to one allele or base or amino acid. As a result, the sequence space (as well as the fitness landscape) is extremely high-dimensional. It is a strange space, however, in that only relatively few discrete values in each dimension are allowed. For proteins, for example, each dimension consists of twenty discrete points representing the twenty amino acids. A typical simple fitness landscape for a trimer in a two-letter code (A and B) is shown in fig. 5. Again, this diagram does not do justice to the high dimensionality of the space. Another useful but misleading representation is where the discrete nature of the sequence alphabet is ignored, resulting in diagrams such as fig. 6. The advantage of this representation is that it provides an intuitive idea of fitness peaks, valleys, and ridges.

Eigen modeled evolution by considering a flow reactor containing self-replicating molecules of RNA, with an influx of mononucleotides and an outflow to keep the concentration constant [11-14]. As with biological RNA, the replication rate is not perfect, but mutations naturally arise. The dynamics can be treated with standard approaches from chemical engineering. The result indicated that rather than having a single "wild-type" genotype, instead there would be a cloud of different genotypes centered in the sequence space around a prototypical sequence. The cloud could represent the ultimate

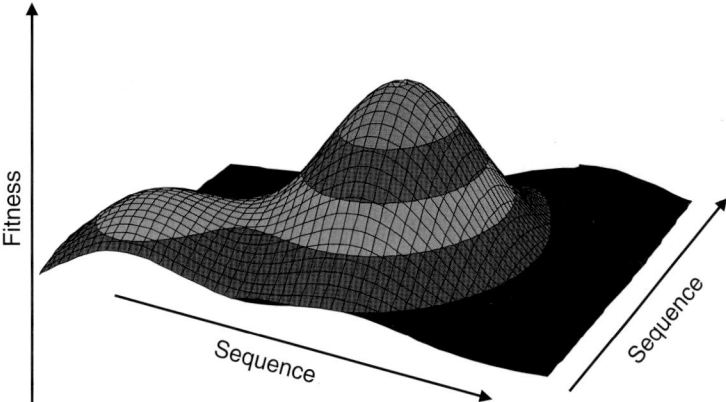

Fig. 6. – Fitness landscape with sequences represented as continuous variables.

steady-state solution—not a transient phenomenon eventually resolved by natural selection. It was then possible to talk about the evolutionary process in terms of changes and competition between these various clouds, which took on the role of species in classical evolutionary theory and thus were named *quasi-species* [13, 14]. Eigen's results demonstrated that in order to consider the fitness of any particular quasi-species, it was necessary to consider the fitness of the prototypical sequence as well as the surrounding sequences. A broader, flatter, but lower fitness peak could out-compete a sharper, narrower albeit higher fitness peak, depending upon the overall mutation rate. Finally, there could be qualitative changes in the evolutionary process brought on by quantitative changes in the mutation rate. Specifically, there was a certain critical mutation rate above which the evolutionary dynamics became random and incoherent, with a loss of the genetic information [14].

Again things change when we consider finite populations, due to the resulting stochastic nature of the evolution as well as the discrete nature of the individuals. On a flat fitness landscape an infinite population would expand to fill the available space. In reality, the cloud of members retains its cohesiveness. The edge of the cloud is characterized by a dilute population of members. Such members are highly unstable with respect to evolution, in that any fluctuations in population that take their number down to zero results in an extinction of this subpopulation from which it cannot recover. The center of the cloud is more resistant to these fluctuations. The result is that there is a tendency to eliminate the outlying members of the population cloud, so that the cloud remains centered on the prototypical sequence. The resulting cloud can then wander in a stochastic manner in the fitness landscape [15, 16].

One of the more important results of this approach towards evolution is the dependence of the evolutionary process on the fitness landscape. Certain characteristics of the landscape have been especially emphasized by Schuster, Fontana, and their co-workers. (For a good review, see [17].) They investigated the fitness landscape for RNA molecules, taking advantage of rapid algorithms for computing the ground-state conformation [18,19]. They found that these molecules had a large degree of neutrality, that is, it was possible to make large changes in the sequence while retaining the same structure. Considering the sequence as genotype and the structure as phenotype, there would be many changes in genotype consistent with a single phenotype. The many-to-few sequence to structure mapping results in large "neutral networks", that is, regions in the sequence space with constant fitness.

The dynamics of the population cloud combined with the large neutral networks can have a large impact on the evolutionary dynamics. The sequence cloud is free to randomly sample the neutral network. The individuals on the tail of the cloud allow the population to sample the fitness landscape at some distance from the prototypical sequence in a large number of different directions. If the tail of this distribution overlaps a region of higher fitness, the whole cloud can adapt in this direction with a small change in sequence. Note that the resulting dynamics (long periods of neutral evolution, punctuated by rare but rapid adaptive change) is exactly what is described by Kimura's neutral theory.

In this model, the properties of the intersection points between various neutral networks become critical. One important property of the fitness landscape is how close the various neutral networks were to each other. For RNA, it is possible to go from one native structure to almost any reasonably common different structure with only a small change of the sequence, a phenomenon known as *shape-space covering* [17]. As a result RNA can evolve quickly for different structures and possibly different functions. This phenomenon seems not to be true of theoretical models of proteins, in that it is more difficult to go from one structure to another [20-22]. As a result, evolutionarily related proteins tend to have the same structure, something called "structural inertia". This may be due to the larger number of amino acids compared with the number of RNA bases, or the relatively small number of sequences that will form a viable protein in any structure.

3. – Applications to proteins

To summarize the previous section:

- We must include the role of stochastic effects resulting from finite population sizes.

  We should be aware of the presence and effect of neutrality in the fitness landscape. One reason for this neutrality is the many-to-few mapping of genotype to phenotype.

- The properties of the fitness landscape such as the size, distribution, and connectiveness of the neutral networks can have a major effect on evolutionary dynamics.

- Techniques drawn from statistical physics are useful in understanding evolution; it is a natural approach to dealing with the general properties of a large number of individuals behaving stochastically.

- Selective pressure does not necessarily result in the organism selecting the highest peak on the fitness landscape. It is not correct to equate evolution with optimization, or even adaptation.

- Even if a feature fulfills some important function, we cannot conclude that feature evolved in an adaptive way to fulfill this function. These features may represent spandrels.

Let us try to apply some of these lessons to understanding protein properties by considering how the sequence translates to phenotype and eventually to fitness.

3˙1. *Maximum foldability and the distribution of protein structures.* – Proteins have three major evolutionary constraints: they must fold to a structure in a reasonable time, the structure they fold to must perform a function, and the folded structure must be stable enough to perform that function reliably while resisting side-reactions such as aggregation and proteolysis. All three of these requirements are complicated. For instance, much experimental and theoretical effort has gone into determining how a protein is able to fold given the vast number of possible conformations, a non-trivial process [23-25]. While a number of simple models have been developed for understanding the qualitative aspects of protein folding, even less is known about the more general properties of protein functionality. Research into protein stability is hampered by our limited understanding of the interactions that determine the folded state. Underlying all of these problems is the fact that proteins are complicated, involving thousands of atoms interacting with the surrounding solvent.

In order to make the system more tractable, we (as others) have looked at simple lattice models of proteins. For instance, we can consider a protein model consisting of a chain of 27 monomers confined to a $3 \times 3 \times 3$ three-dimensional cubic lattice as in fig. 7. The bonds are all of unit length with adjacent residues existing at adjacent sites. It is possible to enumerate all of the possible self-avoiding walks, not counting reflections and rotations: there are a total of 103346 walks for the $3 \times 3 \times 3$ cubic lattice.

The energy function is of the simple contact form

$$E = \sum_{i<j} \gamma_{ij} u(r_{ij}), \tag{3}$$

where $u(r_{ij})$ is equal to one if residues i and j are not adjacent in sequence but are on adjacent lattice sites and zero otherwise, and γ_{ij} represents the energy contribution for contact between residues i and j.

As mentioned above, protein structures remain relatively constant in evolutionary time while sequences are surprisingly plastic. This results in an interesting time-scale separation: protein folding takes place in seconds to minutes, too short for much evolution to occur. As a result, on the folding timescale it is the sequence that determines the folded structure. Conversely, on the evolutionary timescale, sequences change but only in such a way that the folded state remains constant. So on the evolutionary timescale, it is the structure that determines (or at least limits) the sequence.

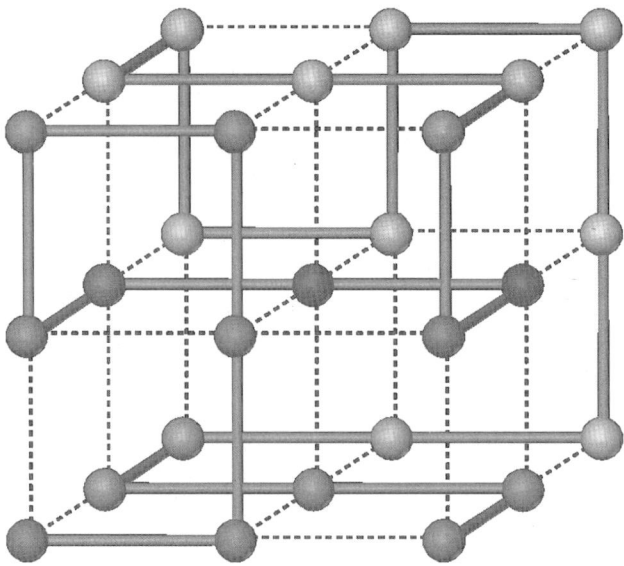

Fig. 7. – Lattice model of proteins.

The challenge in folding is to find the native state while avoiding all of the other free-energy minima. Wolynes and co-workers adapted ideas from spin-glass theory to this problem, conjecturing that there were two possible transitions, one to the folded state, the other to a glassy state from which folding was impossible [26-29]. Using the Random Energy Model [30], it is possible to show that folding could be assisted while freezing to the glassy state discouraged by increasing what we called the *foldability* $\mathcal{F}$, that is, the depth of the native state energy minimum compared with the average of the random states relative to the width in the distribution of random state energies [27-29,31]. Shakhnovich and Dill and their respective co-workers used lattice models to demonstrate that a closely related factor could distinguish between sequences that could and could not fold in numerical simulations [32-38].

As emphasized above, evolution is not equal to optimization, and may even not represent adaptation. Yet for now let us consider how the foldability could be optimized, that is, the limit of infinite selective pressure. In this case it is possible, given a target native-state conformation and an ensemble of random conformations, to solve for the best values of γ_{ij} (that is, the values that maximize the foldability $\mathcal{F}$) in closed form [27,28,31]. While this solution is exact, it is possible to write down an approximate solution by neglecting correlations between the presence of various contacts. The optimal values of the interactions are given by

$$\gamma_{ij}^{\text{opt}} \approx -\frac{u(r_{ij}^T) - \langle u(r_{ij})\rangle}{(1 - \langle u(r_{ij})\rangle)(\langle u(r_{ij})\rangle)}, \qquad (4)$$

where $u(r_{ij}^T)$ is the value of $u(r_{ij})$ in the native-state conformation (that is, is equal to one if and only if this contact is present in the native state) and the averages are over the random conformations. Note that in the case of contacts present in the native state $(u(r_{ij}^T) = 1)$, γ_{ij}^{opt} is approximately inversely proportional to the frequency of this contact in non-native states. This results in the prediction that the long-range contacts between very separated parts of the protein chain should be especially strong, and the propensities for local structure should be weak [31,39].

For each structure, we can define a maximum possible foldability, that is, the foldability with the optimal set of interactions. Some structures have very high maximum foldabilities; other structures had much lower maximum foldabilities. In general, the maximum possible foldability is determined by the number of long-range contacts with small values of $\langle u(r_{ij}) \rangle$:

$$\mathcal{F}_{\text{opt}} \approx \sum_{\text{native-state contacts}} \frac{1 - \langle u(r_{ij}) \rangle}{\langle u(r_{ij}) \rangle}. \tag{5}$$

Given the discussion above about the difference between evolution, optimization, and adaptation, why should we care about $\mathcal{F}_{\text{opt}}$? It is because this parameter is an important characteristic of the fitness landscape in the same way that a mountain can be characterized by its height. Let us consider a more neutralist perspective, that the foldability is not optimal but just has to be high enough with foldability $\mathcal{F}$ greater than some critical value $\mathcal{F}_{\text{crit}}$; all sequences with $\mathcal{F} > \mathcal{F}_{\text{crit}}$ would have fitness equal to one while sequences with $\mathcal{F} < \mathcal{F}_{\text{crit}}$ would have fitness equal to zero. In this case, changes in the sequence that maintain a value of $\mathcal{F} > \mathcal{F}_{\text{crit}}$ would be neutral changes. If $\mathcal{F}_{\text{opt}}$ is large, the interactions that determine the folded state can be far from optimal and still have a foldability larger than $\mathcal{F}_{\text{crit}}$. Conversely, you would have to find near-optimal interactions in order to have $\mathcal{F} > \mathcal{F}_{\text{crit}}$ for a structure with a lower $\mathcal{F}_{\text{opt}}$. As a result, there will be many more sequences that will fold into structures with large $\mathcal{F}_{\text{opt}}$ than would fold into structures with lower $\mathcal{F}_{\text{opt}}$; that is, there would be a strong correlation between the optimizability (value of $\mathcal{F}_{\text{opt}}$ for any structure) and the designability (the number of sequences that would successfully fold into that structure) [31,40,41]. This point is made graphically in fig. 8. Given the random nature of evolution, we would expect that such highly designable structures would be more likely to arise through evolutionary dynamics. In addition, we find that there is a strong correlation between the value of $\mathcal{F}_{\text{opt}}$ and the size of the neutral network [21,22]. This would suggest that such highly optimizable structures would be more resistant to random mutations and more able to evolve to different functions. As $\mathcal{F}_{\text{opt}}$ is correlated with the number of long-range contacts present in the folded state, we would expect proteins that result from evolution to have many such long-range contacts.

Note that this does not imply that structures with many shorter-range contacts fold faster. There is some evidence to the contrary [42], although this analysis is complicated by the domination of secondary-structure interactions in the calculation, interactions that are omitted from the current theory due to their highly cooperative nature. The theory discussed here would be more appropriate for the assembly of already formed (or

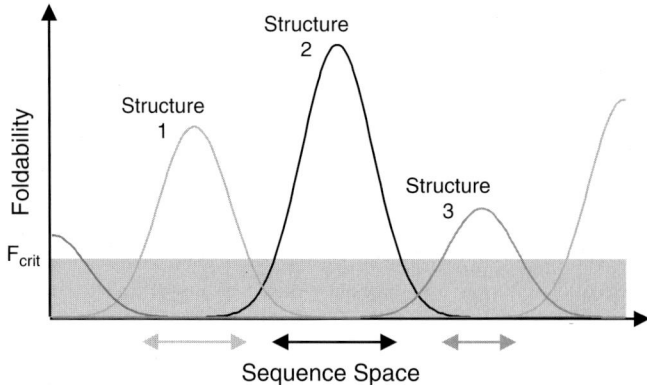

Fig. 8. – Representation of the relationship between $\mathcal{F}_{opt}$ and designability. As the various contact energies are varied away from their optimal values, the foldability of the protein decreases. If $\mathcal{F}_{opt}$ is larger, there is a larger range of interactions that will have foldabilities larger than $\mathcal{F}_{crit}$ as shown in the figure—notice the difference in the range of interactions that will fold into structure 2 compared with the other structures. This bias will increase for larger values of $\mathcal{F}_{crit}$. Given the large dimension of the interaction (as well as sequence) space, small differences in ranges of interaction parameters will result in large differences in the number of sequences folding successfully into one structure or another.

partially formed) secondary structure elements, where the secondary-structure contacts are already formed.

More recently, designability has been calculated by considering all possible lattice proteins constructed by a two-letter amino acid alphabet [43]. While the results are qualitatively similar, both the overall distribution of designabilities as well as the relative designability of different structures show a strong difference with the results with more complete alphabets [44]. This suggests that it may be dangerous extrapolating from such studies to more realistic protein models.

It is interesting to note that the distribution of designabilities changes as the value of $\mathcal{F}_{crit}$ is altered [41, 44]. For low values of $\mathcal{F}_{crit}$, the distribution of designabilities is rather Gaussian, becoming exponential for larger values of this parameter. Finally, for very large values of $\mathcal{F}_{crit}$, the distribution of designabilities becomes a highly stretched exponential of the form $\rho(\text{designability}) \sim \exp[-\alpha \times \text{designability}^\beta]$, with β less than 1.

Recently, we performed simulations of population evolution [45]. In these simulations, we considered 25 residue proteins confined to a 5×5 two-dimensional maximally compact square lattice, with each monomer located at one lattice point. This provides us with 1081 possible conformations represented by the 1081 self-avoiding walks on this lattice, neglecting structures related by rotation, reflection, or inversion. We used a simple lattice potential of the form of eq. (3), modified to take into account the possible amino acids

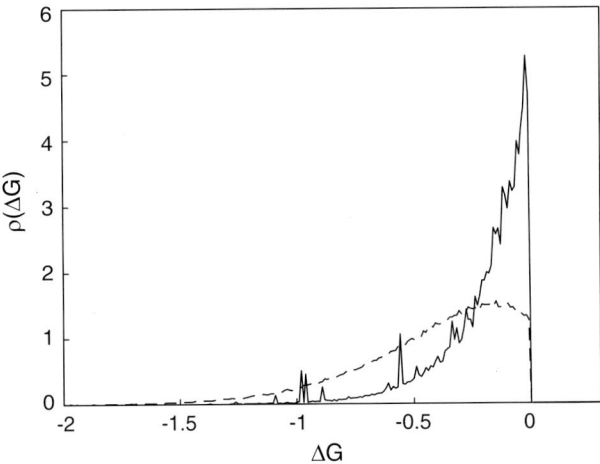

Fig. 11. – The distribution of stabilities for random stable ($\Delta G < 0$) sequences (solid line) compared with the distribution that results from population evolution (dashed line).

subpopulation Φ_1 with the functionality requiring minimal stability became the sole surviving subpopulation. In only one run did the protein subpopulation requiring moderate stability become dominant, and in every run the subpopulation of proteins with a high stability requirement quickly vanished. Because of the natural propensity for proteins to be marginally stable *independent of functionality*, functional mechanisms compatible with marginal stability are at a great evolutionary advantage.

Marginal stability may represent a "spandrel", a naturally occurring tendency that biology can use for its own advantage. If so, this tendency has provided biology with a robust system characterized by easy adaptation of new functionalities. In any case, these evolutionary considerations suggest that it is not necessarily true that proteins are marginally stable because it is required for functionality—rather, proteins may require marginal stability to function because they are marginally stable!

3˙4. *Including explicit functionality*. – We have also neglected the role of other forms of evolutionary pressure such as the need for the folded protein to fulfill its biological function. Being able to fold is crucial but is only one of a long list of requirements a protein must fulfill. The assumption in the work described above is that the requirements of foldability are uncorrelated with these other requirements—two different structures are *a priori* equally likely to be compatible with a given biological activity—and therefore we could neglect functionality when considering protein structures. Is this a valid assumption? How can we bring in functionality into these models? One common aspect of functionality is the ability to recognize and bind some other molecule, including peptides. We developed a model where proteins were evaluated by their ability to bind a specified peptide, as shown in fig. 13 [83]. In this example, the peptide was fixed as QIFW and the protein was allowed to evolve. As shown, we considered 16-mer two-dimensional lattice

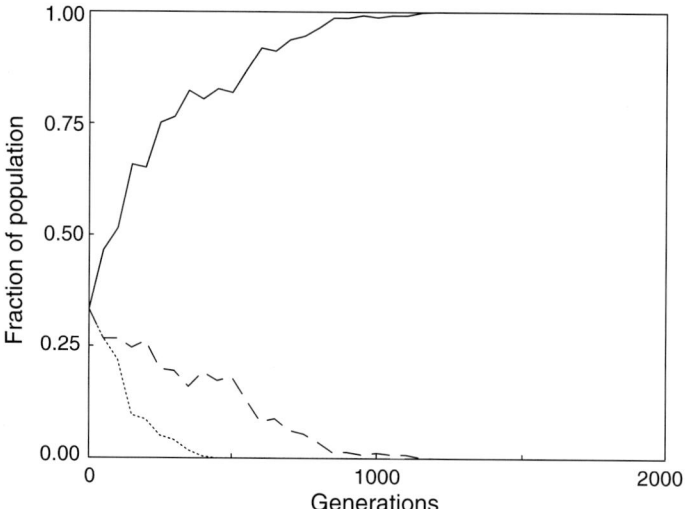

Fig. 12. – Typical evolution run, showing how the subpopulation with a mechanism for functionality requiring minimal stability (solid line) has an evolutionary advantage over mechanisms requiring moderate (dashed line) or high stability (dotted line).

proteins. In order to better represent the thermodynamics, we considered all possible conformations of this protein, including the 69 compact structures that can be fit on a 4×4 square lattice, as well as the 802,016 non-compact conformations. We considered that the protein could be either in an unfolded (non-compact state), in a folded but unbound state, or folded and binding the peptide ligand. We did not consider it possible for an unfolded protein to bind the ligand. There are a total of eight ways of binding the ligand given four sides to the protein and two orientations of the peptide. The energy of all of the various states were calculated using eq. (6), with the Miyazawa-Jernigan potentials used for both intraprotein contacts as well as protein-ligand interactions. It is impossible to determine the fraction of proteins that are binding peptide without knowing the relative concentrations of both. We can, however, easily calculate a relative probability of a protein binding a ligand by considering the weak-binding limit. In this limit, the relative probability is given by

$$(7) \qquad P(\text{ligand bound}) \propto \frac{\sum_{\substack{\text{compact conformations} \\ \text{with bound peptide}}} \exp[-E/kT]}{\sum_{\substack{\text{all compact and non-compact} \\ \text{conformations}}} \exp[-E/kT]}.$$

In previous simulations modeling population evolution, we used neutral evolution with constraints: proteins were either viable or non-viable, and all viable proteins had the same fitness. In this model we considered that the fitness of a protein was proportional to the relative probability of binding the target ligand. A population of 1000 generations was constructed by copying a random sequence. This population was then subjected

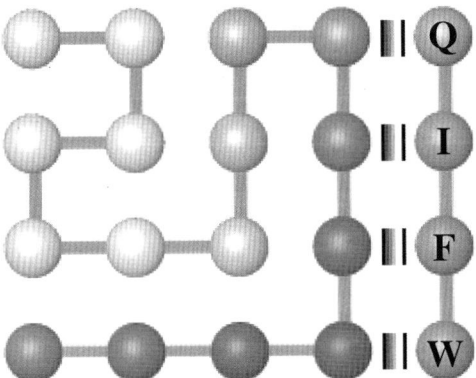

Fig. 13. – Model for representing protein functionality, in this case binding of a QIFW tetrapeptide.

to random mutations at a rate that averaged 20 total mutations in the population per generation. We then computed the relative probability of binding the target ligand, and chose the next generation randomly with replacement where the probability of choosing any particular individual was proportional to the binding probability. Figure 14 shows how the ability of the protein to bind to the ligand changed during the simulation. We also computed an effective number of compact states by computing the relative probability P_i that a compact state would be in any of the 69 possible compact states. η, the effective number of compact states, is then given by

$$(8) \qquad \eta = \left(\sum_{\text{all compact conformations}} \frac{1}{P_i^2} \right)^{-1}.$$

η has the attractive property that it is equal to one if only one state is occupied, and is equal to the total number of states if all are equally occupied. As in the population runs described earlier, we can consider the designability, the fraction of viable sequences that fold into each of the various 69 structures, and compare these values with the occupancy, the fraction of the sequences in the final generation of the evolutionary simulations that folds into that structure. In this case viability for the designability calculations was considered as having a probability of being folded of 50% or more, that is, $\Delta G_{\text{folding}} \leq 0$. This comparison is made in fig. 15 for 5000 different simulations of a population of 1000 proteins evolving for 1000 generations to bind QIFW.

There are three observations we can make. The first is that the designability and the occupancy are generally well correlated. More work is required with different target peptides to better establish this trend. We can also see that there are structures that deviate from the general relationship. Finally, we can tell that the overall distribution of designabilities and the overall distribution of occupancies are quite different. The designabilities follow the trends described earlier, with many rare and a few likely

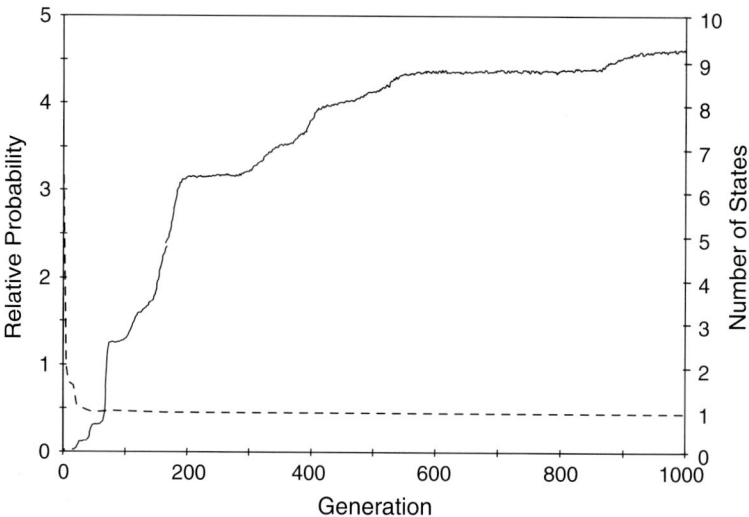

Fig. 14. – Evolution of average relative binding probability and effective number of states during a typical simulation, where fitness was represented by the ability to bind the tetra-peptide QIFW.

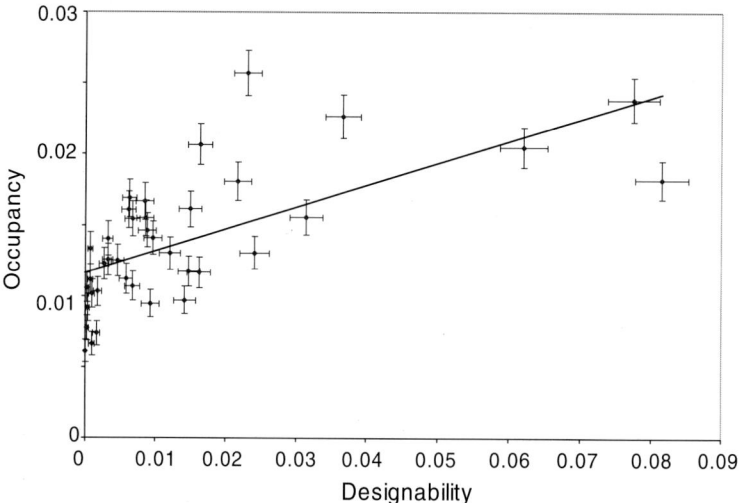

Fig. 15. – Correlation between designability $\mathcal{V}$ and occupancy $\mathcal{O}$. The occupancy represents the frequency of sequences forming any given structure at the end of evolutionary simulations where the fitness corresponded to the ability of the protein to bind the tetra-peptide QIFW. The correlation coefficient between these two quantities is 0.66.

structures. The occupancies are more well balanced, with some proteins resulting from the evolutionary simulation that are effectively never found among random protein sequences. This seems to be at variance with the population simulations described above, which resulted in a more *uneven* distribution of occupancies than designabilities. The difference between these two simulations has to deal with the initial conditions. In the previous study, we allowed the protein population to equilibrate among all of the various possible structures. Sequences that folded into structures that had smaller designabilities would be less resistant to mutations, and would be correspondingly less frequent. In the current simulations, we started with a random set of sequences with low fitness. As the simulation proceeded, the average level of the fitness increased, so that the fitness required for reproducing continually increased. As a result, the volume of the sequence space available to each structure continually decreased. It is interesting to note that, in general, the population very quickly chose a folded native state within tens of generations, as seen in fig. 14. This native state was preserved for the remainder of the simulation. As a result, the probability of each of the folded states was decided when the overall fitness required for reproduction was quite low. As described above, low required values of fitness generally results in a more equitable distribution of proteins among the various structures. This indicates that when we try to explain the current observed distribution of structures, we must consider the fitness landscape that was present when the structures were being determined.

4. – Conclusion

Proteins are the result of a long evolutionary process. In order to understand their properties, we must consider the manner in which they arose. There have been many developments in evolutionary theory that can provide interesting and important perspectives. Increasingly, evolutionary theory has focused on random processes, chance events with consequences that, through becoming fixed in a population, increase the "information content" of the species. The randomness of the processes does not mean that we have to give up attempting to understand the underlying theory. Just as we can understand the general patterns of statistical physics, we can comprehend and model the qualitative and quantitative behavior of evolutionary change. And similarly to statistical physics, attention must shift from fitness to "entropy" or other measures of the number of possibilities open to the system. Random processes will preferentially end up in situations corresponding to many of these possibilities, or alternatively, high entropy. That is why there has been recent attention on the designability of different structures as providing an explanation for why certain structures are so over-represented.

We have described in this paper how other phenomena can be explained as well, such as why proteins that fold under kinetic control would still most likely fulfill the thermodynamic hypothesis, why proteins are marginally stable, and why the functionality of proteins may evolve in such a way as to require marginal stability. We have also looked at the effects of explicit evolutionary pressure for simple functionality. It is important to remember during these analyses that non-intuitive aspects may arise, especially as we often have little intuition considering high-dimensional spaces such as the fitness landscape.

* * *

I would like to acknowledge the co-workers involved in the projects that are described in this paper: N. BUCHLER, S. GOVINDARAJAN, D. POLLOCK, R. RECABARREN, D. TAVERNA, and P. WILLIAMS. I would also like to thank current and past members of the Goldstein lab for important insights and perspectives, and to TODD RAEKER for computational assistance. Financial support was provided by NIH grant LM05770 and NSF equipment grant BIR9512955.

REFERENCES

[1] LI W. H. and GRAUR D., *Fundamentals of Molecular Evolution* (Sinauer, Sunderland) 1991.
[2] VOLKENSTEIN M. V., *Physical Approaches to Biological Evolution* (Springer-Verlag, Berlin) 1994.
[3] PAGE R. D. M. and HOLMES E. C., *Molecular Evolution: A Phylogenetic Approach* (Blackwell Science, Oxford) 1998.
[4] KIMURA M., *Nature (London)*, **217** (1968) 624.
[5] NEI M. and IMAIZUMI Y., *Heredity*, **21** (1966) 183.
[6] KING J. L. and JUKES T. H., *Science*, **164** (1969) 788.
[7] BOWIE J. U., REIDHAAR-OLSON J. F., LIM W. A. and SAUER R. T., *Science*, **247** (1990) 1306.
[8] GOULD S. J. and LEWONTIN R. C., *Proc. R. Soc. London, Ser. B*, **205** (1979) 581.
[9] FISHER R. A., *J. Animal Ecology*, **12** (1943) 54.
[10] WRIGHT S., *International Proceedings of the Sixth International Congress on Genetics*, Vol. **1** (1932) p. 356.
[11] EIGEN M., *Naturwiss.*, **58** (1971) 465.
[12] EIGEN M. and SCHUSTER P., *Naturwiss.*, **64** (1977) 541.
[13] EIGEN M. and MCCASKILL J., *J. Phys. Chem.*, **92** (1988) 6881.
[14] EIGEN M., MCCASKILL J. and SCHUSTER P., *Adv. Chem. Phys.*, **75** (1989) 149.
[15] DERRIDA B. and PELITI L., *Bull. Math. Biol.*, **53** (1991) 355.
[16] HUYNEN M. A, STADLER P. F. and FONTANA W., *Proc. Natl. Acad. Sci. USA*, **93** (1996) 397.
[17] SCHUSTER P. and FONTANA W., *Physica D*, **133** (1999) 427.
[18] ZUKER M. and STIEGLER P., *Nucl. Acids Res.*, **9** (1981) 133.
[19] ZUKER M. and SANKOFF D., *Bull. Math. Biol.*, **46** (1984) 591.
[20] BORNBERG-BAUER E., *Biophys. J.*, **73** (1997) 2393.
[21] GOVINDARAJAN S. and GOLDSTEIN R. A., *Biopolymers*, **42** (1997) 427.
[22] GOVINDARAJAN S. and GOLDSTEIN R. A., *Proteins*, **29** (1997) 461.
[23] LEVINTHAL C., in *Mössbauer Spectroscopy in Biological Systems*, edited by P. DEBRUNNER, J. C. M. TSIBRIS and E. MUNCK (University of Illinois Press, Urbana) 1969, p. 22.
[24] UNGER R. and MOULT J., *Bull. Math. Biol.*, **55** (1993) 1183.
[25] HART W. G. and ISTRAIL S., *J. Comp. Biol.*, **4** (1997) 1.
[26] BRYNGELSON J. D. and WOLYNES P. G., *Biopolymers*, **30** (1990) 171.

[27] GOLDSTEIN R. A., LUTHEY-SCHULTEN Z. A. and WOLYNES P. G., *Proc. Natl. Acad. Sci. USA*, **89** (1992) 4918.
[28] GOLDSTEIN R. A., LUTHEY-SCHULTEN Z. A. and WOLYNES P. G., *Proc. Natl. Acad. Sci. USA*, **89** (1992) 9029.
[29] GOLDSTEIN R. A., LUTHEY-SCHULTEN Z. A. and WOLYNES P. G., in *Proceedings of the 26th Annual Hawaii International Conference on System Sciences*, edited by T. N. MUDGE, V. MILUTINOVIC and L. HUNTER, Vol. **1** (IEEE Computer Society Press, Los Alamitos) 1993, p. 699.
[30] DERRIDA B., *Phys. Rev. Lett.*, **45** (1980) 79.
[31] GOVINDARAJAN S. and GOLDSTEIN R. A., *Biopolymers*, **36** (1995) 43.
[32] SHAKHNOVICH E. I. and GUTIN A. M., *Prot. Eng.*, **6** (1993) 793.
[33] SHAKHNOVICH E. I. and GUTIN A. M., *Proc. Natl. Acad. Sci. USA*, **90** (1993) 7195.
[34] SHAKHNOVICH E. I., *Phys. Rev. Lett.*, **72** (1994) 3907.
[35] SHAKHNOVICH E. I., in *Protein Structure by Distance Analysis*, edited by H. BOHR and S. BRUNAK (IOS Press, Amsterdam) 1994, p. 201.
[36] ŠALI A., SHAKHNOVICH E. I. and KARPLUS M. J., *J. Mol. Biol.*, **235** (1994) 1614.
[37] ŠALI A., SHAKHNOVICH E. I. and KARPLUS M. J., *Nature (London)*, **369** (1994) 248.
[38] CHAN H. S. and DILL K. A., *J. Chem. Phys.*, **100** (1994) 9238.
[39] GOVINDARAJAN S. and GOLDSTEIN R. A., *Proteins*, **22** (1995) 413.
[40] GOVINDARAJAN S. and GOLDSTEIN R. A., *Proc. Natl. Acad. Sci. USA*, **93** (1996) 3341.
[41] BUCHLER N. E. G. and GOLDSTEIN R. A., *J. Chem. Phys.*, **112** (2000) 2533.
[42] BAKER D., *Nature*, **405** (2000) 39.
[43] LI H., HELLING R., TANG C. and WINGREEN N., *Science*, **273** (1996) 666.
[44] BUCHLER N. E. G. and GOLDSTEIN R. A., *Proteins*, **34** (1999) 113.
[45] TAVERNA D. M. and GOLDSTEIN R. A., *Biopolymers*, **53** (2000) 1.
[46] MIYAZAWA S. and JERNIGAN R. L., *Macromol.*, **18** (1985) 534.
[47] LIPMAN D. J. and WILBUR W. J., *Proc. R. Soc. London (Biol.)*, **245** (1991) 7.
[48] LEVITT M. and CHOTHIA C., *Nature (London)*, **261** (1976) 552.
[49] CHOTHIA C., *Nature (London)*, **357** (1992) 543.
[50] ORENGO C. A., JONES D. T. and THORNTON J. M., *Nature (London)*, **372** (1994) 631.
[51] GOVINDARAJAN S., RECABARREN R. and GOLDSTEIN R. A., *Proteins*, **35** (1999) 408.
[52] WANG Z.-X., *Proteins*, **26** (1996) 186.
[53] BLUNDELL T. and JOHNSON M. S., *Protein Sci.*, **2** (1993) 877.
[54] ALEXANDROV N. N. and GŌ N., *Protein Sci.*, **3** (1994) 866.
[55] ZHANG C.-T., *Protein Engin.*, **10** (1997) 757.
[56] ZHANG C.-T. and DELISI C., *J. Mol. Biol.*, **284** (1998) 1301.
[57] LEVINTHAL C., *J. Chim. Phys.*, **65** (1968) 44.
[58] ANFINSEN C., *Science*, **181** (1973) 223.
[59] KIM P. S. and BALDWIN R. L., *Annu. Rev. Biochem.*, **59** (1990) 631.
[60] DILL K. A., *Biochem.*, **29** (1990) 7133.
[61] BERKENPAS M. B., LAWRENCE D. A. and GINSBURG D., *EMBO J.*, **14** (1995) 2969.
[62] THOMAS P. J., QU B. and PEDERSON P. L., *Trends. Biochem. Sci.*, **20** (1995) 456.
[63] MITRAKI A., FANE B., HAASE-PETTINGELL C., STURTEVANT J. and KING J., *Science*, **253** (1991) 54.
[64] BAKER D., SOHL J. L. and AGARD D. A., *Science*, **356** (1992) 263.
[65] HONEYCUTT J. D. and THIRUMALAI D., *Proc. Natl. Acad. Sci. USA*, **87** (1990) 3526.
[66] GUTIN A. M., ABKEVICH V. I. and SHAKHNOVICH E. I., *Proc. Natl. Acad. Sci. USA*, **92** (1995) 1282.
[67] LEOPOLD P. E., MONTAL M. and ONUCHIC J. N., *Proc. Natl. Acad. Sci. USA*, **89** (1992) 8721.

[68] BRYNGELSON J. D., ONUCHIC J. N., SOCCI N. D. and WOLYNES P. G., *Proteins*, **21** (1995) 167.
[69] DILL K. A. and CHAN H. S., *Natl. Struct. Biol.*, **4** (1997) 10.
[70] SOCCI N. D., ONUCHIC J. N. and WOLYNES P. G., *J. Chem. Phys.*, **104** (1996) 5860.
[71] ONUCHIC J. N., LUTHEY-SCHULTEN Z. and WOLYNES P. G., *Annu. Rev. Phys. Chem.*, **48** (1997) 545.
[72] PLOTKIN S. S., WANG J. and WOLYNES P. G., *J. Chem. Phys.*, **106** (1997) 2932.
[73] GOVINDARAJAN S. and GOLDSTEIN R. A., *Proc. Natl. Acad. Sci. USA*, **95** (1998) 5545.
[74] SAVAGE H. J., ELLIOT C. J., FREEMAN C. M. and FINNEY J. L., *J. Chem. Soc. Faraday Trans.*, **89** (1993) 2609.
[75] VOGL T., JATZKE C., HINZ H. J., BENZ J. and HUBER R., *Biochemistry*, **36** (1997) 1657.
[76] RUVINOV S., WANG L., RUAN B., ALMOG O., GILLILAND G. L., EISENSTEIN E. and BRYAN P. N., *Biochemistry*, **36** (1997) 10414.
[77] GIVER L., GERSHENSON A., FRESKGARD P. O. and ARNOLD F. H., *Proc. Natl. Acad. Sci. USA*, **95** (1998) 12809.
[78] PRIVALOV P. L. and KHECHINASHVILI N. N., *J. Mol. Biol.*, **86** (1974) 665.
[79] RASMUSSEN B. F., STOCK A., RINGE D. and PETSKO G. A., *Nature*, **357** (1992) 423.
[80] ZAVODSZKY P., JOZSEF K., SVINGOR A. and PETSKO G. A., *Proc. Natl. Acad. Sci. USA*, **98** (1998) 7406.
[81] TSOU C. L., *Enzyme Engin. XIV*, **864** (1998) 1.
[82] TAVERNA D. M. and GOLDSTEIN R. A., submitted to *Proteins*.
[83] WILLIAMS P. D., POLLOCK D. D. and GOLDSTEIN R. A., *Scientific Visualization and Modelling*, in press.

What evolution can tell us about protein-DNA interactions

L. Mirny

Department of Chemistry and Chemical Biology, Harvard University
Cambridge, MA 02138, USA

M. S. Gelfand

Integrated Genomics Moscow - Moscow, Russia

1. – Introduction

Protein-DNA interactions are central for the regulation of gene expression in the cell. Much progress has been made since the first DNA-binding protein has been isolated [1]. The highest resolution picture of protein-DNA interactions is coming from more than 200 X-ray and NMR solved structures of protein-DNA complexes [2]. As this information was accumulated, structures have been thoroughly examined by the authors. Protein-DNA complexes have been studied by chemical modifications (see [3] for review) and site-specific mutagenesis (*e.g.* [4,5]); and binding motifs and interactions have been classified [6-9]. Recently three groups [10-12] extensively studied representative protein-DNA complexes: chemical and physical properties of the interfaces, their polarity, size, shape and packing.

Although X-ray and NMR structures give us the most detailed picture of protein-DNA interactions, the structures are missing information about the energetics of the interactions and the relative importance of different residues and nucleotides in recognition. Hence, by analyzing protein-DNA complexes alone, one cannot tell why a protein selects one DNA site to bind instead of the others.

By mutating the protein and the DNA site one can identify the relative importance of different residues and nucleotides in protein-DNA recognition. These experiments are labor-intensive, making it impossible to study all possible mutations of a few residues. An enormous number of such mutations, however, have already been tested in the "natural laboratory" by molecular evolution. Families of homologous proteins tell us about mutations that were tolerated by the protein and those that were not. On the DNA side, sites recognized by the same protein, or by its orthologues in closely related organisms are identified by footprinting assays and bioinformatic techniques. Multiple alignments of both footprinted DNA sites [13] and homologous proteins [14-18] are publicly available.

In this study we combine and systematically analyze structures of protein-DNA complexes, footprinted DNA sites, and multiple alignments of DNA-binding proteins. The goal is to identify and understand the primary determinants of specific DNA recognition by proteins.

In the first section we study how conservation of nucleotides in the DNA site is linked to the nucleotides' structural role in the protein-DNA complex. By comparing sites recognized by the same protein, we identify base pairs conserved in evolution. Using structures of protein-DNA complexes we compute the number of interactions every base pair has with the protein and match this number with the conservation of this base pair. The first result of this study is that base pairs that have more interactions with the protein are more conserved in the binding sites. As natural as it is, this result has never been reported before.

Next we study the LacI family of homologous proteins and show that certain residues binding DNA exhibit a very special pattern of conservation: they are conserved within orthologues (that have the same binding specificity) and are variable between paralogues (that have different DNA binding specificities). This kind of pattern can serve as a "signature" of specificity determining residues. We develop a method to find such residues in the families of proteins grouped according to their function. This method is somewhat similar to evolutionary trace analysis [19,20]. Since our method relies on rigorous statistical control, unlike evolutionary trace analysis, it does not require knowledge of protein structure to sort out false-positives. Applied to the LacI family our method identified 12 specificity determinants. When mapped onto the structure, 3 residues are binding DNA and 6 are surrounding the ligand-binding pocket in the ligand-binding domain. Available experimental information supports the critical role of the identified DNA-binding residues in determining the specificity of DNA recognition.

2. – DNA's point of view

2˙1. *Results.* – For our analysis we select all bacterial transcription factors for which sufficient footprinted sites in the DPI database [13] and high-resolution X-ray or NMR structures are *both* available. Unfortunately, only five proteins satisfy these criteria: Crp, PurR, TrpR, Ihf and MetJ. For each structure we compute the number of contacts n_i each base pair i has with the protein, *i.e.* the number of heavy atoms that are at a distance less than $4.0\,\text{Å}$ from a protein atom. To focus on sequence-specific interactions

TABLE I. – *Correlation between the number (n) of interactions with the proteins a base pair has and its variability (S) in footprinted sites.*

Protein	Number of sites	Corr. coeff. (r)	Association (γ)
Crp	49	−0.62	**−0.98**
PurR	23	−0.50	**−0.77**
TrpR	4	−0.60	**−0.63**
Ihf	27	−0.72	**−0.64**
MetJ	16	+0.10	**−0.25**
MetJ SELEX holo	75	−0.59	**−0.80**
MetJ SELEX apo	56	−0.50	**−0.80**

of the DNA with the protein, we exclude atoms belonging to the sugar-phosphate DNA backbone, because they do not depend on the DNA sequence. We also compute the number of hydrogen bonds (including water-mediated) and the number of hydrophobic interactions each base pair has with the protein. Hydrogen bonds are computed using NUCPLOT/HBPLUS [21]. Two groups are said to have a hydrophobic interaction if both have CHARMM [22] group-charge less than 0.3 and they are separated by less than 3.5 Å.

For aligned footprinted sites we compute variability (information contents) [23] at each position as

$$S_i = - \sum_{x=A,C,G,T} f_i(x) \log f_i(x), \qquad (1)$$

where $f_i(x)$ is the frequency of nucleotide x in position i at the site. Next we compute the correlation between **S** and **n**. We use both the traditional linear correlation coefficient r [24] and a 2×2 association measure γ [25]. The 2×2 measure is used to compute the association between categorical variables. To use it, we classify positions as being variable ($S_i > S_{\text{cut}}$) vs. conserved ($S_i \leq S_{\text{cut}}$) and as strongly involved ($n_i > n_{\text{cut}}$) vs. slightly involved ($n_i < n_{\text{cut}}$) in interactions with the protein. To eliminate ambiguity in setting the cut-off S_{cut} and n_{cut} we use medians of **S** and **n** accordingly. This way we obtain a 2×2 variability-involvement frequency table ρ,

$$\begin{aligned}
\rho_{11} &= \text{number of positions with } S_i > S_{\text{cut}} \text{ and } n_i > n_{\text{cut}}, \\
\rho_{12} &= \text{number of positions with } S_i \leq S_{\text{cut}} \text{ and } n_i > n_{\text{cut}}, \\
\rho_{21} &= \text{number of positions with } S_i > S_{\text{cut}} \text{ and } n_i \leq n_{\text{cut}}, \\
\rho_{22} &= \text{number of positions with } S_i \leq S_{\text{cut}} \text{ and } n_i \leq n_{\text{cut}}.
\end{aligned} \qquad (2)$$

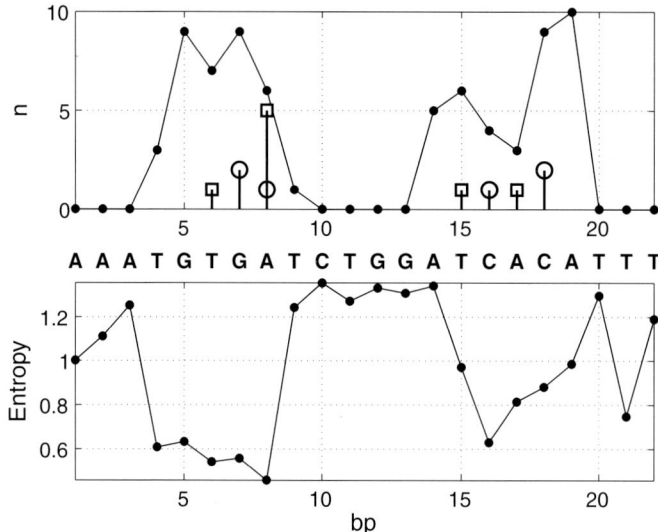

Fig. 1. – Crp site. Top: the thin line shows the number of interactions n base pairs have with the protein. The number of hydrogen bonds formed by the base pair (including water-mediated bonds) is shown by large circles. The number of hydrophobic interactions between a base pair and a protein is shown by large squares. Bottom: the variability (entropy) in footprinted DNA sites. The "consensus" (most frequent) nucleotides are shown by letters above the plot.

Then the association between S and n is measured as [25]

$$\gamma = \frac{\rho_{11}\rho_{22} - \rho_{12}\rho_{21}}{\rho_{11}\rho_{22} + \rho_{12}\rho_{21}}. \tag{3}$$

Table I summarizes the results for all the five proteins. Strikingly, for all proteins except MetJ a strong negative correlation is observed. This indicates that base pairs that have more interactions with the protein n are more important for recognition, and hence have lower variability S. This transparent result has never been reported before. Importantly, all types of interactions were counted together. We did not discriminate between hydrogen bonds, hydrophobic or electrostatic interactions. When any single type of interaction is taken into account the correlation is much lower (see [26] for details).

Crp. Figure 1 presents S_i and n_i for the complex of catabolite gene activator protein (CAP) with its site. CAP is a homodimer. The binding site of each domain can be seen as the region of high n_i and low S_i in the figure. Interestingly, the "right" site is slightly less conserved and, correspondingly, has less tight interactions with the protein. Most of the interactions are formed by ARG180, ARG185 and GLU181 in both chains. They form both hydrogen bonds and hydrophobic interactions (by C_β, C_γ atoms interacting with the CH_3 group of T). Neither the hydrogen bonding pattern, nor the hydrophobic pattern alone correlate with the observed conservation S. The total number of all types

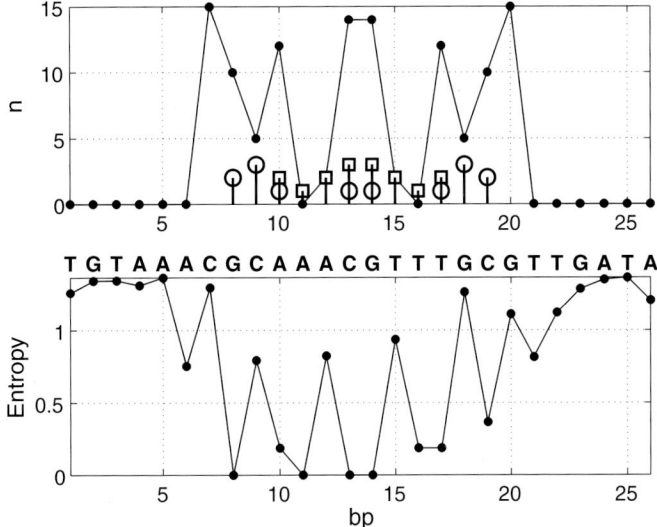

Fig. 2. – PurR site. Notation as in fig. 1.

of interactions n, however, exhibits a strong correlation with S (the strongest among the proteins studied here.)

PurR. For purine repressor both S and n are very symmetric (see fig. 2). However, the perfect symmetry of n is the result of the X-ray structure that was built assuming the two-fold symmetry of the molecule [27]. The correlation between S and n is high with a few exceptions, *e.g.* base pairs AT in positions 11 and 16 are very conserved, but have no interactions with the protein. Most other positions show a regular trend: S decreases as n increases. On the protein side, residues that have most of the contacts with the bases are THR14, ARG24, LEU52, ALA49 and ALA53. As in the case of Crp, both hydrogen bonding and hydrophobic interactions are involved and only their combination exhibits correlation with S.

TrpR. Only four natural footprinted sites are available for TrpR, leading to a poor profile of S. In spite of this problem, correlation with n is significant (see fig. 3). Both n and S are symmetric exhibiting the distinct pattern of highly conserved $A_6 C_7 T_8$ and $A_{17} G_{18} T_{19}$. C_7 and G_{18} are the nucleotides that interact the most with the protein. Both half-sites have lots of hydrophobic interactions with the protein and very few hydrogen bonds. Other conserved base pairs are $G \cdot C_4$ and $C \cdot G_{21}$. Each pair has 7 interactions with the protein and a single hydrogen bond. However, mutations that eliminate this hydrogen bond have a very modest effect on the stability of the complex [28]. Perhaps, other types of interactions are determining specific recognition by Trp.

Comparison with sites obtained by SELEX lead to the correlation $r = -0.43$. However, the motif obtained by SELEX is asymmetric and only the half-site is conserved (GNACTAG motif). This inconsistency with the natural sites could result from differ-

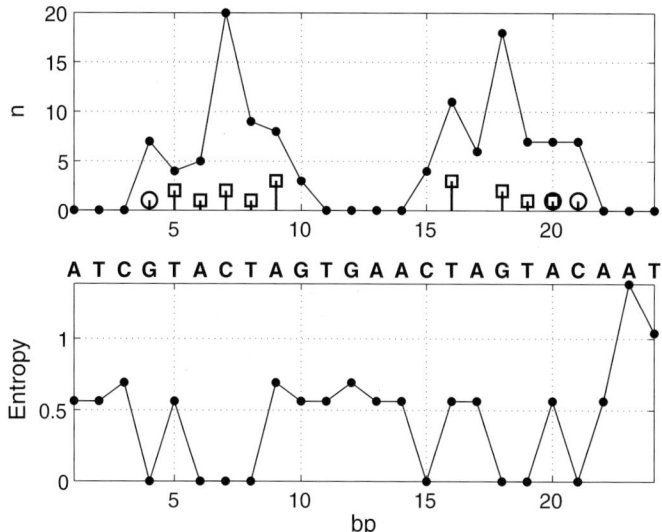

Fig. 3. – Trp site. Notation as in fig. 1.

ent modes of binding observed in Trp repressor, which exhibits both dimer and tandem binding [28, 29].

Ihf. Integration host factor (IHF) is known to bend DNA 160° at the binding site. The site consists of two regions: a 5′ region with no clear consensus and a 3′ region with a significant but very small consensus. In accord with this data, the X-ray structure of the IHF complex shows very few (if any) protein-DNA contacts in the 5′ region and tight protein-DNA interactions in the 3′ region [30]. Our analysis brings quantitative support to these observations. Figure 4 shows the number of protein-DNA interactions and variability of the base pairs in the IHF site. Our results indicate that conservation in the 3′ region can be very well explained by direct protein interactions with the DNA. Two peaks in n correspond to the regions where two proline residues (one from each protein chain) intercalate the DNA. Four arginines, ARG59 and ARG62 (from both A and B chains), are forming almost as many interactions with the bases as intercalating prolines. Most of the other interactions in these regions are formed by LYS65 (chains A&B), ILE72 (chain A), ASN63 (chains A&B) and GLY61 (chains A&B). While arginines are involved in direct and water-mediated hydrogen bonding, prolines and isoleucine are forming hydrophobic interactions with the bases. Two out of three hydrogen bonds with the bases, however, are formed by non-conserved G at position 29 and non-conserved C at position 32. Position 29 is occupied by G only in 15% of the sites and position 32 is occupied by C in 19% indicating that hydrogen bonding of these nucleotides does not lead to strong specificity. Another hydrogen bond and several non-bonded interactions are formed by ARG46(B) with base pairs at positions 20–22. These interactions are also apparently non-specific as base pairs at these positions are not conserved. In summary,

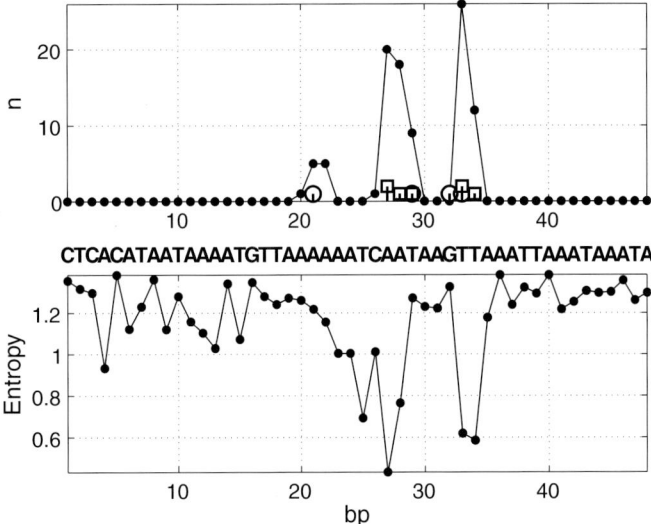

Fig. 4. – Ihf site. Notation as in fig. 1.

a 0.72 correlation is observed in the IHF site, while the hydrogen binding pattern alone cannot explain the observed conservation. In contrast, hydrophobic interactions seem to correlate with the pattern of conservation.

MetJ is binding to arrays of two-to-five adjacent copies of an eight base-pair "metbox" sequence. Naturally occurring operators differ from the consensus sequence to a greater extent as the number of metboxes increases. This makes the motif obtained from the individual eight base-pair sites very weak exhibiting no correlation with the number of direct protein-DNA complexes. However, the conservation pattern of SELEX-derived sites does correlate with the number of interactions between the base pairs and the protein.

In summary, we showed that in the five different bacterial transcription factors the number of interactions a base pair has with the protein strongly correlates with conservation of this base pair. The origin of this correlation is clear: some of the direct interactions between the nucleotides and the protein are stabilizing the complex; then mutations of a more interacting base pair are more destabilizing and are eliminated in evolution. For the same reason residues that have more interactions in the protein (buried residues) are more conserved. Although this result for residues has been known for decades, a similar result for base pairs in protein-DNA complexes is reported here for the first time. Another result concerns the role of hydrogen bonds that are widely believed to dominate in determining the specificity and stability of protein-DNA complexes. Our results, on the contrary, indicate that hydrogen bonds alone cannot explain the pattern of conservation in the site. Only when hydrogen bonds, hydrophobic and other interactions are taken together, does this number correlate with patterns of conservation.

The nature of protein-DNA interactions is very complex and involves hydrogen bonds, hydrophobic and electrostatic interactions and effects of "indirect readout" related to water extrusion, and local DNA bending and twisting. Surprisingly, such a simple parameter as the number of direct interactions (that does not take into account even the different strength of interactions) is able to explain the patterns of conservation in the DNA binding sites. This result makes us believe that more complex models of protein-DNA energetics would be able to predict binding motifs for DNA-binding proteins.

3. – Proteins' point of view

3`1. Results

3`1.1. Conservation of DNA-binding residues. The examination of known protein DNA-complexes reveals several residues binding DNA bases. How conserved are these residues in protein evolution? To address this question we focus on a large LacI family of homologous DNA-binding proteins. All of them are bacterial transcription factors regulating the expression of proteins involved in the sugar/nucleotide metabolism.

Figure 5 presents the multiple alignment of the DNA-binding domains in the LacI family. For each residue we computed the variability as

$$(4) \qquad S_i = -\sum_{x=1}^{20} f_i(x) \log f_i(x),$$

where x is a type of amino acid and $f_i(x)$ is its frequency in the position i of the multiple alignment. Variability computed this way shows that some DNA-binding residues are very conserved, while others are not (see fig. 6).

In order to understand the origin of this high variability we split the LacI family into subgroups of orthologous proteins. We start from *E. coli* homologues of LacI. For each of them we find orthologues in a close bacterial genome by the bidirectional-hit method [31]. We discard orthologues, when a bidirectional hit is absent or weak (see [32] for details). By this method we obtain a family of 54 proteins grouped into 15 sub-families of orthologous proteins. All found orthologues are aligned by ClustalW [33] Importantly, most of the residues appearing variable across the *whole* family are conserved *within* every orthologous sub-family. This suggests that such residues can serve as specificity determinants. In fact, orthologous proteins from relatively close genomes are believed to have the same cellular function [31]. Hence, orthologous transcription factors are likely to regulate the same genes and bind similar sites on the DNA. Hence, we expect that residues determining DNA-binding specificity are conserved within a sub-family of orthologous transcription factors (*e.g.*, within PurR proteins from *E. coli*, *H. influenzae* and *V. cholerae*). Moreover, specificity determinants must differ among paralogous proteins as they are binding different sites (and regulate different genes) in the same organism (*e.g.*, PurR and GalR in *E. coli*). Some variable DNA-binding residues exhibit exactly this pattern of variation: they are conserved within every single orthologous sub-family and are different in different sub-families.

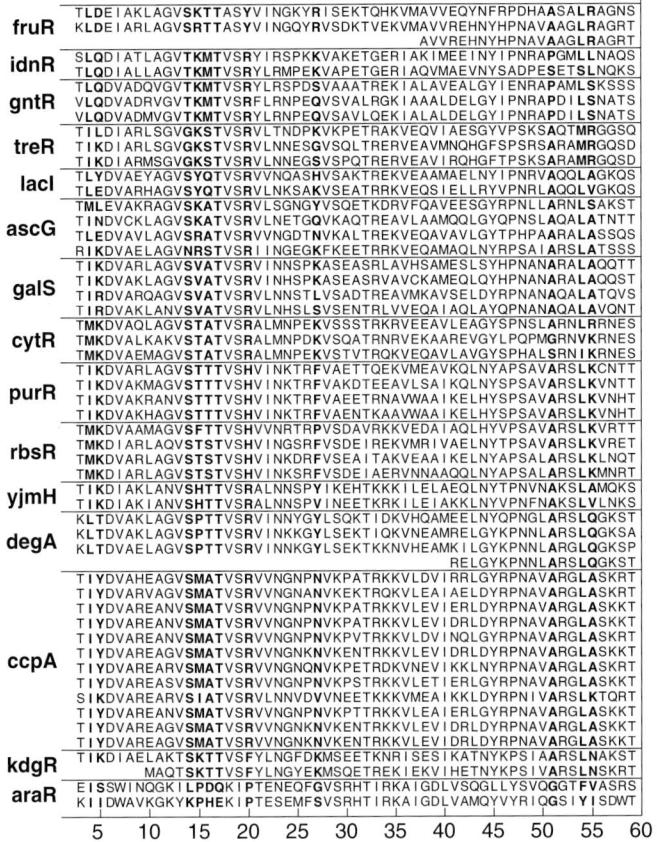

Fig. 5. – DNA-binding domain of LacI proteins grouped into orthologous sub-families. Sub-families are separated by lines, the name of each sub-family is shown on the left. Residues binding the nucleotides (non-backbone DNA atoms) are shown in bold. Note that some of them are very conserved (*e.g.*, S14, T17) while some are not. Residues in positions 15, 16 and 55 are specificity determinants as they are conserved within most sub-families and are different between sub-families.

This pattern of variability can be considered a "signature" of the specificity-determining residues. Based on this idea, we developed a method to search for the specificity-determining residues in protein families.

3˙1.2. Specificity determinants of LacI/GalR family. In order to identify residues with the pattern described above, we use *mutual information* as a measure of association between a residue type x and a sub-family index y:

$$(5) \qquad I_i = \sum_{x,y} f_i(x,y) \log \frac{f_i(x,y)}{f_i(x)f(y)},$$

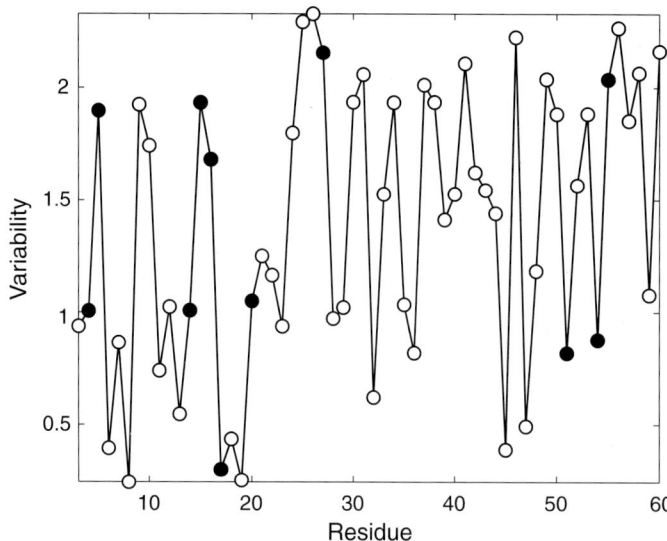

Fig. 6. – Variability (entropy) in the DNA-binding domain of the LacI family. Nucleotide-binding residues are shown by filled circles.

where $f_i(x)$ is the frequency of residue type x in the position i of the multiple alignment, $f(y)$ is the fraction of proteins belonging to orthologous sub-family y, and $f_i(x,y)$ is the frequency of residue type x in the proteins of sub-family y in the position i of the multiple alignment. Summation is over all types of residues $x = 1 \cdots 20$ and over all sub-families y (for LacI $y = 1 \cdots 15$). Mutual information has several important properties: 1) it is always positive; 2) it equals zero if and only if x and y are statistically independent; and 3) a large value of I_i indicates a strong association between x and y. The variability and composition of position i in the multiple alignment influence I_i as well. Hence, we cannot rely on the value of I_i as an indicator of association; instead we estimate the statistical significance of I_i. We start from the zero-hypothesis of no association between x and y. Next we compute $P(I_i)$ the probability of observing I_i under this zero-hypothesis. Positions in the multiple sequence alignment that exhibit low $P(I_i)$ are likely to be specificity determinants.

We use shuffling to compute $P(I)$. For each position i we take a column of $\mathbf{x}$ and randomly shuffle this vector. Next we compute mutual information for shuffled $\mathbf{x}_{\text{sh}}$ and original $\mathbf{y}$: $I_{\text{sh}} = I(\mathbf{x}_{\text{sh}}, \mathbf{y})$. The procedure is repeated 10^3 times for each i. At each i we get a distribution $f(I_{\text{sh}})$, that turns out to be Gaussian even at the tails (data not shown). From this distribution one can compute the mutual information expected under the zero-hypothesis as $I_i^0 = E[f(I_{\text{sh}})]$. However, I_i^0 is systematically lower than I_i, since sequences within an orthologous sub-family are more similar to each other than between sub-families. Importantly, this bias is systematic and is the same for all positions in the protein. To compensate for this bias we make a linear transformation of the mutual

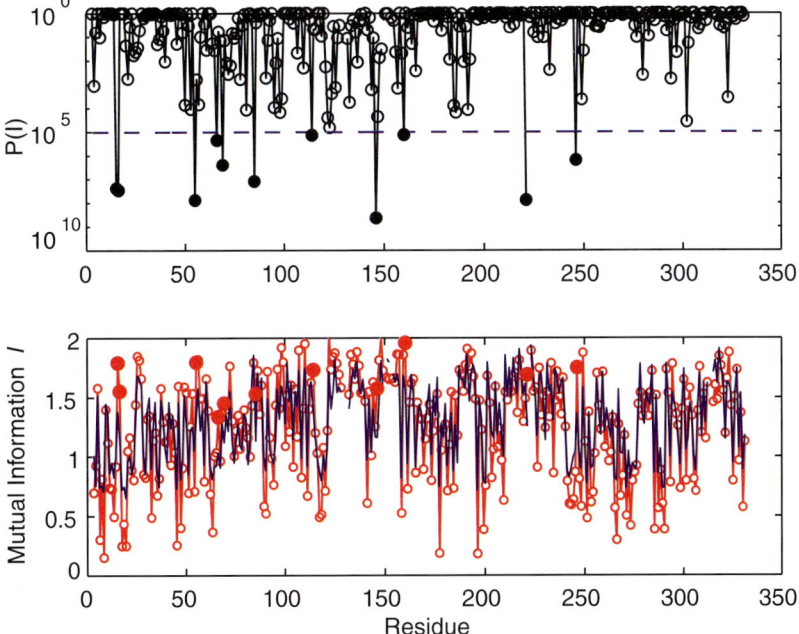

Fig. 7. – Identification of specificity determinants in the LacI family. Top: $P(I)$ probability to observe mutual information I by chance. Positions with $P(I) < 10^{-5}$ (below the broken line) are shown by filled circles. These positions are specificity determinants. Bottom: observed (red) and expected (blue) mutual information. Correlation 0.97. Note few specificity determinants (filled red circles) with mutual information much higher than expected.

information obtained by shuffling: $I'_{sh} = aI_{sh} + b$. Coefficients a and b are chosen to minimize the squared deviation between I_i and I_i^0, i.e. $\sum_i (I_i - I_i^0)^2 \to \min$. Then from $f(I'_{sh})$ we compute $P(I_i)$. Details and derivation of the method are published elsewhere [32].

Figure 7 presents the mutual information I, the expected mutual information I^0 and the probability $P(I)$ computed for the LacI family. This plot reveals several important results: 1) The correlation between observed I_i and expected I_i^0 is very high ($\rho = 0.97$); this indicates that i) the model used to compute expected mutual information is accurate and ii) the vast majority of positions in the LacI family exhibit no functional association. 2) Very few positions have low $P(I)$ (i.e., I_i much greater than I_i^0) indicating that these positions have a strong association with the function of the protein and are probably specificity determinants.

Positions with strong function association ($P(I) < P_{\text{cut-off}} = 10^{-5}$) are: 15, 16, 55, 66, 69, 85, 114, 123, 146, 160, 221, 246. (The numbering is according to the 3D structure of PurR, PDB code 1wet.) To understand the role of these residues we map them onto the structure of PurR.

Figure 8 presents the structure of the PurR-DNA complex with specificity-determining residues shown by space-fill. Examination of the structure brings us to the following conclusions. 1) Only three residues THR15, THR16 and LYS55 out of 12 are located in the DNA-binding domain. All three are deeply buried in the DNA grooves forming a net of interactions with the bases. 2) Another 7 specificity-determining residues, SER69, ALA66, CYS123, ASP146, ASP160, PHE221, and GLY246, are located in the ligand-binding pocket or close to it (within 8.5 Å from the ligand; see the insert). Such a structural location indicates that these residues are involved in ligand recognition. Since different orthologues have different ligands, these residues change from sub-family to sub-family, but stay the same within sub-families. PHE221 is of special interest as its aromatic rings directly interact with the aromatic ligand of PurR (hypoxanthine or guanine). The other functionally linked residues, ALA66 CYS85 and LYS114, are located far from the DNA and the ligand and are either "false positives" or have some special role in allosteric regulation. For example LYS114 of one PurR chain is located next to LYS114 of the other chain and may be important for correct dimerization. In summary, the structural location of identified residues supports the view that they serve as specificity determinants in the LacI family. This includes the specificity of DNA recognition and ligand binding specificity.

The role of positions 15, 16 and 55 in specific DNA recognition is evident from a series of mutant experiments [34,35]. When TYR15 and GLN16 of LacI were mutated to residues observed at these positions in the paralogues (malI, rafR, cytR etc.) the mutants were preferentially binding corresponding operators of these proteins (malI, rafR, cytR etc.). Similarly, when GalR was mutated to have LacI's residues in positions 15 and 16 it was specifically binding sites of LacI. That is strong experimental evidence that positions 15 and 16 are responsible for determining DNA-binding specificity in proteins of this family. Although residue in 55 is binding DNA in the minor groove, this residue was shown to be critical for DNA recognition by PurR [35].

To the best of our knowledge residues we selected in the ligand-binding domain (except for 146) have not been the subject of mutagenic studies. Although mutations of several other residues were shown to interfere with ligand binding, it is not clear how they influence specificity of ligand recognition. Our analysis suggests several possible experiments to test the specificity of ligand binding.

First, one can make single mutations of the putative specificity-determining residues and study the binding affinity and specificity of the mutants. Second, as in experiments with DNA-binding residues, one can "transplant" some or all of the outlined residues from, say, LacI to PurR and measure the selective binding of LacI ligand *vs.* PurR ligand by the mutant. The main question is whether specificity can be re-designed by changing this small set of theoretically predicted residues. Third, since most of these proteins are involved in relatively simple and well-understood transcription regulation processes one can make an *in vivo* study, *e.g.*, PurR protein with LacI DNA specificity and vice versa. If successful, this kind of "chimeric" protein can be used to re-design the network of cellular regulation.

In summary, we studied protein-DNA interactions and the evolution of DNA-binding

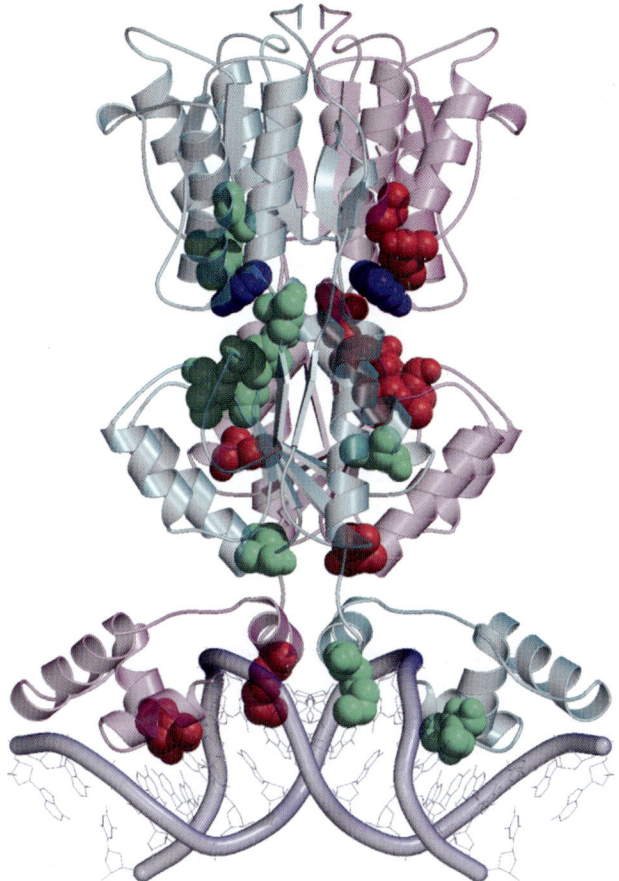

Fig. 8. – Structure of PurR-DNA complex. Two protein chains are shown by semi-transparent ribbons in green and pink. Ligands (guanine) are shown by blue space-fill. Residues identified as specificity determinants are shown by space-fill and colored according to their chain (green and red). Note (1) three residues deeply buried into the DNA, (2) a set of residues in the ligand binding site.

proteins by analyzing their sequences and structures. First, we found that base pairs that have more interactions with the protein are more conserved in evolution. We also showed that, in contrast to the prevalent view, hydrogen bonds are not the main players in protein-DNA recognition. Only when taken together can hydrogen bonds, hydrophobic and electrostatic interactions explain differing conservation of different base pairs in the DNA site. In the second part of this work we focused on LacI/PurR family of bacterial transcription regulators. We showed that certain residues responsible for DNA recognition exhibit *no conservation* among homologues. These residues are conserved among orthologous proteins that bind the same site and are different in paralogous proteins that bind different sites. Based on this idea we developed a method to identify such residues in the multiple alignment of proteins grouped according to their specificity. Using this method we found 12 specificity-determining residues in the LacI/PurR family. Structural location and available experimental information strongly support the role of these residues as functional determinants. The method is general and can be applied to any family of proteins grouped according to their function.

* * *

The work reported in this lecture was supported by the WILLIAM F. MILTON Fund and the Harvard Society of Fellows.

REFERENCES

[1] GILBERT W. and MULLER-HILL B., *Proc. Natl. Acad. Sci. USA*, **58** (1967) 2415.
[2] BERMAN H., ZARDECKI C. and WESTBROOK J., *Acta Crystallogr. D Biol. Crystallogr.*, **54** (1998) 1095.
[3] LARSON C. and VERDINE G., *The Chemistry of Protein-DNA Interactions*, in *Bioorganic Chemistry: Nucleic Acids* (Oxford University Press) 1996, pp. 324-346.
[4] FIELDS D., HE Y., AL-UZRI A. and STORMO G., *J. Mol. Biol.*, **271** (1997) 178.
[5] BROWN B. and SAUER R., *Proc. Natl. Acad. Sci. USA*, **96** (1999) 1983.
[6] HARRISON S., *Nature*, **353** (1991) 715.
[7] PABO C. and SAUER R., *Annu. Rev. Biochem.*, **61** (1992) 1053.
[8] WINTJENS R. and ROOMAN M., *J. Mol. Biol.*, **262** (1996) 294.
[9] SAUER R. and HARRISON S., *Curr. Opin. Struct. Biol.*, **6** (1996) 51.
[10] JONES S., VAN H., BERMAN H. and THORNTON J., *J. Mol. Biol.*, **287** (1999) 877.
[11] NADASSY K., WODAK S. and JANIN J., *Biochemistry*, **38** (1999) 1999.
[12] PABO C. and NEKLUDOVA L., *J. Mol. Biol.*, **301** (2000) 597.
[13] ROBISON K., MCGUIRE A. and CHURCH G., *J. Mol. Biol.*, **284** (1998) 241.
[14] BATEMAN A., BIRNEY E., DURBIN R., EDDY S., HOWE K. and SONNHAMMER E., *Nucleic Acids Res.*, **28** (2000) 263.
[15] CORPET F., SERVANT F., GOUZY J. and KAHN D., *Nucleic Acids Res.*, **28** (2000) 267.
[16] KRAUSE A., STOYE J. and VINGRON M., *Nucleic Acids Res.*, **28** (2000) 270.
[17] DODGE C., SCHNEIDER R. and SANDER C., *Nucleic Acids Res.*, **26** (1998) 313.
[18] HENIKOFF J., GREENE E., PIETROKOVSKI S. and HENIKOFF S., *Nucleic Acids Res.*, **28** (2000) 228.
[19] LICHTARGE O., BOURNE H. and COHEN F., *J. Mol. Biol.*, **257** (1996) 342.
[20] LICHTARGE O., YAMAMOTO K. and COHEN F., *J. Mol. Biol.*, **274** (1997) 325.
[21] LUSCOMBE N., LASKOWSKI R. and THORNTON J., *Nucleic Acids Res.*, **25** (1997) 4940.
[22] Chapter CHARMM: The Energy Function and Its Parameterization with an Overview of the Program.
[23] STORMO G., SCHNEIDER T. and GOLD L., *Nucleic Acids Res.*, **14** (1986) 6661.
[24] DEGROOT M., *Probability and Statistics* (Addison-Wesley Pub. Co, Reading, Mass.) 1996.
[25] GOODMAN L. and KRUSKAL W., *Measures of Association for Cross Classifications* (Springer Series in Statistics, Springer-Verlag, New York) 1979.
[26] MIRNY L. and GELFAND M., to be published.
[27] SCHUMACHER M., CHOI K., ZALKIN H. and BRENNAN R., *Science*, **266** (1994) 763.
[28] GRILLO A., BROWN M. and ROYER C., *J. Mol. Biol.*, **287** (1999) 539.

[29] LAWSON C. and CAREY J., *Nature*, **366** (1993) 178.
[30] RICE P., *Curr. Opin. Struct. Biol.*, **7** (1997) 86.
[31] TATUSOV R., GALPERIN M., NATALE D. and KOONIN E., *Nucleic Acids Res.*, **28** (2000) 33.
[32] MIRNY L. and GELFAND M., to be published.
[33] THOMPSON J., HIGGINS D. and GIBSON T., *Nucleic Acids Res.*, **22** (1994) 4673.
[34] LEHMING N., SARTORIUS J., KISTERS-WOIKE B., VON W.-B. and MULLER-HILL B., *EMBO J.*, **9** (1990) 615.
[35] GLASFELD A., KOEHLER A., SCHUMACHER M. and BRENNAN R., *J. Mol. Biol.*, **291** (1999) 347.

Two models of amino acid conservation in proteins

N. V. Dokholyan and E. I. Shakhnovich

Department of Chemistry, Harvard University
12 Oxford Street, Cambridge, MA 02138, USA

1. – Introduction

The growing amount of biological information [1-6] makes it possible to study the common features that many biomolecules (proteins and DNA) share. This informational explosion has already been widely utilized in understanding patterns of DNA and protein sequences. For example, study of the GenBank sequences of various organisms led to a plausible theory of expansion of dimeric tandem repeats in non-coding DNA sequences [7, 8] and discovery of long-range correlations in the nucleotide patterns in non-coding DNA [9, 10]. The availability of complete genome sequences and mRNA expression data for all genes resulted in the *MobyDick* algorithm of identification of DNA sequence motifs that control gene expression [11].

Recently, it has been conjectured [12-14] that study of the conserved amino acids in families of structurally similar proteins can shed light on the functionally, kinetically and thermodynamically important amino acids in proteins. The basic belief behind the majority of such studies is that evolution optimizes, to a certain extent, the properties of proteins, so that they become more stable, and have better folding and functional properties. Here we use the "optimization" hypothesis of molecular evolution to understand the impact of evolution on the thermodynamic stability of proteins. The main question we would like to address is to what extent we can use this hypothesis to predict thermodynamically important residues (that result in the more stable proteins) using the data provided by molecular evolution in the form of sequence alignments.

Recent studies of Mirny and Shakhnovich [15] identified the presence of universally conserved amino acids across the families of proteins sharing the same fold. These conserved residues have been linked to protein stability, kinetic properties or function. Various experiments [16-25] have identified some of the conserved residues to have predicted specific roles.

The link between conserved amino acids and their role in proteins makes it crucial to model evolution. It is important to realize, however, the complexity of evolutionary processes that may result in a two-tier group of proteins [4,5] (see fig. 1). The first group, called the group of homologous proteins or homologs, is a set of protein sequences that have at least 25% sequence similarity and are structurally similar. The second group, called the group of structurally homologous proteins or analogs, is a set structurally similar proteins that may have less than 25% sequence similarity. Analogs include several families of homologs and generally constitute a larger set of proteins than homologs. We conjecture that such organization of structurally similar proteins is the result of the separation of the evolutionary time scale. On a time scale τ_0, a set of mutations occur that do not affect those amino acids that play thermodynamical, kinetical and/or functional roles. As a result, there is little variation in sequence at the important sites of proteins. If a mutation occurs at the thermodynamically, kinetically and/or functionally important sites, it usually substitutes amino acids with close physical properties so that the core, nucleus and/or functional site are not disrupted and the protein folds into its family fold, is stable in this fold, and preserves its function. At this time scale, a family of homologs is born.

Rarely, at time scale τ, correlated mutations occur that modify *several* amino acids at the core, nucleus and/or functional site, so that the stability and kinetics of proteins are not altered. Such a set of mutations can drastically modify the sequence of the protein. However, within the time scale τ_0, a family of homologs is born within which there is conservation of (already new) amino acids in the specific (important) sites of homologous proteins. Although there are alternations in the specific sites of the proteins at the time scale τ, these sites are more preserved than the rest of the sequence, making themselves a target for identifying thermodynamically, kinetically and/or functionally important amino acids.

We propose here two models to illustrate our view. The first one, the Z-score model, is based on the design of a set of structurally identical sequences by the Z-score minimization [26-28]. The idea is to find the similarities in the sequences of such a set and to recover those residues that are conserved across this set. The protein folding theory [29,30] suggests that Z-score minimization is equivalent to maximizing the energy gap between misfolded or unfolded conformations and the native state of the protein. It has been pointed out that such maximization results in stable and fast-folding proteins. Thus, by designing sequences that have the same fold, we attempt to mimic evolution in diversifying protein sequences for the same fold family. In addition, the Z-score model is a dynamical model, *i.e.* there is an implicit time scale that allows one to follow the evolution of sequences during the design procedure (see subsect. 3˙1).

The second model, the mean-field model, is proposed to capture the conservatism of amino acids in protein folds that is the result of evolutionary pressure to preserve the stability of the proteins. The mean-field model is based on the idea that mutation rates are a function of an effective field acting on this amino acid from the rest of the protein.

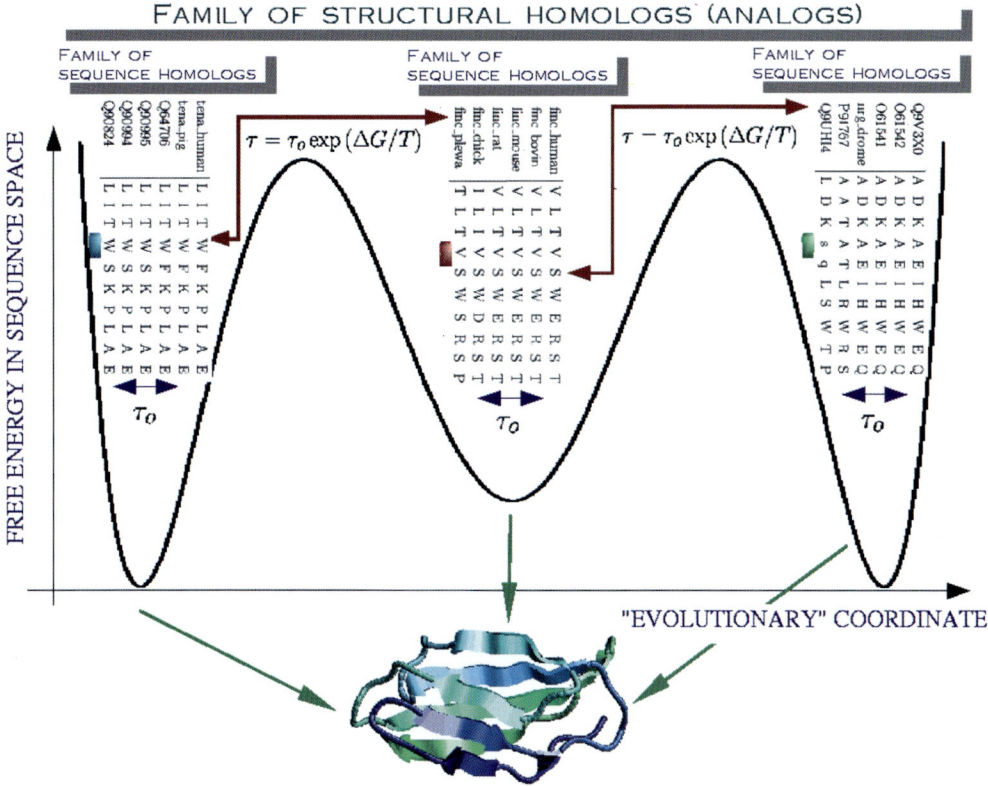

Fig. 1. – A schematic representation of the evolutionary processes that result in conservation patterns of amino acids. For a given family of folds, *e.g.* Ig in this diagram, there are several alternative minima (3) in the hypothetical free-energy landscape in the sequence space as a function of the "evolutionary" reaction coordinate (*e.g.*, time). Each of these minima is formed by mutations in protein sequences at time scales, τ_0, that do not alter the protein's thermodynamically and/or kinetically important sites, forming families of homologous proteins. Transitions from one minimum to another occur at time scales $\tau = \tau_0 \exp[\Delta G/T]$. At time scale τ mutations occur that would alter several amino acids at the important sites of the proteins in such a way that the protein properties are not compromised. At time scale τ the family of analogs is formed. In three minima we present three families of homologs (1ten, 1fnf, and 1cfb) each comprised of six homologous proteins. We show 10 positions in the aligned proteins: from 18 to 28. It can be observed that at position 4 (marked by blocks) in each of the families presented in the diagram amino acids are conserved within each family of homologs, but vary between these families. This position corresponds to position 21 in Ig fold alignment (to 1ten) and is conserved (see figs. 4(a) and 7(a)).

These rates are assumed to stay fixed on the time scale of evolution of a single family of homologs. We propose this model to understand to what degree stability is important in conserving amino acids. By comparing the outcomes of the proposed models we might be able to answer questions of the evolutionary time scale and the degree to which sequences, belonging to the same fold family, are separated from each other. In the next sections we describe the methods and models we use.

2. – Methods

2˙1. Protein model. – We use the C_β representation of proteins in which each pair of amino acids is in contact if their C_β's (C_α in the case of Gly) are within the distance 7.5 Å [31]. We use the Miyazawa-Jernigan (MJ) [32] matrix of pair potentials to represent the interaction between each pair of 20 amino acids. The total potential energy of the protein can be written as follows:

$$(1) \qquad E = \frac{1}{2} \sum_{i \neq j}^{N} U(\sigma_i, \sigma_j) \Delta_{ij},$$

where N is the length of the protein, σ_i is an amino acid at the position $i = 1, \ldots, N$. $U(\sigma_i, \sigma_j)$ is the corresponding element of the MJ matrix of pairwise interactions between amino acids σ_i and σ_j. Δ_{ij} is the element of the contact matrix, that is defined to be 1 if the contact between amino acids i and j exists (*i.e.* the distance between these amino acids in the native (ground) state is smaller than 7.5 Å), and 0, if the above contact does not exist:

$$(2) \qquad \Delta_{ij} \equiv \begin{cases} 1, & |r_i^{NS} - r_j^{NS}| \leq 7.5 \text{ Å}, \\ 0, & |r_i^{NS} - r_j^{NS}| > 7.5 \text{ Å}, \end{cases}$$

where r_i^{NS} is the position of the i-th residue when the protein is in the native conformation.

2˙2. The 6-letter potential. – Due to the similarities in properties of the 20 types of amino acids, one can classify these amino acids into 6 distinct groups: aliphatic $\{AVLIMC\}$, aromatic $\{FWYH\}$, polar $\{STNQ\}$, positive $\{KR\}$, negative $\{DE\}$, and special (reflecting their special conformational properties) $\{GP\}$. We construct the potential of interaction, $U_6(\hat{\sigma}_i, \hat{\sigma}_j)$, between the six groups of amino acids, $\hat{\sigma}$, by computing the average interaction between these groups, *i.e.*

$$(3) \qquad U_6(\hat{\sigma}_i, \hat{\sigma}_j) = \frac{1}{N_{\hat{\sigma}_i} N_{\hat{\sigma}_j}} \sum_{\sigma_k \in \hat{\sigma}_i, \sigma_l \in \hat{\sigma}_j} U_{20}(\sigma_k, \sigma_l),$$

where σ denotes amino acids in 20-letter representation and $U_{20}(\sigma_k, \sigma_l)$ is the 20-letter matrix of interaction MJ; $\hat{\sigma}$ denotes amino acids in 6-letter representation. $N_{\hat{\sigma}}$ is the number of actual amino acids of type $\hat{\sigma}$, *e.g.* for the aliphatic group $N_{\hat{\sigma}} = 6$. The 6-letter interaction potential for MJ 20-letter potential is given in table I.

TABLE I. – *A 6-letter potential derived (see subsect. 2˙2) for MJ 20-letter potential. The symbols "l", "r", "p", "+", "−" and "s" denote 6 distinct corresponding groups of amino acids: aliphatic $\{AVLIMC\}$, aromatic $\{FWYH\}$, polar $\{STNQ\}$, positive $\{KR\}$, negative $\{DE\}$, and special (reflecting their special conformational properties) $\{GP\}$.*

	\multicolumn{6}{c}{6-letter MJ potential}					
	l	r	p	+	−	s
l	−0.31	−0.39	−0.22	0.01	−0.41	−0.12
r	−0.39	−0.27	−0.32	−0.02	−0.28	−0.12
p	−0.22	−0.32	−0.41	−0.25	0.07	−0.29
+	0.01	−0.02	−0.25	−0.10	−0.18	−0.18
−	−0.41	−0.28	0.07	−0.18	0.01	−0.05
s	−0.12	−0.12	−0.29	−0.18	−0.05	0.04

2˙3. *The measure of the information context of the sequences.* – In both the Z-score and mean-field models, to study the information context of the sequences, we compute the sequence entropy, $S_X(k)$, at each position, k, of the sequence,

$$(4) \qquad S_X(k) = -\sum_\sigma P_X(\sigma_k) \ln P_X(\sigma_k),$$

where $P_X(\sigma_k)$ is the probability that we observe an amino acid σ_k at the k-th position. Subscript $X = Z$ or MF denotes the model: Z-score or mean-field correspondingly. The summation is taken over all possible values of σ_k.

The effect of switching to a 6-letter representation of amino acids from the 20-letter representation on the sequence entropy, $S_6(k)$, is that all values $S_6(k)$ are typically smaller than that of $S_{20}(k)$. For a M-letter alphabet with all letters equally represented, i.e. $P_X(\sigma_k) = 1/M$, the entropy is equal to $\ln M$. Thus, we expect that the difference between the typical values of $S_{20}(k)$ and $S_6(k)$ is approximately $\ln(20/6) \approx 1.2$. The case when all letters of a M-letter alphabet are equally presented corresponds to the maximal value of the entropy (see appendix A for derivation), *i.e.*

$$(5) \qquad S_M(k) \le \ln M.$$

2˙4. *The entropy of the protein fold families.* – Theoretical predictions from statistical-mechanical analysis can be compared with data on real proteins. In order to determine conservatism in real proteins, we assume that the space of sequences that fold into the same protein structure presents a two-tier system, where homologous sequences are grouped into families and there is no recognizable sequence homology between families despite the fact that they fold into closely related structures [15, 33, 34].

Using the database of protein families with close sequence similarity (HSSP database [4]), we compute frequencies of amino acids at each position, k, of aligned sequences, $P_m(\sigma_k)$, for a given, m-th, family of proteins. We average these frequencies across all N_s families sharing the same fold that are present in FSSP database [5]:

$$P_{\text{acr}}(\sigma_k) = \frac{1}{N_s} \sum_{m=1}^{N_s} P_m(\sigma_k). \tag{6}$$

Next, we determine the sequence entropy, $S_{\text{acr}}(k)$, at each position, k, of structurally aligned protein analogs:

$$S_{\text{acr}}(k) = -\sum_{\sigma} P_{\text{acr}}(\sigma_k) \ln P_{\text{acr}}(\sigma_k). \tag{7}$$

3. – Two models

3`1. *Designing by Z-score minimization.* – We start with the sequence of a given protein and perform a Monte Carlo search for the mutation that energetically favors interactions in such a sequence. The Monte Carlo design algorithm is based on the minimization of the so-called Z-score, defined as

$$Z = \frac{E^* - \langle E \rangle}{\sigma(E)}, \tag{8}$$

which corresponds to the minimization of the energy gap between the native state, E^*, of the selected sequence and the average energy of the set of random structures (decoys) for a given sequence, $\langle E \rangle$ (see fig. 2). The protein folding theory [29, 30] suggests that Z-score minimization is equivalent to maximizing energy gap between misfolded or unfolded conformations and the native state of the protein. It has been pointed out that such maximization results in stable and fast-folding proteins. The energy gap must be "significant", meaning that E^* must deviate from $\langle E \rangle$ by more than one standard deviation σ: $E^* < \langle E \rangle - \sigma$. Many researchers have pointed out (see, *e.g.*, review [35]) that minimization of the Z-score corresponds to the stabilization of the protein in its native state. The minimization of the Z-score is performed at some design temperature, T_{des}.

The design proceeds as follows: i) we select an amino acid σ_i at a random position $1 \leq i \leq N$; ii) we substitute this amino acid by σ_i' with probability p,

$$p = \begin{cases} 1, & \text{if } \delta Z < 0, \\ \exp\left[-\frac{\delta Z}{T}\right], & \text{if } \delta Z > 0, \end{cases} \tag{9}$$

where $\delta Z = Z(\sigma_i') - Z(\sigma)$ is the difference between the Z-scores of the mutated and the original proteins. We design each of $N_s = 100$ sequences by running the simulations for N_m Monte Carlo steps.

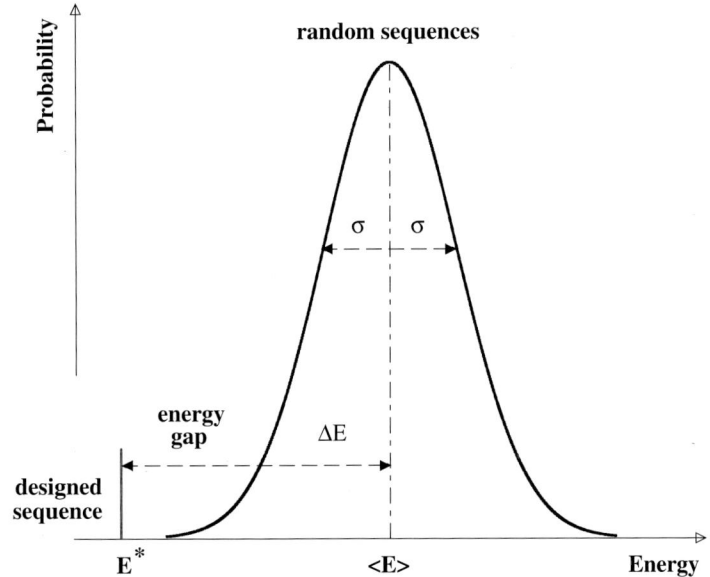

Fig. 2. – A schematic representation of the probability distribution of energies of the random sequences vs. the designed sequence.

Computation of $\langle E \rangle$ and $\sigma(E)$ is straightforward:

$$\langle E \rangle = \frac{1}{2} \sum_{i \neq j} U(\sigma_i, \sigma_j) f_{ij}, \tag{10}$$

and

$$\sigma^2(E) = \langle (E - \langle E \rangle)^2 \rangle = \frac{1}{2} \sum_{i \neq j} f_{ij}(1 - f_{ij}) U(\sigma_i, \sigma_j) + \mathcal{O}(f_{ij}^2), \tag{11}$$

where f_{ij} is the frequency of a contact between monomers i and j, *i.e.*

$$f_{ij} = \langle \Delta_{ij} \rangle. \tag{12}$$

We estimate frequencies by making two assumptions: 1) the distribution, $P(\ell = |i - j|; i, j)$ of the contact distances, $\ell = |i - j|$, between various amino acids at the positions i and j is universal among globular proteins; and 2) the actual frequency of contacts between various amino acids, i and j, is only a function of the absolute value of the length of contacts, $|i - j|$, and is equal to the distribution of the contact lengths, *i.e.*

$$f_{ij} = f_{|i-j|} = P(\ell). \tag{13}$$

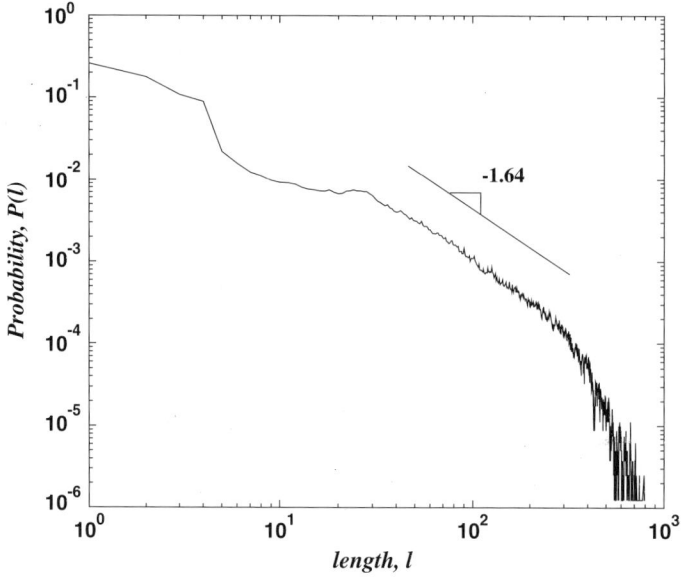

Fig. 3. – Double-logarithmic plot of the distribution, $P(\ell)$, of contacts of length $\ell = |i - j|$ in the ensemble of approximately 10^3 representative globular proteins in (PDB) [2,3]. The contact between residues positioned i-th and j-th along the protein chain is defined by eq. (2) using the C_β-representation of proteins. The parallel line in the range of length $20 < \ell < 200$ indicates the power-law behavior of $P(\ell)$ in this region, $P(\ell) \sim \ell^{-1.64}$. The region $5 < \ell < 20$ is specific to proteins and has been discussed in detail in [36].

Both assumptions are motivated by the fact that the variety of protein structures known to date samples adequately the conformational space of proteins under study and the variety is large enough that all the information about the secondary structure peculiarities of individual proteins are averaged out.

So, in order to estimate frequencies, f_{ij}, according to eq. (13) we compute the distribution of contacts of length $\ell = |i-j|$ in the ensemble of approximately 10^3 representative globular proteins in Protein Data Bank (PDB) [2,3]. The distribution, shown in fig. 3, is obtained using C_β-representation of proteins. The contacts are defined by eq. (2).

The estimation of frequencies is one of the key ingredients to protein design. An alternative approach to that proposed above is to assume that the set of conformational decoys is the set of all possible random coil states of a homopolymer collapsed at the temperatures below theta-point temperature, $T < T_\theta$—these are the states that decoy random heteropolymers explore at the folding transition temperature. Thus, we can determine the frequencies of contacts in an ensemble of random heteropolymers by taking the time average of a contact matrix element Δ_{ij} in the possible conformations of a homopolymer at $T < T_\theta$.

To compute the frequencies of contacts for a homopolymer of length N, we use discrete molecular dynamics simulations [37-39]. We model a homopolymer by N beads on a string with the interaction distances scaled to 7.5 Å. (See [38] for a detailed description of the model and the algorithm.) We run the simulation at the temperature T_θ (ϵ parameter [38] is set to -1) for 10^7 time units([1]). After 10^7 time units of simulations we compute the frequency f_{ij} of each of the $N(N-1)/2$ contacts in our homopolymer.

There are two principal drawbacks of the second method: 1) the probability of occurrence of stable elements of the structure in homopolymers resembling secondary structure in proteins is so low that the distribution of contact lengths, $P(\ell)$, in homopolymers (not shown) drastically differs from that shown for real proteins in fig. 3. 2) The model of a homopolymer used in the simulations strongly differs (*e.g.* in flexibility) from real proteins. In fact, the problem of building an appropriate model for the chain flexibility is so important that small variations in it result in drastically different kinetics from a realistic one (*e.g.*, appearance of the intermediate states) [40]. We find that both of these drawbacks make the "homopolymer" approach of estimating the frequencies very inefficient, so we omit it in further discussion.

After we obtain N_s number of designed sequences, we compute the probability of an amino acid σ_k to be in the k-th position, $P_Z(\sigma_k)$, as the frequency of occurrence of this amino acid,

$$P_Z(\sigma_k) = \frac{N(\sigma_k)}{N_s}, \tag{14}$$

where $N(\sigma_k)$ is the total number of occurrences of an amino acid σ_k at the position k. Next, using eq. (4) we compute the sequence entropy, $S_Z(k)$.

3˙2. *Mean-field model*. – We develop a model that provides a rationale for conservatism patterns caused by selection for stability. Our model is of *equilibrium* evolution that maintains stability and other properties achieved at an earlier, prebiotic stage. To this end, we propose that stability selection accepts only those mutations that keep the energy of the native protein, E, below a certain threshold E_0 necessary to maintain an energy gap [26,41-43]. The requirement to maintain an energy threshold for the viable sequences makes the equilibrium ensemble of sequences analogous to a microcanonical ensemble. In analogy with statistical mechanics, a more convenient and realistic description of the sequence ensemble is a canonical ensemble, whereby strict requirements on energy of the native state are replaced by a "soft" evolutionary pressure that allows energy fluctuations from sequence to sequence but makes sequences with high energy in the native state unlikely. In the canonical ensemble of sequences, the probability of finding a particular sequence, $\{\sigma\}$, in the ensemble follows the Boltzmann distribution [26,42-44]

$$P(\{\sigma\}) = \frac{\exp[-E\{\sigma\}/T]}{Z}, \tag{15}$$

([1]) In the discrete molecular dynamics algorithm, the time unit is the average time between subsequent collisions.

where T is the effective temperature of the canonical ensemble of sequences that serves as a measure of evolutionary pressure and $Z = \sum_{\{\sigma\}} \exp[-E\{\sigma\}/T]$ is the partition function taken in sequence space.

Next, we apply a mean-field approximation that replaces all multiparticle interactions between amino acids by the interaction of each amino acid with an effective field Φ acting on this amino acid from the rest of the protein. This approximation presents $P(\{\sigma\})$ in a multiplicative form as $\prod_{k=1}^{N} p(\sigma_k)$ of probabilities to find an amino acid σ at position k. $p(\sigma_k)$ also obeys the Boltzmann statistics

$$p(\sigma_k) = \frac{\exp[-\Phi(\sigma_k)/T]}{\sum_\sigma \exp[-\Phi(\sigma_k)/T]}. \tag{16}$$

The mean-field potential $\Phi(\sigma_k)$ is the effective potential energy between amino acid σ_k and all amino acids interacting with it, *i.e.*

$$\Phi(\sigma_k) = \sum_{i \neq k}^{N} U(\sigma_k, \sigma_i) \Delta_{ki}. \tag{17}$$

The potential Φ is similar in spirit to the protein profile introduced by Bowie *et al.* [45] to identify protein sequences that fold into a specific 3D structure.

For each member, m, of the fold family (FSSP database [5]), presented in fig. 1, we compute the mean-field probability, $p_m(\sigma_k)$, using eq. (16). This probability, $p_m(\sigma_k)$, for each fold family member corresponds to the frequency of amino acids, σ_k, at positions k, for a given family of homologs. Then, we compute the average over all members of the fold family mean-field probability,

$$p_{\text{MF}}(\sigma_k) = \frac{1}{N_s} \sum_{m=1}^{N_s} p_m(\sigma_k). \tag{18}$$

This quantity corresponds to the $P_{\text{acr}}(\sigma_k)$ presented in [15]. Equations (16)-(18), along with a properly selected energy function, U, make it possible to predict the probabilities of all amino acid types and the sequence entropy $S_{\text{MF}}(k)$ at each position k,

$$S_{\text{MF}}(k) = -\sum_\sigma p_{\text{MF}}(\sigma_k) \ln p_{\text{MF}}(\sigma_k) \tag{19}$$

from the native structure of a protein. The summation is taken over all possible values of σ.

If the stability selection were a factor in evolution of proteins and our model captures it, then we should observe correlation between predicted mean-field–based sequence entropies, $S_{\text{MF}}(k)$, and actual sequence entropies $S_{\text{acr}}(k)$ in real proteins. Thus, the question is: "Can we find such T, so that the predicted conservatism profile $S_{\text{MF}}(k)$ matches the real one $S_{\text{acr}}(k)$?"

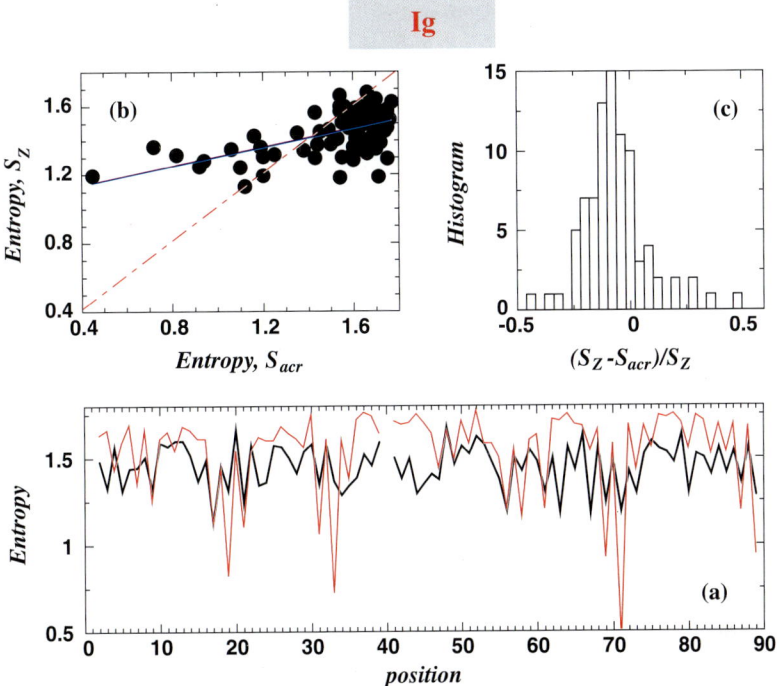

Fig. 4. – (a) The values $S_Z(k)$ (black line) and $S_{\text{acr}}(k)$ (red line) for all positions, k, for the Ig-fold. The lower the values of $S_Z(k)$ the more conservative amino acids are at these positions. (b) The scatter plot of $S_Z(k)$ vs. observed $S_{\text{acr}}(k)$. The linear regression correlation coefficients are shown in table II. The blue line is the linear regression that has a slope different than 1 (red line), corresponding to the $S_Z(k) = S_{\text{acr}}(k)$ relation. (c) The histogram of the relative differences between $S_Z(k)$ and $S_{\text{acr}}(k)$.

By varying the values of the temperature T in the range $0.1 \leq T \leq 4.0$, we minimize the distance, $D^2 \equiv \sum_{k=1}^{N}(S_{\text{MF}}(k) - S_{\text{acr}}(k))^2$, between the predicted and observed conservatism profiles. We exclude from this sum such positions in structurally aligned sequences that have more than 50% gaps in the structural (FSSP) alignment. We denote by T_{sel} the temperature that minimizes D.

4. – Results

We study three folds: Immunoglobulin fold (Ig), Oligonucleotide-binding fold (OB), and Rossman fold (R). We compute the correlation coefficient [46] between values of $S_X(k)$ ($X = \{Z, \text{MF}\}$), obtained at T_{des} and T_{sel}, and $S_{\text{acr}}(k)$ for all three folds. The results are summarized in table II. The plots of $S_X(k)$ and $S_{\text{acr}}(k)$ vs. k as well as their scatter plots are shown in figs. 4-9.

TABLE II. – *The values of the correlation coefficient r for the linear regression of $S_X(k)$ ($X = \{Z, \mathrm{MF}\}$) vs. S_{acr} for Ig, OB, and R folds and the corresponding optimal values of the temperature $T = T_{\mathrm{des}}$ and T_{sel}, and the number of mutations per designed sequence, N_m, for the Z-score model.*

Fold	N_s	Representative protein		Correlation coefficient				
		PDB code [2,3]	N	$S_Z(k)$ vs. $S_{\mathrm{acr}}(k)$	T_{des}	N_{mut}	$S_{\mathrm{MF}}(k)$ vs. $S_{\mathrm{acr}}(k)$	T_{sel}
Ig	51	1TEN	89	0.56	0.65	100	0.63	0.34
OB	18	1MJC	69	0.45	0.5	100	0.69	0.19
R	166	3CHY	128	0.48	0.5	200	0.71	0.25

4˙1. *Z-score model*. – We find that the correlation between $S_Z(k)$ and $S_{\mathrm{acr}}(k)$ strongly depends on the number of mutations, N_m, we introduce during the design of a protein. This fact is in accord with our view (see fig. 1) of protein evolution. On a short time scale, $\tau_0 \sim 10^2$ Monte Carlo steps, mutations rarely alter amino acids with specific important properties such as participation in stabilization of proteins, in the nucleation processes

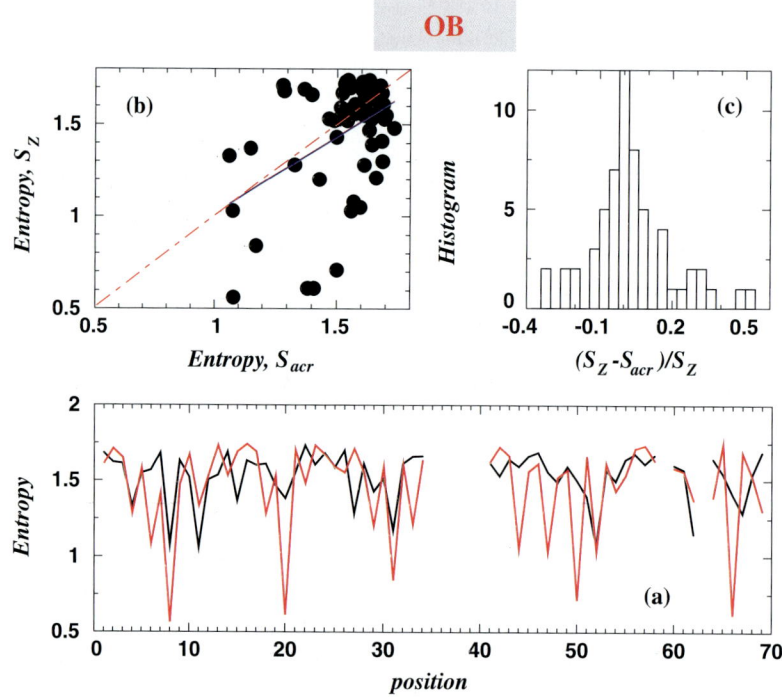

Fig. 5. – (a)-(c) The same as fig. 4 but for the OB-fold.

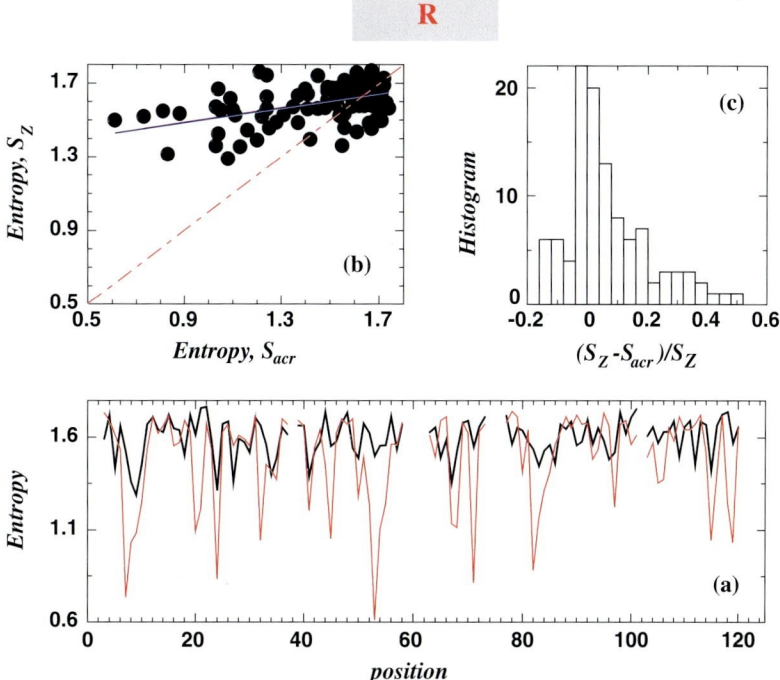

Fig. 6. – (a)-(c) The same as fig. 4 but for R-fold.

in folding kinetics, and/or function(s) of proteins. These mutations diversify the family, m, of homologs, $\mathcal{M}_h^m$. On a larger scale, $\tau \gg \tau_0$, correlated mutations modify the core and/or nucleus site(s) of the proteins without compromising their stability, folding rates and function(s). Thus, at the time scale τ evolution moves from one family of homologs to another, diversifying the underlying family of analogs, $\mathcal{M}_a$, $\bigcup_m \mathcal{M}_h^m \subseteq \mathcal{M}_a$. The ensemble of analogs is still much smaller than the ensemble, $\mathcal{M}_0$, of all possible sequences ($\mathcal{M}_a \subseteq \mathcal{M}_0$), which is of the size 6^N (in a 6-letter alphabet)—for $N = 100$ residue protein this number is of the order of 10^{78}.

It is important that for the number of mutations that are of the order of protein length we find correlation between entropies of the designed sequence, $S_Z(k)$, and the empirically observed one, $S_{\text{acr}}(k)$ (see table II). This correlation depends on the input random number, indicating that the selected sequences constitute a family of homologs, $\mathcal{M}_h^m$, that is closer to or more distant from an original sequence family of homologs, $\mathcal{M}_h^0$ (both $\mathcal{M}_h^m$ and $\mathcal{M}_h^0$ belong to a given family of analogs, $\mathcal{M}_a$).

Temperature becomes a sensitive parameter when the number of mutations, N_m, is of the order of protein length. Temperature affects the average value of the entropy over all possible positions. So, we use this parameter to adjust the average value of the $\langle S_Z \rangle$ to the average value of $\langle S_{\text{acr}} \rangle$.

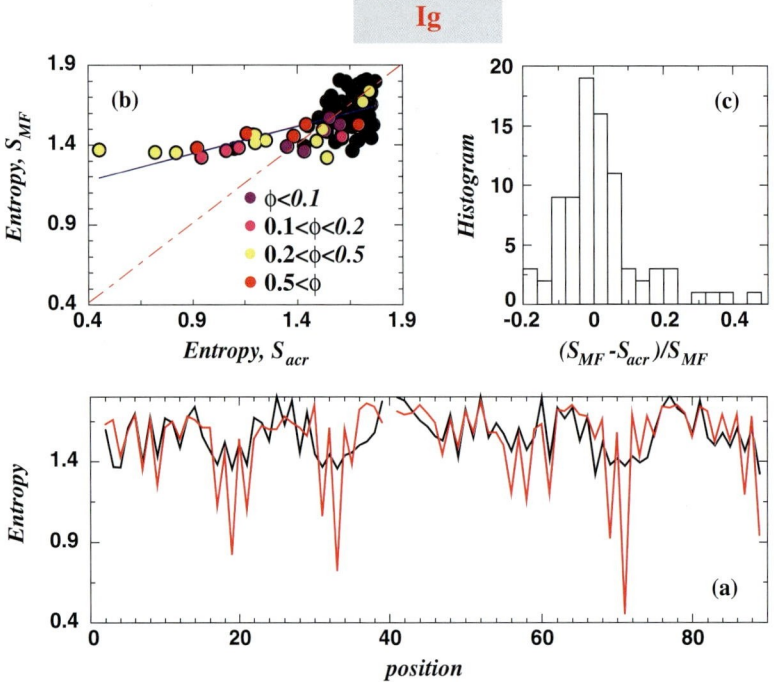

Fig. 7. – (a) The values $S_{\mathrm{MF}}(k)$ (black line) and $S_{\mathrm{acr}}(k)$ (red line) for all positions, k, for the Ig-fold. The lower the values of $S_{\mathrm{MF}}(k)$ the more conservative amino acids are at these positions. (b) The scatter plot of predicted $S_{\mathrm{MF}}(k)$ vs. observed $S_{\mathrm{acr}}(k)$. The linear regression correlation coefficients are shown in table II. The blue line is the linear regression that has a slope different than 1 (red line), corresponding to the $S_{\mathrm{MF}}(k) = S_{\mathrm{acr}}(k)$ relation. (c) The histogram of the relative differences between $S_{\mathrm{MF}}(k)$ and $S_{\mathrm{acr}}(k)$. In (b) we assign colors to data points corresponding to amino acids with the specific range of ϕ-values [47]: red, if $0.5 < \phi < 1$, yellow, if $0.2 < \phi < 0.5$, magenta, if $0.1 < \phi < 0.2$, violet if $\phi < 0.1$, and black if ϕ-values are not determined.

Here we present the results for the selected ensembles of the designed sequences, $\mathcal{M}_{\mathrm{h}}^{m}$, after being optimized during N_m mutations. Besides the correlation between $S_Z(k)$ and $S_{\mathrm{acr}}(k)$ that we find (see table II and figs. 4–6), more importantly, the profiles of $S_Z(k)$ and $S_{\mathrm{acr}}(k)$ are in a visible concert with each other.

4·2. *Mean-field model*. – The correlation between $S_{\mathrm{MF}}(k)$ and $S_{\mathrm{acr}}(k)$ is remarkable for all three folds and indicates that our mean-field model is able to select the conserved amino acids in protein fold families. It is fully expected that the correlation coefficient is smaller than 1. The reason for this is that computation of $S_{\mathrm{MF}}(k)$ takes into account the evolutionary selection for stability only and it does not take into account possible additional pressure to optimize kinetic or functional properties.

The additional evolutionary pressure due to the kinetic or functional importance of

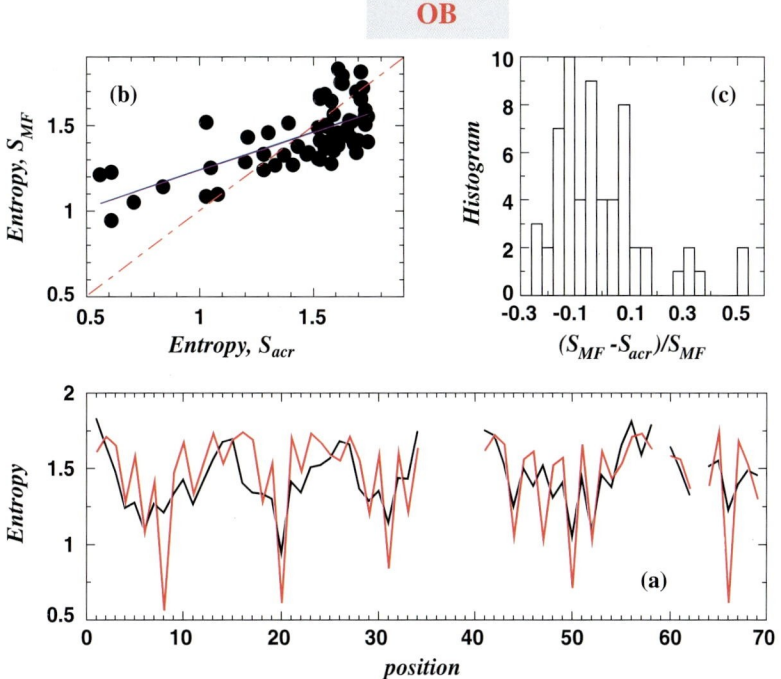

Fig. 8. – (a)-(c) The same as fig. 2 but for the OB-fold.

amino acids results in pronounced deviations of S_{MF} from S_{acr} for few amino acids that may be kinetically or functionally important. A number of amino acids whose conservatism is much greater than predicted by our model form a group of "outliers", from the otherwise very close correspondence between S_{MF} and S_{acr}. To demonstrate that some of those amino acids are important for folding kinetics and as such they can be under additional evolutionary pressure, we color data points on the S_{MF} vs. S_{acr} scatter plot according to the range of ϕ-values [48] which the corresponding amino acids fall into. The thermodynamic and kinetic roles of individual amino acids were studied extensively i) by Hamill et al. [47] for the TNfn3 (1TEN) protein, and ii) by López-Hernández and Serrano [18] for the CheY protein. We use the ϕ-values for individual amino acids obtained in [47,18]. We observe that i) for TNfn3 protein most of the points in fig. 7(b) that belong to the outlier group have ϕ-values ranging from 0.2 to 1, and ii) for CheY protein most of the points (for which ϕ-values are known) in fig. 9(b) that belong to the outlier have ϕ-values ranging from 0.3 to 1. Figures 7(b) and 9(b) demonstrate that the presence of additional evolutionary pressure due to the kinetic importance of amino acids results in stronger conservatism of specific positions than predicted by mean-field theory.

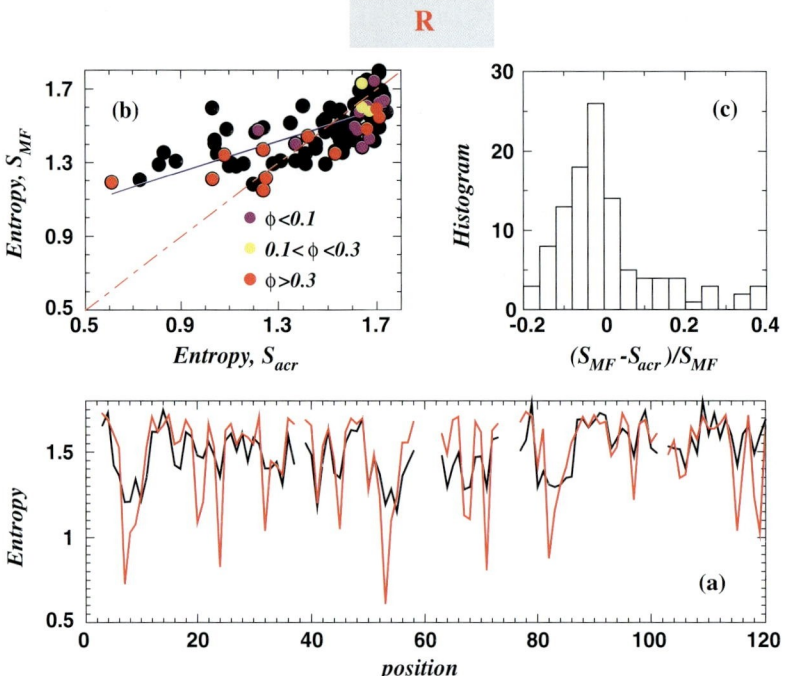

Fig. 9. – (a)-(c) The same as fig. 2 but for R-fold. In (b) we assign colors to data points corresponding to amino acids with the specific range of ϕ-values [18]: red, if $0.3 < \phi < 1$, yellow, if $0.1 < \phi < 0.3$, violet if $\phi < 0.1$, and black if ϕ-values are not determined.

5. – Conclusion

To conclude, we present two models that attempt to explain sequence conservation caused by the most basic and universal evolutionary pressure in proteins to maintain stability. Using the Z-score model, we show that separation of two basic time scales in evolution is a plausible scenario for the sequence heterogeneity of structurally homologous proteins. The two basic time scales are τ_0 and $\tau \gg \tau_0$; i) at time scales, τ_0, most mutations that occur in the protein sequences do not alter the protein's thermodynamically, kinetically, and/or functionally important sites, forming families of homologous proteins; ii) at time scales $\tau \gg \tau_0$ mutations occur that would alter several amino acids at the important sites of the proteins in such a way that the properties of proteins are not compromised. At time scale τ the family of analogs is formed.

The mean-field model predicts sequence entropy very well for the majority of amino acids, but not for all of them. The amino acids that exhibit considerably higher conservatism than predicted from stability pressure alone are likely to be important for function and/or folding. Comparison of the "base-level" stability conservatism $S_{\mathrm{MF}}(k)$ with $S_{\mathrm{acr}}(k)$—actual conservatism profile of a protein fold—allows one to identify functionally and kinetically important amino acid residues and potentially gain specific insights into the folding and function of a protein.

* * *

We thank R. S. DOKHOLYAN for careful reading of the manuscript and S. V. BULDYREV, A. V. FINKELSTEIN, A. YU. GROSBERG, and L. A. MIRNY for helpful discussions. Mean-field model was developed in [14] with L. A. MIRNY. NVD is supported by NIH postdoctoral fellowship (GM20251-01). EIS is supported by NIH grant RO1-52126.

APPENDIX A.

The maximal value of the sequence entropy

To estimate the maximal value of the sequence entropy, $S(k)$, we use the method of Lagrange multipliers to get rid of a constrain set by the fact that

$$\text{(A.1)} \qquad \sum_\sigma P(\sigma) = 1.$$

We compute the maximal value of the unconstrained entropy,

$$\text{(A.2)} \qquad S_\lambda(k;\{\sigma\}) = -\sum_{\sigma=1}^{M} P(\sigma) \ln P(\sigma) + \lambda \left(\sum_{\sigma=1}^{M} P(\sigma) - 1 \right),$$

where λ is the Lagrange multiplier.

The extremum of $S_\lambda(k;\{\sigma\})$ is set by the set of equations

$$\text{(A.3)} \qquad \frac{\partial S_\lambda(k;\{\sigma\})}{\partial P(\sigma)} = -\ln P(\sigma) + \lambda - 1 = 0.$$

The solutions to eq. (A.3) is

$$\text{(A.4)} \qquad P(\sigma) = \exp[\lambda - 1] = \frac{1}{M};$$

the latter equality comes from eq. (A.1). The second derivative of $S_\lambda(k;\{\sigma\})$ with respect to $P(\sigma)$ is always negative, so the extremal value of $S_\lambda(k;\{\sigma\})$ is its maximal value. The maximal value of the sequence entropy can be computed using eqs. (A.2)–(A.4) and is equal to $\ln M$.

REFERENCES

[1] BENSON D. A., BOGUSKI M. S., LIPMAN D. J., OSTELL J., OUELLETTE B. F. F., RAPP B. A. and WHEELER D. L., *Nucl. Acid Res.*, **27** (1999) 38.
[2] BERNSTEIN F. C., KOETZLE T. F., WILLIAMS G. J. B., MEYER E. F. jr., BRICE M. D., RODGERS J. R., KENNARD O., SHIMANOUCHI T. and TASUMI M., *J. Mol. Biol.*, **112** (1977) 535.
[3] ABOLA E. E., BERNSTEIN F. C., BRYANT S. H., KOETZLE T. F. and WENG J., *Protein Data Bank*, in *Crystallographic Databases-Information Content, Software Systems, Scientific Applications*, edited by F. H. ALLEN, G. BERGERHOFF and R. SIEVERS (Data Commission of the International Union of Crystallography, Cambridge) 1987, pp. 107-132.
[4] DODGE C., SCHNEIDER R. and SANDER C., *Nucl. Acid Res.*, **26** (1998) 313.
[5] HOLM L. and SANDER C., *J. Mol. Biol.*, **233** (1993) 123.
[6] KABSCH W. and SANDER C., *Biopolymers*, **22** (1983) 2577.
[7] DOKHOLYAN N. V., BULDYREV S. V., HAVLIN S. and STANLEY H. E., *Phys. Rev. Lett.*, **79** (1997) 5182.
[8] DOKHOLYAN N. V., BULDYREV S. V., HAVLIN S. and STANLEY H. E., *J. Theor. Biol.*, **202** (2000) 273.
[9] PENG C.-K., BULDYREV S. V., GOLDBERGER A. L., HAVLIN S., SCIORTINO F., SIMONS M. and STANLEY H. E., *Nature*, **356** (1992) 168.
[10] BULDYREV S. V., GOLDBERGER A. L., HAVLIN S., PENG C.-K., SIMONS M. and STANLEY H. E., *Phys. Rev. E*, **47** (1993) 4514.
[11] BUSSEMAKER H. J., LI H. and SIGGIA E. D., *Proc. Natl. Acad. Sci. USA*, **97** (2000) 10096.
[12] PTITSYN O. B., *J. Mol. Biol.*, **278** (1998) 655.
[13] MIRNY L. A., ABKEVICH V. I. and SHAKHNOVICH E. I., *Proc. Natl. Acad. Sci. USA*, **95** (1998) 4976.
[14] DOKHOLYAN N. V., MIRNY L. A. and SHAKHNOVICH E. I., cond-mat/0007084 (2000).
[15] MIRNY L. A. and SHAKHNOVICH E. I., *J. Mol. Biol.*, **291** (1999) 177.
[16] LORCH M., MASON J., CLARKE A. and PARKER M., *Biochemistry*, **38** (1999) 1377.
[17] SCHINDLER T., PERL D., GRAUMANN P., SIEBER V., MARAHIEL M. and SCHMID F., *Proteins: Struc. Func. Genet.*, **30** (1998) 401.
[18] LÓPEZ-HERNÁNDEZ E. and SERRANO L., *Folding & Design*, **1** (1996) 43.
[19] RUSSELL R., SASIENI P. and STERNBERG M., *J. Mol. Biol.*, **282** (1998) 903.
[20] WELCH M., OOSAWA K., AIZAWA S. and EISENBACH M., *Biochemistry*, **33** (1994) 10470.
[21] BELLSOLELL L., CRONET P., MAJOLERO M., SERRANO L. and COLL M., *J. Mol. Biol.*, **257** (1996) 116.
[22] WILCOCK D., PISSABARRO M. T., LÓPEZ-HERNÁNDEZ E., SERRANO L. and COLL M., *Acta. Cryst. D*, **54** (1998) 378.
[23] VILLEGAS V., MARTÍNEZ J. C., AVILÉS F. X. and SERRANO L., *J. Mol. Biol.*, **283** (1998) 1027.
[24] VAN NULAND N. A. J., CHITI H., TADDEI N., RAUGEI G., RAMPONI G. and DOBSON C., *J. Mol. Biol.*, **283** (1998) 883.
[25] VAN NULAND N. A. J., MEIJBERG W., WARNER J., FORGE V., SCHEEK R., ROBBILARD G. and DOBSON C., *Biochemistry*, **37** (1998) 622.
[26] SHAKHNOVICH E. I. and GUTIN A. M., *Proc. Natl. Acad. Sci. USA*, **90** (1993) 7195.
[27] ABKEVICH V. I., GUTIN A. M. and SHAKHNOVICH E. I., *Folding & Design*, **1** (1996) 221.
[28] SHAKHNOVICH E. I., *Curr. Opinion Struc. Biol.*, **7** (1997) 29.
[29] BRYNGELSON J. D. and WOLYNES P. G., *Proc. Natl. Acad. Sci. USA*, **84** (1987) 7524.

[30] ABKEVICH V. I., GUTIN A. M. and SHAKHNOVICH E. I., *Proteins: Struc. Func. Genet.*, **31** (1998) 335.
[31] JERNIGAN R. L. and BAHAR I., *Curr. Opinion Struc. Biol.*, **6** (1996) 195.
[32] MIYAZAWA S. and JERNIGAN R. L., *J. Mol. Biol.*, **256** (1996) 623.
[33] ROST B., *Folding & Design*, **2** (1997) S19.
[34] TIANA G., BROGLIA R. and SHAKHNOVICH E. I., *Proteins: Struc. Func. Genet.*, **39** (2000) 244.
[35] SHAKHNOVICH E. I., *Folding & Design*, **3** (1998) R45.
[36] BEREZOVSKY I. N., GROSBERG A. Y. and TRIFONOV E. N., *FEBS Lett.*, **466** (2000) 283.
[37] ZHOU Y. and KARPLUS M., *Proc. Natl. Acad. Sci. USA*, **94** (1997) 14429.
[38] DOKHOLYAN N. V., BULDYREV S. V., STANLEY H. E. and SHAKHNOVICH E. I., *Folding & Design*, **3** (1998) 577.
[39] DOKHOLYAN N. V., BULDYREV S. V., STANLEY H. E. and SHAKHNOVICH E. I., *J. Mol. Biol.*, **296** (2000) 1183.
[40] BORREGUERO J. M., DOKHOLYAN N. V., BULDYREV S., STANLEY H. E. and SHAKHNOVICH E. I., in preparation (2000).
[41] SALI A., SHAKHNOVICH E. I. and KARPLUS M., *J. Mol. Biol.*, **235** (1994) 1614.
[42] RAMANATHAN S. and SHAKHNOVICH E. I., *Phys. Rev. E*, **50** (1994) 1303.
[43] FINKELSTEIN A. V., GUTIN A. and BADRETDINOV A., *Proteins*, **23** (1995) 142.
[44] PANDE V. S., GROSBERG A. YU. and TANAKA T., *Phys. Rev. E*, **51** (1995) 3381.
[45] BOWIE J. U., LUTHY R. and EISENBERG D., *Science*, **253** (1991) 164.
[46] PRESS W. H., FLANNERY B. P., TEUKOLSKY S. A. and VETTERLING W. T., *Numerical Recipes* (Cambridge University Press, Cambridge) 1989.
[47] HAMILL S. J., STEWARD A. and CLARKE J., *J. Mol. Biol.*, **297** (2000) 165.
[48] ITZHAKI L. S., OTZEN D. E. and FERSHT A. R., *J. Mol. Biol.*, **254** (1995) 260.

PROTEIN DESIGN AND INTERACTION

Physical selection of protein structures

A. V. FINKELSTEIN

Institute of Protein Research, Russian Academy of Sciences
142290, Pushchino, Moscow Region, Russian Federation

In this lecture we will discuss the general rules established for three-dimensional structures of globular proteins, and will try to find a general reason underlying the empirically observed rules.

An overview of protein structures has shown long ago that the majority of them fit a limited set of folding patterns [1-5], though many proteins, very similar in their folds, seem to have no common ancestry or function. This observation forms the basis of modern classifications of protein folds. Technically, such classifications have been supported by the possibility of computer sorting of protein structures [6].

SCOP [7] and CATH [8] (fig. 1) are now the most popular classifications of protein folds. Though different in details, they coincide in their main features. These classifications give a convenient survey of the three-dimensional (3D) protein structures. Figure 1 presents such an overview, with examples of the most popular protein folds. Actually, the classified folds refer to relatively small proteins and to protein domains, while large proteins (which are more typical of eukaryotes than of prokaryotes) consist of a few such domains. It is noteworthy that the domain architectures demonstrate no evolution from bacteria to mammals.

Figure 1 draws our attention to the most common protein folds. This can give the impression that all the proteins can fit the small set of common folding patterns. Actually, this is not exactly so, and the "80/20 law" is fully applicable to protein structures. In its initial form, this law reads: "80% of beer is consumed by only 20% of the population". As for proteins, the law reads "80% of protein families belong to only 20% of observed folds". The remaining 20% of proteins are scattered over the "unusual" folds, which form 80% of the observed folds.

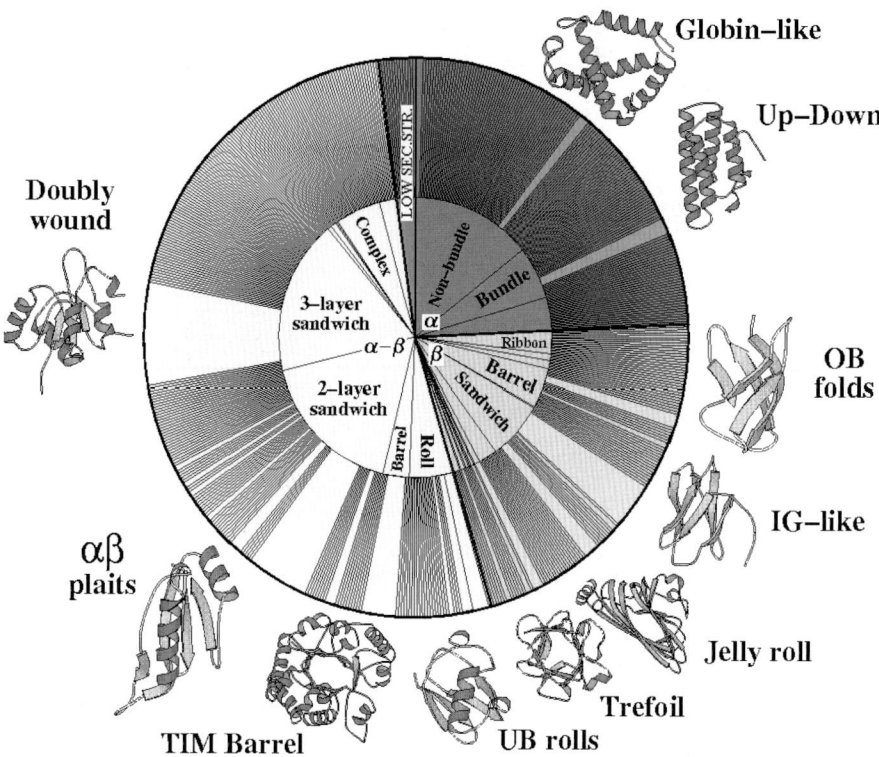

Fig. 1. – "CATHerine plot" showing the distribution of non-homologous structures of protein globules amongst different structural classes ("α", "β", "α-β" and "Low Secondary Structure" proteins), different architectures within the classes ("Non-bundle", "Bundle", etc.) and different fold families (within the architectures) in the CATH database [7]. The angle subtended for a given segment reflects the proportion of structures within a given class (sector), architecture (inner circle) or fold family (outer circle). The most populated "superfolds" are indicated in paler shades and illustrated with a drawing of a representative from the family. These drawings stress the arrangement of secondary structures, α-helices and β-strands (arrows), forming the frameworks of protein molecules. Reproduced, with a minor modification, from [7].

Why do most proteins fit the limited set of folds? And why not all of them (as the DNA chains do)? And—since the common ancestry and common functions do not seem to be enough to explain the fold similarity—can this similarity be explained by the necessity to satisfy the general physical principles of folding of stable protein structures?

Before starting to answer these questions, we have to ask, what is the structural level where the similarity of different protein structures is revealed?

More than twenty years ago it became clear [1-3] that this is the level of "folding patterns" (or simply "folds"), which is situated between two traditional structural levels, the secondary structure of chain pieces and the detailed atomic 3D structure of the whole protein. The folding pattern can be described as the crude mutual

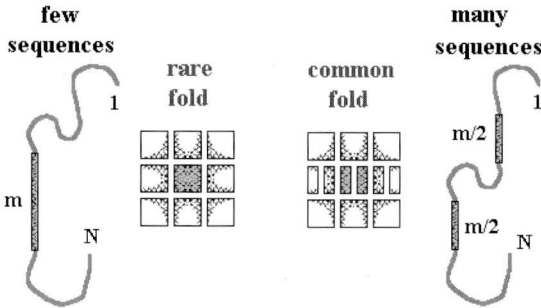

Fig. 7. – Multilayer packing with the internal α-helix is much less common than the multilayer packing with the internal β-sheet (cf. fig. 1). The following reason for this observation is suggested. The internal secondary-structure segment must consist only of non-polar residues (otherwise, the globule would explode), and the length of this segment is dictated by the globule's diameter, which is large, since the multilayer packing requires a large globule. The α-helix consists of twice more residues than the extended β-strand of the same length. The probability of occurrence of one long entirely hydrophobic sequence of m residues in *a given chain position* ($\sim p^m$, where $p \sim 1/2$ is the average fraction of non-polar residues in protein chains) is small, but it is the same as the probability of two twice shorter hydrophobic blocks in *two given chain positions* ($\sim (p^{m/2}) \times (p^{m/2}) = p^m$). However, one block can be placed in the N-residue chain in only $\sim N$ different ways, while two blocks can be positioned in $\sim N^2/2$ different ways. Therefore, if $Np^m \ll 1$ (which is always the case for the compact multilayer packings [5]) there are much more sequences compatible with two internal β-strands than with one internal α-helix.

energetically, and such a protein could not be stable. Therefore very large compact globules of a "normal" amino acid composition must be unstable. Thus, for the sake of stability, they must be divided into sub-globules called domains (which we indeed usually see in large proteins).

Figures 4 and 1 show one more peculiar feature of the common folding patterns: namely, the β-sheet is often surrounded by α-helices, while the opposite (fig. 6) is extremely rare.

This phenomenon can be explained from the same reasoning ("usual" *vs.* "rare" sequences). Let us compare the pair of β-strands, situated in the middle of the globule, with the α-helix, also situated in the middle (fig. 7). If we fix the chain positions of these two β-strands, on the one hand, and of the α-helix, on the other, the fully hydrophobic sequences (necessary, for the sake of stability, for both these competing structures) will have the same number of residues of the same type and thus absolutely the same odds to occur (see legend to fig. 7). However, the fully hydrophobic sequence aimed for the internal α-helix consists of one continuous piece, while the sequence aimed for the pair of internal β-strands consists of two fully hydrophobic pieces. These two fully hydrophobic pieces can have much more positions in the chain than the single piece. Therefore, there are much more sequences compatible with two internal β-strands than with one internal α-helix.

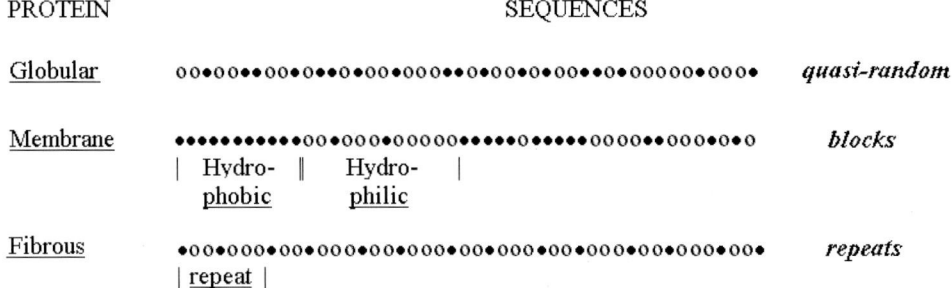

Fig. 8. – Typical alteration of hydrophobic (•) and polar (o) residues in the sequences of the water-soluble globular protein of the membrane and of the fibrous proteins.

In other words, there are much more random sequences, which can stabilize a folding pattern with two internal β-strands than the sequences which can stabilize a fold with a single internal α-helix.

And this advantage of odds (estimated for random sequences!) is exactly what we see in protein folds. Thus, we see that a consideration of what can be more readily formed by random sequences can help to understand the folds of real globular proteins. It seems that, though the proteins are certainly the products of natural selection, this selection has much more space to operate when the fold is compatible with many sequences, and much less when it is compatible with only a few.

Of course, it may seem strange to speak about proteins where each chain is coded by a naturally selected gene and to borrow the arguments from the random sequences. However, let us look at protein sequences. Statistical analysis shows that amino acid sequences of globular proteins, and we are now speaking about them, look almost "random" [13], in the sense that they have neither blocks, typical of membrane proteins, nor repeats, typical of fibrous proteins (fig. 8). The distribution of secondary structures over the chains of globular proteins also looks almost "random", compatible with a random clustering of the non-polar groups making the hydrophobic surfaces of these structures [5, 14]. And what does "to look like a random chain" mean? This means to look like *most* of the chains. Thus, speaking of the structures that can be formed by random sequences, we actually consider the structures compatible with many various sequences... And, up to now, we have obtained some reasonable results!

Meanwhile, we have dealt only with packings of secondary structures. Let us go further and try to apply the same considerations to the loops, *i.e.*, to the connections of secondary structures by the protein chain.

As we have seen, the protein folding patterns are often very beautiful. Moreover, they often resemble the patterns of pottery ornaments (fig. 9). And, according to the neat idea of Jane Richardson who had discovered this resemblance [2], this is because both lines—the ornament and the protein chain—solve the same problem: how to surround a volume by a continuous self-avoiding line.

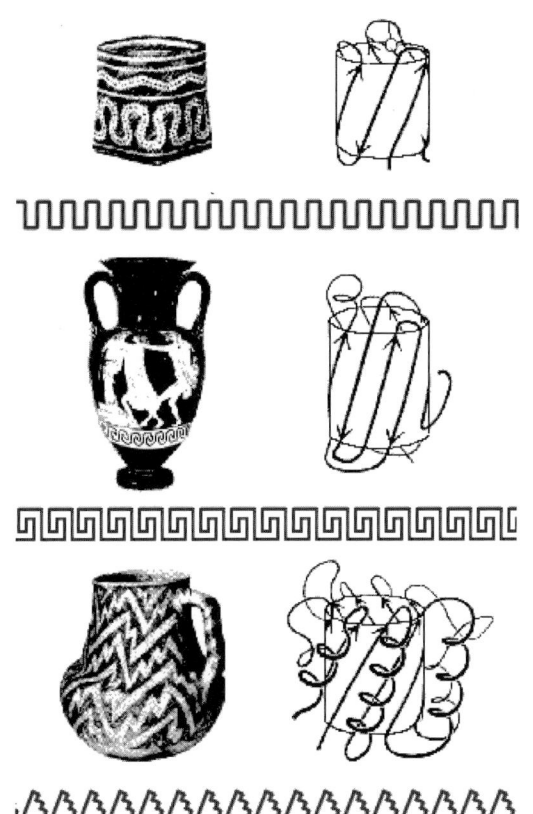

Fig. 9. – Protein folding patterns and pottery ornaments. Two solutions of the problem of the surrounding of a volume by a continuous self-avoiding line. At the top: the "meander motif", typical of some β-proteins; in the middle: the "Greek key" motif, also typical of some β-proteins (see, e.g., IG-like proteins or Jelly rolls in fig. 1); at the bottom: the "lightning motif", typical of some α/β proteins (e.g., TIM barrels). Reproduced from the cover of Nature, **268** (1977) 5620, where a paper of J. Richardson on protein folding patterns [2] has been published.

The self-avoidance effect is achieved in proteins mainly by three means:

1) The secondary-structure regions form a layer (or layers) around the core of a globule (or around two cores; the latter is typical of α/β proteins); the loops do not come inside the core.

2) The loops slide over the surface of the core(s) and do not cross each other.

3) The usual connection of parallel β-strands is the right-handed connection, i.e., all the connections have the same handedness and therefore, taken together, form the super-helices.

The first peculiarity is already clear to us. That is, we understand why the layered

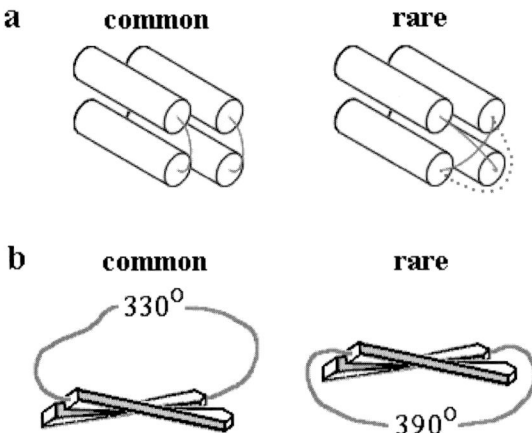

Fig. 10. – Rare and common elements of the protein folding patterns. (a) The overlapping (as well as bypassing) loops, connecting the secondary structures, are relatively very rare in proteins. (b) The left-handed connection of the parallel β-strands is rare in proteins, while their right-handed connection is common.

construction of the secondary-structure packing is energetically favorable, and why the loops in the core (the "structural defects!") are energetically unfavorable.

If we want to understand the other peculiarities (fig. 10), following the same line of fold shape confrontation with fold stability (and, further, with the number of fold-stabilizing sequences), we have to ask: what is energetically wrong with the crossed loops? what is energetically wrong with the left-handed connections?

Indeed, what is energetically wrong with the crossed loops (fig. 10a)? When we say "crossed", we do not mean that one loop runs into another, we only mean that one loop covers the other! Covers, clasps to the core, and screens its polar backbone from water: *i.e.*, the covered loop loses some of its hydrogen bonds with water. And we will again need a "special" sequence to heal the lost bonds and to maintain the stability of the globule with the crossed loops . . . .

However, here we must feel ourselves embarrassed. Indeed, the loop covered by another one loses only one, may be two hydrogen bonds—much less than a loop piercing the entire globule, which has been discussed earlier. The lost one-two hydrogen bonds cost only three-five kilocalories per mole [5]. This is very much less than the total energy of interactions within the globule, amounting to hundreds of kilocalories (according to protein melting data [15]). Moreover, this is much less than the usual margin of stability of the native globule (*i.e.*, the free energy difference between the folded and unfolded protein under physiological conditions) which, according to the same experiments, amounts to about 10 kcal/mol [15]. Then, why does an "energy defect" of only 3–5 kcal/mol virtually exclude the crossing of loops in the native protein globules? And one more question: why the upper loop cannot make an additional bend (fig. 10a) and thus avoid covering and screening of the lower loop from water? Maybe, this is prevented by limited loop

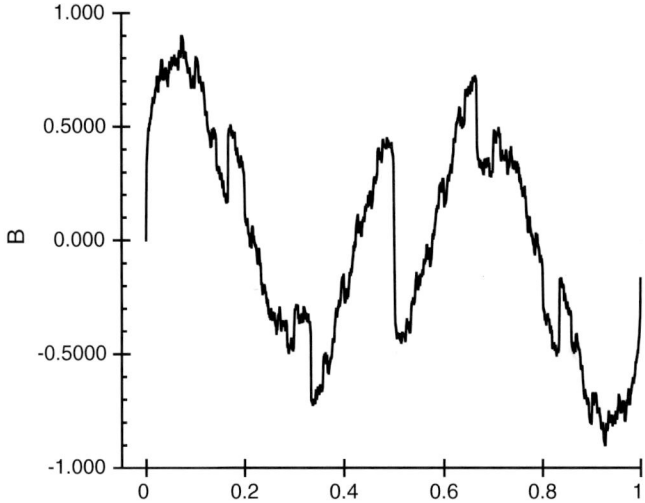

Fig. 3. – A schematic example of a corrugated free-energy landscape, the horizontal axis being the configuration space and the vertical axis the free energy.

In the rest of this lecture we will call J the control parameters of the systems. Average over J will be denoted by a bar (*e.g.*, $\bar{F}$). In the cases I will consider here a quenched disorder is present: the variables J parametrize the quenched disorder.

The presence of many valleys may be understood if we consider the free energy (or the energy at zero temperature) as a function of the configuration space. Different valleys can be associated to different minima of the free energy. If many minima are present we say that the free-energy landscape is corrugated. Of course we cannot make a graph of a corrugated landscape because the configurational space has many dimensions (the number of dimensions goes to infinity with the volume of the system). In fig. 3 we show a schematic example of a corrugated free-energy landscape, just to give a visual impression.

While it is easy to construct a corrugated landscape in random systems, also systems which have no quenched disorder may have a corrugated landscape. It may be amusing to note that the function shown in fig. 3 is not a random function, as the careful reader may have already noticed, but it is given by the simple expression

(11) $$F(x) = \sum_{p=2,\infty} \frac{\sin(2\pi p x)}{p},$$

where the sum runs over all the prime numbers.

The aim of the theory is to make precise statements on the form of the free-energy landscape and on its statistical properties. A crucial tool is the definition of the overlap (or the distance) among two valleys and the study of its probability distribution.

4. – The overlap and its probability distribution

4˙1. Spin glasses. – Let us firstly consider the case of spin glasses. The simplest example is given by Ising spin glasses [2-4]. In this case the Hamiltonian is given by

$$H = -\sum_{i,k} J_{i,k}\sigma_i\sigma_k - \sum_i h_i\sigma_i, \tag{12}$$

where $\sigma = \pm 1$ are the spins. The variables J are random couplings (*e.g.*, Gaussian or ± 1) and the variables h_i are the magnetic fields, which may be point dependent.

Let us consider two different models for spin glasses:

– The Sherrington-Kirkpatrick model (infinite range): all N points are connected: $J_{i,k} = O(N^{-1/2})$. Eventually N goes to infinity.

– Short-range models: i belongs to a L^D lattice. The interaction is nearest neighbour (the variables J are either zero or of order 1) and eventually L goes to infinity at fixed D (*e.g.*, $D = 3$).

We have already remarked that states may be separated making a comparison among them. To this end it is convenient to consider their mutual overlap. Given two configurations, we define their overlap:

$$q[\sigma,\tau] = \frac{1}{N}\sum_{i=1,N}\sigma(i)\tau(i). \tag{13}$$

The overlap among the states is defined as

$$q(\alpha,\gamma) = \frac{1}{N}\sum_{i=1,N}m_\alpha(i)m_\gamma(i) \approx q[\sigma,\tau], \tag{14}$$

where σ and τ are two generic configurations which belong to the states α and γ, respectively.

We define $P_J(q)$ as the probability distribution of the overlap q at given J, *i.e.* the histogram of $q[\sigma,\tau]$, where σ and τ are two equilibrium configurations. Using eq. (9), one finds that

$$P_J(q) = \sum_{\alpha,\gamma} w_\alpha w_\gamma \delta(q - q_{\alpha,\gamma}), \tag{15}$$

where in a finite-volume system the delta-functions are smoothed. If there is more than one state, $P_J(q)$ is not a single delta-function:

$$P_J(q) \neq \delta(q - q_{\mathrm{EA}}). \tag{16}$$

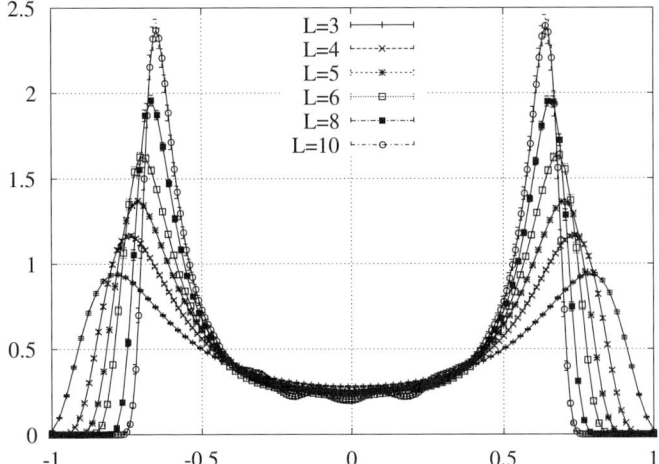

Fig. 4. – The function $P(q)$ after average over many samples in four dimensions for systems of size L^4 ($L = 3 \ldots 10$).

If this happens we say that the replica symmetry is broken: two identical replicas of the same system may stay in a quite different state.

There are many models where the function $P_J(q)$ is non-trivial. Analytic studies have been done in the case of the SK model, where one can prove rigorously that the function $P_J(q)$ is non-trivial. In the finite-dimensional case no theorem has been proved and in order to answer the question whether the function $P_J(q)$ is trivial we must resort to some numerical simulations [5] or to experiments.

In this case numerical simulations present ample evidence that in three and four dimensions the function $P_J(q)$ is non-trivial and strongly depends on J. It seems quite reasonable that when the system becomes infinite there is a peak which evolves toward a delta-function which corresponds to the contribution coming from two configurations σ and τ which belongs to the same state (see for example fig. 4, taken from [6]).

4˙2. *Glasses*. – Let us consider for simplicity a system of N identical particles in a box. If we had two replicas of this system (x and y), we could define their overlap as

$$(17) \qquad q = \sum_{i,k=1,N} f(x_i - y_k),$$

where $f(x)$ is a function which goes to zero very fast with the distance and is near to 1 at distances much smaller that the interatomic distance. However, in glasses no quenched disorder is present, so the system is invariant under translations and it is more convenient to consider the set of all configurations translated by an arbitrary amount. The natural definition of the overlap would be in this case

$$q = \max_t \sum_{i,k=1,N} f(x_i - y_k + t). \tag{18}$$

In the same spirit we could define a distance as

$$d^2 = \min_t \min_\pi \sum_{i=1,N} (x_i - y_{\pi(i)} - t)^2, \tag{19}$$

where the minimum is done over all the $N!$ permutations π.

We expect that in the cases of glasses the overlap (or the distance) has a similar behaviour to that obtained in the case of spin glasses, with different form of the function $P(q)$.

5. – Off-equilibrium dynamics

We have seen that in the glass phase replica symmetry is broken and we expect that there are characteristic violations of the fluctuation dissipation theorem in the off-equilibrium dynamics in the aging regime [7-9]. Indeed in the aging regime the fluctuation dissipation theorem cannot be applied and generalized dissipation relations are valid. The form of these generalized dissipation relations is fixed by the theory and the resulting predictions are confirmed by numerical simulations.

The general problem that we face is to find what happens if the system is carried in a slightly off-equilibrium situation. There are two ways in which this can be done.

- We rapidly cool the system starting from a random (high-temperature) configuration at time zero and we wait a time much larger than the microscopical one. The system orders at distances smaller than a coherence distance $\xi(t)$ (which eventually diverges when t goes to infinity) but remains always disordered at distances larger than $\xi(t)$.

- A second possibility consists in forcing the system in an off-equilibrium state by gently *shaking* it. This can be done for example by adding a small time-dependent magnetic field, which should be however strong enough to force a large scale rearrangement of the system [10].

In the first case we have the phenomenon of ageing. This effect may be *emphasized* if we define a two-time correlation function and a two-time relaxation function (we cool the system at time 0) [7,8]. The correlation function is defined as

$$C(t, t_w) \equiv \frac{1}{N} \sum_{i=1}^{N} \langle \sigma_i(t_w) \sigma_i(t_w + t) \rangle, \tag{20}$$

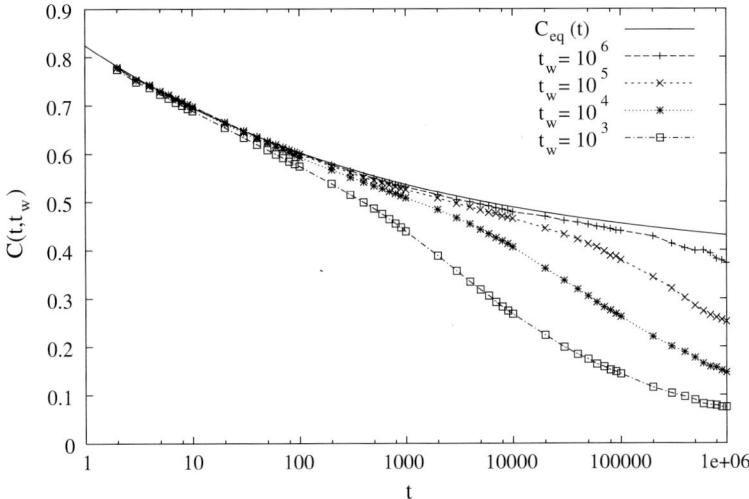

Fig. 5. – The correlation function for spin glasses as a function of time t at different t_w.

which is equal to the overlap $q(t_w, t_w + t)$ among a configuration at time t_w and one at time $t_w + t$. The relaxation function $S(t, t_w)$ is just given by

$$\beta^{-1} \lim_{\delta h \to 0} \frac{\delta m(t + t_w)}{\delta h}, \tag{21}$$

where δm is the variation of the magnetization when we add a magnetic field δh starting from time t_w. More precisely, we can introduce the time-dependent Hamiltonian

$$H = H_0 + \theta(t - t_w) \sum_i h_i \sigma_i. \tag{22}$$

The relaxation function is thus defined as

$$\beta S(t, t_w) \equiv \frac{1}{N} \sum_{i=1}^{N} \left\langle \frac{\partial \sigma_i(t_w + t)}{\partial h_i} \right\rangle. \tag{23}$$

We can distinguish two situations:

– For $t \ll t_w$ we stay in the "quasi-equilibrium" regime [11], $C(t, t_w) \simeq C_{\text{eq}}(t)$, where $C_{\text{eq}}(t)$ is the equilibrium correlation function; in this case $q_{\text{EA}} \equiv \lim_{t \to \infty} \lim_{t_w \to \infty} C(t, t_w)$.

– For $t = O(t_w)$ or larger we stay in the aging regime. If simple aging holds, $C(t, t_w) \propto \mathcal{C}(t/t_w)$. A plot of the correlation function for spin glasses at different t_w is shown in fig. 5.

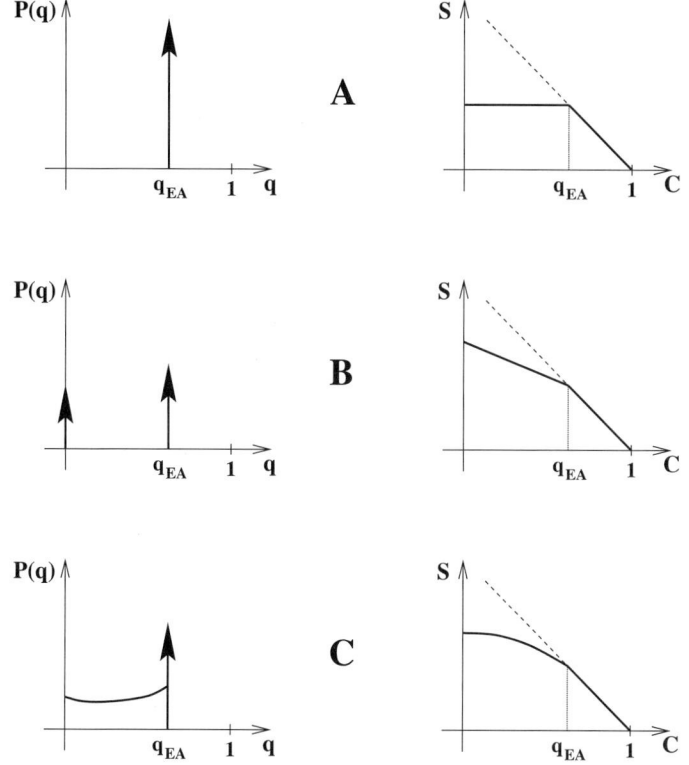

Fig. 6. – Three different forms of the function $P(q)$ and the related function $S(q)$.

In the equilibrium regime, if we plot parametrically the relaxation function of the correlation, we find that

(24) $$\frac{\mathrm{d}S}{\mathrm{d}C} = -1,$$

which is a compact way of writing the fluctuation-dissipation theorem.

Generally speaking, the fluctuation-dissipation theorem is not valid in the off-equilibrium regime. In this case one can use general arguments to derive a relation between statics properties and the form of the function $S(C)$ measured in off-equilibrium [7-9]:

(25) $$\frac{\mathrm{d}S}{\mathrm{d}C} = -\int_0^C \mathrm{d}q P(q) \equiv X(C).$$

In fig. 6 we show three main different kinds of dynamical response $S(C)$, that correspond to different shapes of the static $P(q)$ (which in the case of spin glasses at zero magnetic field should be replaced by $P(|q|)$). Case A corresponds to systems where replica

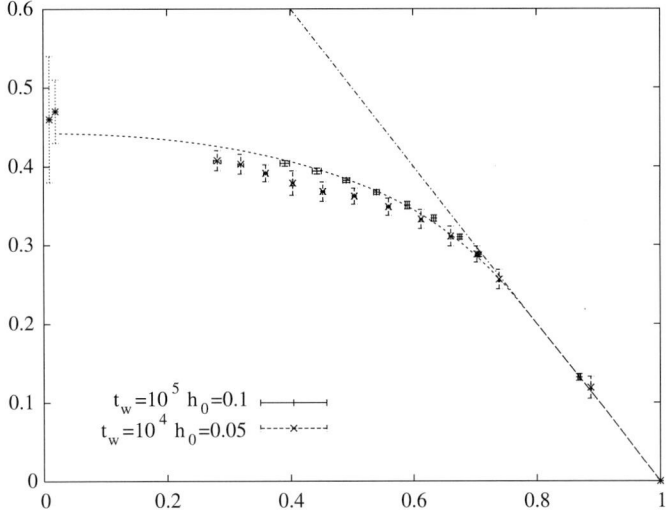

Fig. 7. – Relaxation function *vs.* correlation in the Edwards-Anderson (EA) model in $D = 3$ $T = 0.7 \simeq \frac{3}{4}T_c$ and theoretical predictions from eq. (25).

symmetry is not broken, case B to one step replica symmetry breaking, which should be present in structural glasses, and case C to continuous replica symmetry breaking, which is present in spin glasses.

The validity of these relations has been intensively checked in numerical experiments (see for example fig. 7).

In spin glasses the relaxation function has been experimentally measured many times in the aging regime, while the correlation function has not been measured: it is a much more difficult experiment in which one has to measure thermal fluctuations. Fortunately enough measurements of both quantities for glasses and spin glasses are now in progress. It would be extremely interesting to see if they agree with the theoretical predictions.

Similar measurements have been done also in the case of glasses [12].

6. – The inherent structure picture for glasses

A quite old statement is that *the glass is a frozen liquid*, which I prefer to reformulate as: *a low-temperature liquid is an unfrozen glass*.

The general idea is the following: the phase space of the systems can be approximately decomposed into valleys (or inherent structures) labelled by a, such that the barrier between these valleys is large in the liquid (in the low-temperature region where the viscosity is large the barriers become infinite at T_K [13]).

For simplicity I will assume that valleys do maintain their identity when changing the temperature. Under this assumption there is a one-to-one correspondence between valleys and inherent structures, *i.e.* minima of the Hamiltonian H. The properties of the liquid are therefore connected to the properties of the system near the minima of the Hamiltonian.

The partition function can be approximately written as

$$Z = \sum_a \exp\left[-\beta N f_a(\beta)\right], \qquad (26)$$

where $f_a(\beta)$ is the free-energy density of the a-th valley. As we shall see later, in the infinite-volume limit the previous sum is dominated by those valleys having a given free-energy density $f = f^*(\beta)$.

The number of relevant valleys at $f = f^*(\beta)$ (*i.e.* $\mathcal{N}^*(\beta)$) is supposed to exponentially increase with the size (N) of the system:

$$cN^* \approx \exp\left[N\Sigma^*(\beta)\right]. \qquad (27)$$

It can be argued that the configurational entropy or complexity $\Sigma^*(\beta)$ is approximately near to ΔS: indeed it was suggested long time ago that the configurational entropy Σ^* goes to zero at the phase transition point [13].

Recently using the techniques stemming from replica theory we have added new ingredients to this old scenario:

- We have constructed a microscopic realization of the previous ideas in the mean-field models [14, 15]. In other words there are soluble infinite-range models in which there is an exponentially large number of valleys and their properties can be computed analytically.

- If we assume that *minimal* corrections are present in finite dimensions to mean-field predictions, the mean-field results may be extended to three-dimensional glasses.

- The replica method [2] gives tools to do the appropriate computations, both in the mean-field approach and in short-range models.

As outcome of these advances a many-valley picture with detailed predictions on the landscape has been constructed and the properties of glasses can be computed in the framework of replicated liquid theory [16-19].

7. – The form of the free-energy landscape

Let us suppose that we can define a temperature free-energy functional ($F[\rho]$) as a function of the density $\rho(x)$. It is natural to assume that the different valleys are in a first approximation associated to different local minima of this functional, the free energy evaluated at the local minima being the free energy of the corresponding valley.

I will summarize which are in mean field the properties of the local minima of the free energy that have been computed explicitly in some long-range microscopic models. Although these results have been obtained in some particular models, one can show that they are valid for a quite large class of models (at least in the mean-field approximation [20]).

- For $T > T_c$ there is one relevant minimum with $\rho(x) = $ const. There are many solutions of the equation

$$\frac{\delta F}{\delta \rho} = 0, \tag{28}$$

which are not minima (there are some negative eigenvalues of the Hessian $\mathcal{H}(x,y) = \partial^2 H/\partial x\,\partial y$).

- For $T = T_c$ the negative eigenvalues of the Hessian become positive or zero. There is an exponentially large number of minima, connected by flat regions. The quantity T_c is the critical temperature of the mode-coupling theory.

- For $T_c > T > T_K$ the number of relevant minima becomes proportional to $\exp[N\Sigma^*(\beta)]$, where $\Sigma^*(\beta)$ is the configurational entropy or complexity. The minima are separated by barriers that diverge with the number of particles N (in mean-field theory), however the barriers are finite in real life (*i.e.* beyond mean-field theory).

- For $T_K > T$ the number of relevant minima is finite. The minima are separated by barriers that diverge with N.

For $T_c > T > T_K$ there is a dual description: the system may be described as a *liquid* and also as a *solid* (with an exponentially large number of different solid structures). This duality tells us that all the information needed to study the glass can be extracted from the liquid phase.

8. – Some analytic and numerical results

Analytic and numerical computations have been done in many case. I will present here only the results for the Kob Andersen Lennard Jones binary mixture described in the introduction at density $\rho = 1.2$, with the only aim to show that the formalism here presented may be used to produce quantitative results.

In order to compute the free energy (or the entropy of the valleys) a simple and useful approximation is to assume that the potential near each minimum of the Hamiltonian is approximately a harmonic one and therefore the free energy of a valley can be computed using a harmonic approximation.

The relevant quantities can be either computed analytically (using liquid theory) or extracted by numerical simulations in the liquid phase.

Both the analytic and the numerical results of [19] for the configurational entropy are shown in fig. 8 (from [19]). The correctness of the numerical estimate has been confirmed later by the more accurate computations of [21].

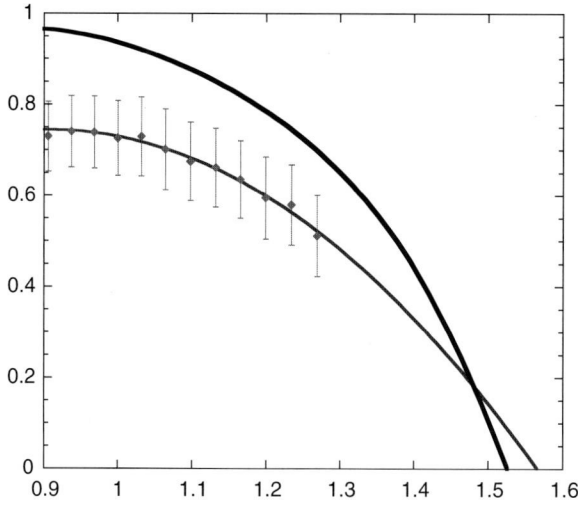

Fig. 8. – Analytic and numerical complexity as a function of $T^{-0.4}$.

These microscopic analytic computations support the proposal that there is a thermodynamic liquid-glass transition characterized by vanishing of the complexity and that at low temperature below T_K the system stays only in a few valleys.

9. – Conclusions

It should be clear from the previous discussion that we start to have a very precise idea on the glass transition which starts to be experimentally tested. Numerical simulations are in very clear agreement with the theoretical picture.

There are many points which we would like to understand better. Let us mention a few of them:

- We would like to have definite numerical and theoretical information on the validity of the conjecture that at the phase transition point there is no divergent length. This could be done looking more carefully to the four-particle correlations.

- It is crucial to have an estimate of the barriers between valleys, which is however theoretically lacking.

- One would like to have experimental systematic measurements of the violations of the fluctuation dissipation theorem in off-equilibrium dynamics with the same kind of accuracy obtained in the computer simulations.

- It would be important to pin down the deep reasons for the distinction between strong glasses.

The solutions of the previous puzzles seems not to be out of reach, so that we hope that in a few years many of the previous points will be clarified.

REFERENCES

[1] MARINARI E., PARISI G., RICCI-TERSENGHI F., RUIZ-LORENZO J. and ZULIANI F., *J. Stat. Phys.*, **98** (2000) 973.
[2] MÉZARD M., PARISI G. and VIRASORO M. A., *Spin Glass Theory and Beyond* (World Scientific, Singapore) 1987.
[3] BINDER K. and YOUNG A. P., *Rev. Mod. Phys.*, **58** (1986) 801.
[4] FISCHER K. H. and HERTZ J. A., *Spin Glasses* (Cambridge University Press, Cambridge) 1991.
[5] For a recent review see MARINARI E., PARISI G. and RUIZ-LORENZO J. J., *Spin Glasses and Random Fields*, edited by A. P. YOUNG (World Scientific, Singapore) 1998.
[6] MARINARI E. and ZULIANI F., *J. Phys. A*, **32** (1999) 7447.
[7] CUGLIANDOLO L. F. and KURCHAN J., *Phys. Rev. Lett.*, **71** (1993) 173; *J. Phys. A*, **27** (1994) 5749.
[8] FRANZ S. and MÉZARD M., *Europhys. Lett.*, **26** (1994) 209.
[9] FRANZ S., MÉZARD M., PARISI G. and PELITI L., *Phys. Rev. Lett.*, **81** (1998) 1758; *The response of glassy systems to random perturbations: a bridge between equilibrium and off-equilibrium*, cond-mat/9903370.
[10] CUGLIANDOLO L. F., KURCHAN J. and PELITI L., *Phys. Rev. E*, **55** (1997) 3898.
[11] FRANZ S. and VIRASORO M., *J. Phys. A*, **33** (2000) 891.
[12] PARISI G., *Phys. Rev. Lett.*, **79** (1997) 3660.
[13] For example see SASTRY S., DEBENEDETTI P. and STILLINGER F. H., *Nature*, **393** (1998) 554; BHATTACHARYA K. K., BRODERIX K., KREE R. and ZIPPELIUS A., cond-mat/9903120; ANGELANI L., PARISI G., RUOCCO G. and VILIANI G., *Phys. Rev. Lett.*, **81** (1998) 4648.
[14] KIRKPATRICK T. R. and THIRUMALAI D., *Transp. Theor. Stat. Phys.*, **24** (1995) 927 and references therein.
[15] CRISANTI A. and SOMMERS H.-J., *J. Phys. I*, **5** (1995) 805.
[16] MÉZARD M. and PARISI G., *Phys. Rev. Lett.*, **82** (1998) 747; *J. Chem. Phys.*, **111** (1999) 1076.
[17] MÉZARD M., *Physica A*, **265** (1999) 352.
[18] COLUZZI B., MÉZARD M., PARISI G. and VERROCCHIO P., *J. Chem. Phys.*, **111** (1999) 9039.
[19] COLUZZI B., PARISI G. and VERROCCHIO P., *J. Chem. Phys.*, **112** (2000) 2933; *Phys. Rev. Lett.*, **84** (2000) 306.
[20] FRANZ S. and PARISI G., *Physica A*, **261** (1998) 317.
[21] KOB W., SCIORTINO F. and TARTAGLIA P., *Phys. Rev. Lett.*, **83** (1999) 3214.

units, is produced in our model by the chain section of the parent conformation placed entirely in either internal or shell regions of the globule. In particular, the probability to have an uninterrupted succession of some k of B-monomer units in the sequence is equal to the probability that the ideal polymer chain (that is, the chain in the parent conformation) has a loop of k monomer units entirely confined in the B region with ends on the separation surface. Similarly, the probability to have an uninterrupted succession of some k of A-monomer units in the sequence is equal to the probability that the ideal parental conformation has a loop of k monomer units confined within the shell A-region, again with ends on the separation surface. Importantly, these loops are independent of each other. Therefore, the above-defined probabilities $P_B(k)$ and $P_A(k)$ provide a complete description of the statistics of protein-like sequences.

We will derive later analytical expressions for probability distributions $P_B(k)$ and $P_A(k)$. It is useful, however, to begin with simple physical arguments yielding

$$(2) \qquad P_{A,B}(k) \simeq \begin{cases} k^{-3/2}, & \text{for } 1 < k < \left(\dfrac{d_{A,B}}{a}\right)^2; \\ \left(\dfrac{d_{A,B}}{a}\right)^{-3} \exp\left[-\lambda_{A,B} k\right], & \text{for } k > \left(\dfrac{d_{A,B}}{a}\right)^2. \end{cases}$$

The upper asymptotic form is valid for short polymer loops, when neither the curvature of the separating boundary nor the overall globule shape play any role. In this regime, $P(k)$ is simply the probability for a random walk to start at the planar wall and to return to it for the first time after k steps. This is the classical probabilistic "first return" problem, for which the $\sim k^{-3/2}$ answer is well known [11]. This scaling is valid for loop sizes $a\sqrt{k}$ smaller than the relevant characteristic length scale, $d_H = R^*$ for the B-loops inside the inner sphere, or $d_P = R - R^*$ for the A-loops in the spherical shell. The second asymptotic form in eq. (2) indicates that for long polymer loops the function $P(k)$ decays exponentially (see below for the calculation of $\lambda_{H,P}$). It is easier to explain this in terms of polymer statistics than random walks: to confine a polymer chain of k monomer units in a cavity costs some entropy ΔS, and at $a\sqrt{k} \gg d$ this entropy goes linearly with k, making the probability, $e^{\Delta S}$, exponential in k.

The result (eq. (2)) is in fact sufficient to explain qualitatively the correlations in protein-like sequences and, in particular, computational data shown in fig. 4. Indeed, according to our discussion, protein-like sequences can be thought of as produced by first generating a random number k_1 from the probability distribution $P_B(k)$ and taking k_1 of B-monomers in a row, then generating k_2 from the probability distribution $P_A(k)$ and taking k_2 of A-monomers, then generating k_3 using $P_B(k)$ and connecting k_3 of B-monomers to the previously assembled tail, and thus continuing until the entire chain is completed. The mathematical scheme, in which a section of random walk is generated with a power law probability distribution, is called a Levy flight [12]. Thus, we can conclude that the long-range correlations in the primary sequences of protein-like copolymers are described by Levy flight statistics. Furthermore, for the $k^{-3/2}$ behavior of $P(k)$, Levy flight is known to have peculiar properties. Since averaged block length diverges for such distribution, the value of D_L is controlled by the longest block, yielding $D_L \sim L^\alpha$ with $\alpha = 1$, which agrees very well with our computational data for modest values of L.

We understand also that this power law should be valid within the range of length scales dictated by the first line of eq. (2). At larger scales, correlations decay exponentially and we expect crossover to classical behavior with $\alpha = 1/2$. In this regard, it is important to look closer at the crossover values of k. This requires to determine the separation radius R^*. In order to achieve the 1 : 1 composition of the sequence, the value of R^* must be chosen in such a way that the volumes of internal B and shell A regions are the same, which requires $R^{*3} = R^3 - R^{*3}$, or $R^* = 2^{-1/3} R \approx 0.8R$. The volume fraction of polymer units in a globule, ρ, is controlled by the energy of interactions of monomer units. It is clear that $\rho = Na^3/(4/3)\pi R^3$, or $R \approx 0.6aN^{1/3}\rho^{-1}$. Therefore, internal A-loops remain in the power law long-range correlation regime up to the length $k < 0.24 N^{2/3}\rho^{-2}$, while for outer B-loops this crossover occurs somewhat earlier: $k < 0.015 N^{2/3}\rho^{-2}$. Thus, we predict that there should be at least about a decade of length scales in which B-loops still exhibit a long-range correlation, while A-loops already cross over to the classical short-range behavior.

Let us now derive the exact analytical expressions for the probability distributions $P_B(k)$, $P_A(k)$ and for the dispersion D_L. It is convenient to use the random walk terminology to describe parent conformation. In this language, for instance, $P_B(k)$ is the probability that the random walker, after entering a sphere of the radius R^*, will exit it after k steps without ever touching the boundary at the intermediate steps. To emphasize the analogy with the first return problem [11], we can say that this is the probability that the walker will arrive to the boundary *for the first time* after k steps.

To address it, we recall that the statistical weight of all random walk trajectories starting at the point $\vec{r}_0$ and arriving after k steps at the point $\vec{r}$, $G(\vec{r}, k|\vec{r}_0)$, obeys the diffusion equation [2]

$$(3) \qquad \frac{\partial G(\vec{r}, k|\vec{r}_0)}{\partial k} = \frac{a^2}{6} \Delta G(\vec{r}, k|\vec{r}_0) + \delta(k)\delta(\vec{r} - \vec{r}_0),$$

where a^2 is the mean-square length of one step (the squared size of one monomer unit along the chain).

To introduce the condition of first return, we have to say that the walker never touches the boundary, which is achieved by imposing the boundary condition

$$(4) \qquad G(\vec{r}, k|\vec{r}_0)|_{|\vec{r}|=R^*} = 0.$$

The probability distribution of the "first return times" in terms of G is then given as the time-dependent flux of diffusing particles through the absorbing wall:

$$(5) \qquad P_B(k) = \left| \oint d\sigma \frac{a^2}{6} \frac{\partial G}{\partial r} \bigg|_{r=R^*} \right|,$$

where $\partial/\partial r$ means the component of gradient normal to the surface, integration is performed over the closed separating surface, and the absolute value is written to avoid thinking about the direction of the flux. The normalization condition $\int P_B(k)\,dk = 1$ is guaranteed by the fact that all diffusing particles eventually leave through the surface. As regards $\vec{r}_0$, it should be taken within a distance of order a from the separating R^* surface.

The problem thus formulated, including eqs. (3)–(5), is easy to solve: we write G in terms of the bilinear expansion $G = \sum_n e^{k\lambda_n}\psi_n(\vec{r})\psi_n(\vec{r}_0)$ over the eigenfunctions ψ_n satisfying $(a^2/6)\Delta\psi_n = \lambda_n\psi_n$ with the boundary condition (4). Upon spherical integration in (5), all angular-dependent harmonics vanish, and we arrive at

$$(6) \qquad P_B(k) = \frac{\pi a^2}{3R^* r_0} \sum_{n=1}^{\infty} n(-1)^{n+1} \exp\left[-\frac{a^2}{6}\left(\frac{n\pi}{R^*}\right)^2 k\right] \sin\left(n\pi \frac{r_0}{R^*}\right).$$

The distribution $P_A(k)$ can be derived similarly, except that now we have to take care of the boundary condition at the outer surface of the globule. To this end, we argue that this condition must be taken in the form

$$(7) \qquad \nabla_{\vec{r}} G(\vec{r}, k|\vec{r}_0)|_{r=R} = 0.$$

Indeed, formally this condition ensures the constant density of monomer units throughout the globule for large values of k, as well as breaking of correlations as soon as the polymer chain is "reflected" by a globular boundary. Physically, this boundary condition reflects the fact that there is always a "sticky layer" (or depletion layer) formed self-consistently along the internal surface of the globule due to the effective attraction of monomer units to the outer region where polymer density is depleted and the excluded-volume effect is reduced. As long as we are not interested in the structure of the surface layer of the globule, we can just replace this layer by the effective boundary condition (7).

After calculations, for $P_A(k)$ we obtain

$$(8) \qquad P_A(k) = \frac{a^2 R^*}{6(R-R^*)^2 r_0} \sum_{n=0} \exp\left[-\frac{a^2}{6}\left(\frac{\zeta_n}{R-R^*}\right)^2 k\right] \frac{\sin\left(\zeta_n((r_0-R^*)/(R-R^*))\right)}{\zeta_n - \sin\zeta_n \cos\zeta_n},$$

where ζ_n satisfies $R\zeta_n \cos\zeta_n - (R-R^*)\sin\zeta_n = 0$.

It is not very easy, but a technical matter to check that indeed eqs. (6) and (8) have asymptotic behavior in accord with eq. (2).

Let us now turn to the dispersion D_L introduced in the beginning of this paper (see fig. 2). From eq. (1) it follows that

$$(9) \qquad D_L^2 = \langle h_L \rangle + 2 \sum_{m=1}^{L} (L-m) p_{BB}(m) - \langle h_L \rangle^2,$$

where $p_{BB}(m)$ is the probability to find two hydrophobic units at the distance m along the chain. The value of $p_{HH}(m)$ can be expressed in terms of the function $G(\vec{r}, m|\vec{r}_0)$ as follows.

The partition function of a polymer chain forming a globule is given by the solution of eq. (3) with the boundary condition (7). In this case $G(\vec{r}, m|\vec{r}_0)$ is equal to the probability that the ideal polymer chain trajectory arrives after m steps at point $\vec{r}$ if it starts at point $\vec{r}_0$, the normalization condition $\int d^3 r G(\vec{r}, m|\vec{r}_0) = 1$ being satisfied. As the globule density is constant and all B-units are inside the sphere of radius R^* with volume $V^* = 4\pi(R^*)^3/3$, the pair probability can be easily calculated:

$$(10) \qquad p_{BB}(m) = \frac{1}{V} \int_{V^*} d^3 r_0 \int_{V^*} d^3 r G(\vec{r}, m|\vec{r}_0).$$

Finally, we have

$$(11) \qquad D_L^2 = \frac{\langle h_L \rangle^2}{L} + \frac{\sqrt{6}a}{R} \sum_{n=1}^{\infty} \left(\frac{\phi(\xi_n R^*/R)}{\xi_n \sin \xi_n} \right)^2 \frac{L(1 - e^{-\alpha_n}) - 1 + e^{-L\alpha_n}}{(1 - e^{-\alpha_n})(e^{\alpha_n} - 1)},$$

where $\alpha_n = (\xi_n a/R)^2/6$, $\phi(\xi) = \xi \cos \xi - \sin \xi$, ξ_n satisfies the equation $\phi(\xi_n) = 0$.

Equation (11) gives Levy flight asymptotic form $D_L \sim L$ for $L \ll 0.3(R/a)^2$. In the whole range of L there is no universal power law, but rather a smooth decrease of the effective exponent in the expression for D_L from $\alpha \approx 1$ for small values of L to $\alpha \approx 1/2$ for large values of L.

The analytical function D_L given by eq. (11) is shown in fig. 4 (first curve). One can see a good agreement of simulations with analytical theory. Small deviations are probably due to the incomplete relaxation to equilibrium of the globule of fig. 2a; it is known that equilibration from "crumpled globule" to the globule with Gaussian statistics requires extremely long times [13].

In conclusion, we have shown that the protein-like copolymers generated according to the coloring procedure proposed earlier [3-6] do exhibit long-range correlations in the primary sequences. These correlations can be described by Levy-flight statistics. This result, first observed in computer experiments, is confirmed by exact and general analytical calculation.

It is now interesting to test this result for the real globular protein sequences simplified to a two-letter AB-representation. The dilemma which is emerging in this connection is the following:

– either our schematic picture of the globular protein shown in fig. 2b (with hydrophobic core, polar envelope and ideal chain statistics inside the globule) is correct, and then we should find the signs of Levy-flight statistics in real protein sequences;

– or, in the case of negative result of this search, some features should be modified in our schematic representation of globular proteins.

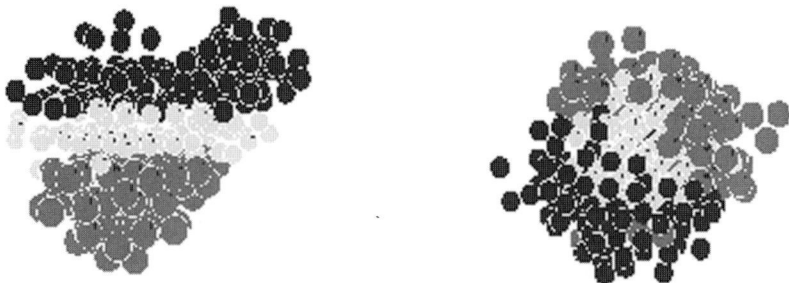

Fig. 5. – The "parent" conformation of the membrane-protein–like copolymer (left part) and the conformation obtained after equilibration in computer simulation (right part).

4. – AB-copolymers which mimic membrane proteins

Another way for coloring monomeric units inside the dense homopolymer globule mimics some properties of real membrane proteins. It is well known that real membrane proteins are located inside the membrane in such a way that some part (about 30% from the whole number) of the amino acid units (mainly the hydrophobic and uncharged ones) are located inside the bilipid layer of the membrane while the other amino acid units are located in water environment inside and outside the cell. In our rough model we assigned the type B to monomeric units which lie inside a cross-section of a parent globule by a narrow flat layer. So, the B-part of a parent conformation has the form of a narrow disk. We have taken 30% of all links to be of B-type. We present, in the left part of fig. 5, the snapshot of an original conformation of just prepared AB-copolymer globule. We marked both hemispheres of outer A-links (70% from the whole amount) of the original globule into two different colors (black and grey) to see whether the "parent" microsegregated structure can be re-established after the equilibration procedure.

We have studied the conformations of such AB-copolymer chain for the following values of interaction energies $(\varepsilon_{AA}, \varepsilon_{AB}, \varepsilon_{BB}) = (-1, -1, -2)$, *i.e.* there is always an attraction between monomer units, but attraction between the B-units is stronger. We have performed the procedure described above for the chains of $N = 256$ units using for simulations the Monte Carlo (MC) method and the bond fluctuation model [14]. This sequence design scheme was repeated many times, and the results were averaged over ~10^6 MC steps and different initial configurations.

We have indeed found that such a chain shows the effect of stability of a "parent" microsegregated structure (a typical conformation obtained after the procedure of decollapse and following collapse of that chain is shown on the right part of fig. 5). The spherical B-core is formed instead of original disk-like B-core that is, of course, natural due to the isotropy of the interaction potential. But one can see definitely that the grey units have much more contacts with each other than with the black units and vice versa, *i.e.* the grey and black units are segregated from each other. It can also be seen from fig. 6 that upon decreasing temperature the number of contacts between all monomeric

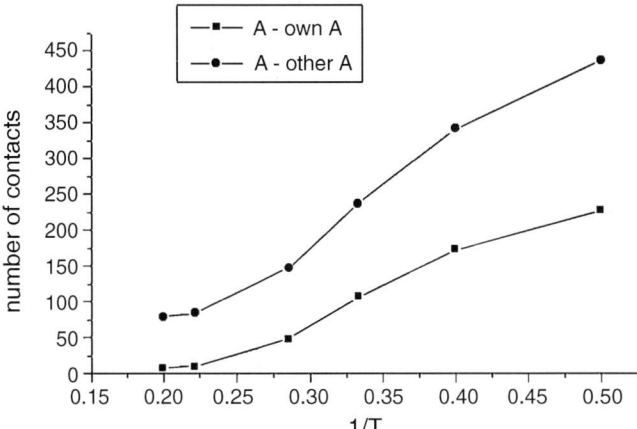

Fig. 6. – Temperature dependence of the number of contacts between A-monomer units which belong to different (circles) and to the same (squares) hemispheres.

units increases because of the transition into the globular state, but the number of contacts between A-links from one hemisphere increases faster than that between A-links from different hemispheres. In other words, we can say that the protein-like copolymer "inherited" (or "memorized") some important structural features of the "parent" globule which were then reproduced in the other conditions.

We have found the explanation of this effect. We have analyzed histograms of length of A-units sequences and found out that those connecting the same hemisphere (we called them "loops") have small length, about 1 unit in average, and those connecting different hemispheres ("bridges") are longer, about 5 units long (fig. 7). Therefore "loop" sections in their majority are too short to play the role of bridges after refolding.

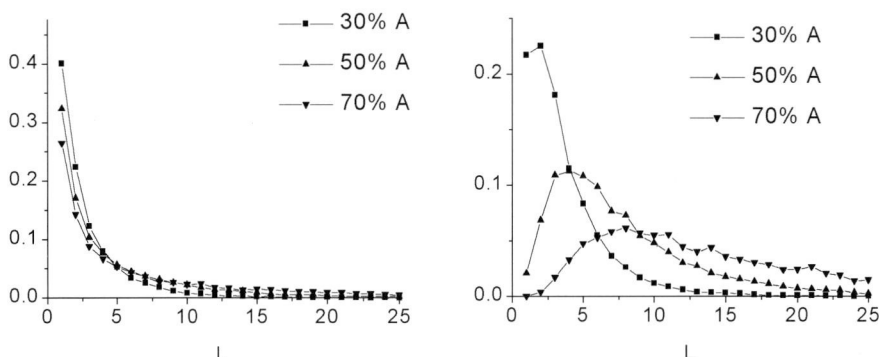

Fig. 7. – Histogram of distribution of the lengths of "loop" A-units sequences (a) and "bridge" B-units sequences (b).

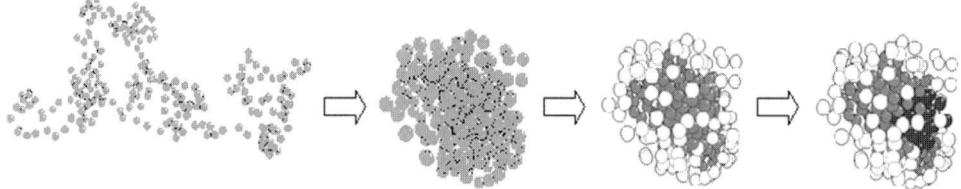

Fig. 8. – The "coloring" procedure for the ABC-copolymer modeling proteins with an active enzymatic center.

5. – ABC-copolymers modeling proteins with active enzymatic center

As another way of preparation of the primary structure of a copolymer chain, we have studied the ABC-copolymer prepared in the course of a "triple coloring" of some particular homopolymer globule in the following way: we assigned the type A to the surface monomeric units, the type B to the inner monomeric units (as has been previously done for protein-like copolymers), and the type C to those inner monomeric units which lie inside a small sphere whose center does not coincide, however, with the center of mass of the parent homopolymer globule (see fig. 8).

Our idea was to prove whether such a "parent" conformation can be recombined in the course of the equilibration procedure for different sets of interaction parameters including the restoration of the originally given distance between the centers of the B-core and the C-core. We intended to prove whether the position of C-links inside the primary sequence together with a specially chosen interaction potential can lead to stable reconstruction of the spatial conformation of the whole chain.

However, there is a strong dependence on the way in which we perform the equilibration (see fig. 9). In that figure the right branch represents a quick decrease of temperature during the coil-globule transition, and the left branch represents the slow annealing procedure (smooth decreasing of temperature). A quick temperature decrease brings us to the frozen non-equilibrium conformation. However, we see that using annealing procedure we are able to achieve a real globule conformation (the global minimum of energy, not local ones).

We performed computer simulations for the chain of $N = 256$ monomeric units using attractive interaction potentials for B- and C-links (the attraction for C-link was taken stronger that that for B-links). We have found in our computer experiment that such ABC-copolymers normally restore their original structure with B- and C-cores although we have not succeeded up to now in determining the interaction potential which would allow us to get the center of the C-core at the same distance from the center of the B-core as in the original conformation.

Nevertheless, we have definitely found the effect of restoration of the active center after the following procedure: we switch off the attraction between C-links and let them "dissolve" inside the dense B-core; and after the switching of the attraction between C-links we observe the restoration of the C-core again.

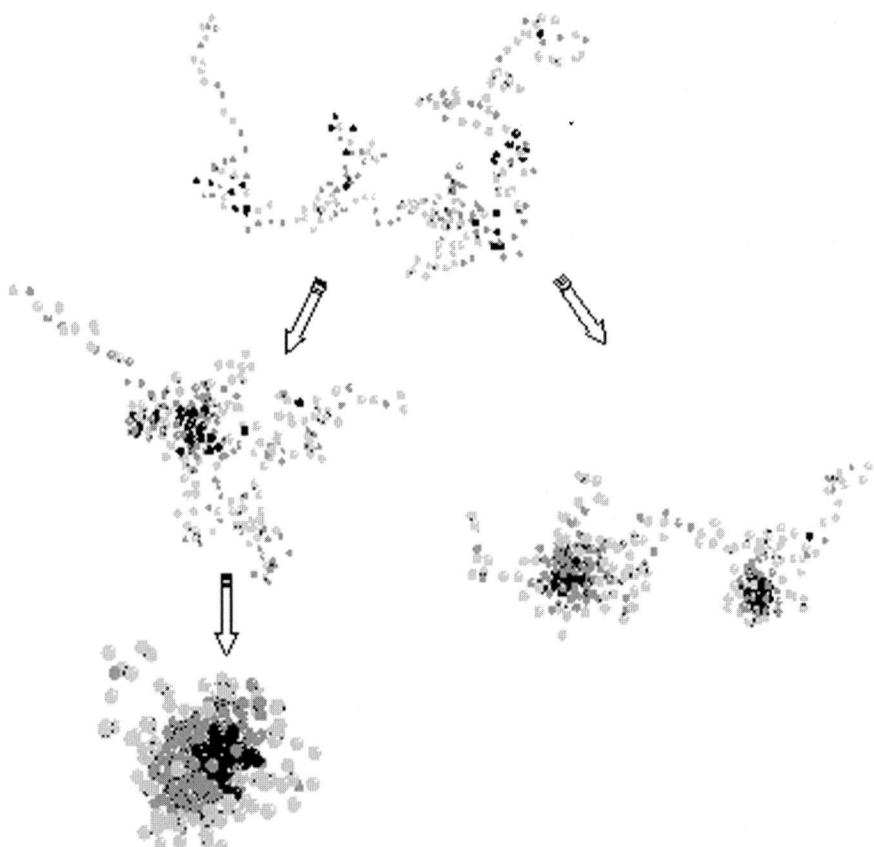

Fig. 9. – Annealing procedure. Equilibrated globule on the left side ($T = 1.0$), frozen conformation on the right side ($T = 1.4$).

We have compared the properties of ABC-copolymers with designed (as described above) primary sequence with the properties of quasi-random ABC-copolymers. The quasi-random sequence was prepared in the following way. At the first stage we prepared the protein-like AB-copolymer according to the original scheme described in the introduction. Then we have "recolored" some of the internal B-units into C-units randomly, *i.e.* the BC-sequence is a purely statistical one. The attraction between C-units in this case was also taken to be stronger than that between B-units. We have found that ABC-copolymers with our designed primary sequence form larger clusters of active units than this quasi-random ABC-copolymer.

To investigate the influence of the primary sequence on the formation of the selected core in the globule, we have considered the case of equally favorable attraction for B-B, B-C and C-C contacts. In this case there is no difference between B- and C-links from the energetic point of view. We have found a pronounced difference between the histograms

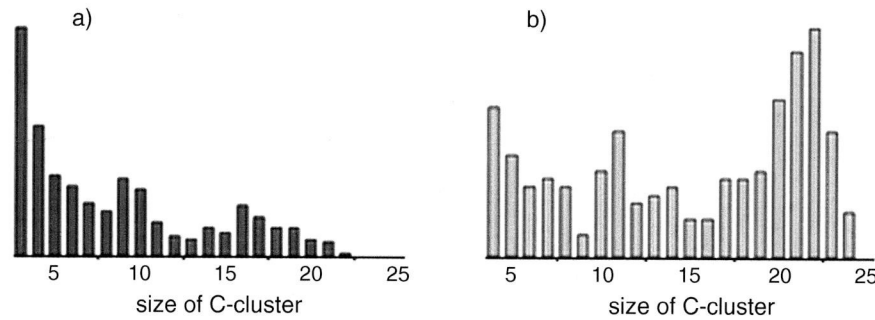

Fig. 10. – Histograms (non-normalized) of the C-units cluster size for quasi-random (a) and designed (b) primary sequences of B- and C-units. (B- and C-units are equivalent from the viewpoint of attraction energies.)

of the C-cluster size for designed (fig. 10b) and quasi-random ABC-sequences (fig. 10a) in globular state. One can see from fig. 10 that in the same spatial conformation (at the same temperature) there are rather small C-clusters which dominate for the case of quasi-random sequence, while for the designed sequence the large C-clusters which include almost all C-links have much larger probability (there were totally 25 C-units in the chain).

6. – Adsorption-tuned AB-copolymers

Let us now generalize the idea which was introduced above. Actually, the primary structure of protein-like copolymer was generated via a kind of "coloring procedure" (in two "colors", A and B) for a homopolymer chain in the globular state. However, the special primary sequence can be obtained not only from globular conformation; *any specific polymer chain conformation can play the role of a "parent" one.*

The simplest example of this kind is connected with the conformation of a homopolymer chain adsorbed on the plane surface. Let us "color" the links in direct contact with the surface in some typical "instant snapshot" conformation (see fig. 11). This corresponds to the assumption that the surface catalyses some chemical transformation of the adsorbed links. Then we will end up with AB-copolymer for which the sequence design was performed in the "parent" adsorbed state. After desorption such AB-copolymer will have special functional properties: it will be *"tuned to adsorption"*.

Indeed, we have performed Monte Carlo computer experiments along the lines of the sequence design scheme outlined above [15] for a chain of 32 monomer units. In the conformations of the adsorbed homopolymer chain 8 units which are closest to the surface were identified and denoted as B-units, others were designated as A-units. Then we studied the adsorption behavior of the AB-copolymer chain obtained in this way on the plane surface induced by the increase of attraction of B-units to this surface and compared it with the behavior of the corresponding random and random-block copolymers (for details, see [15]).

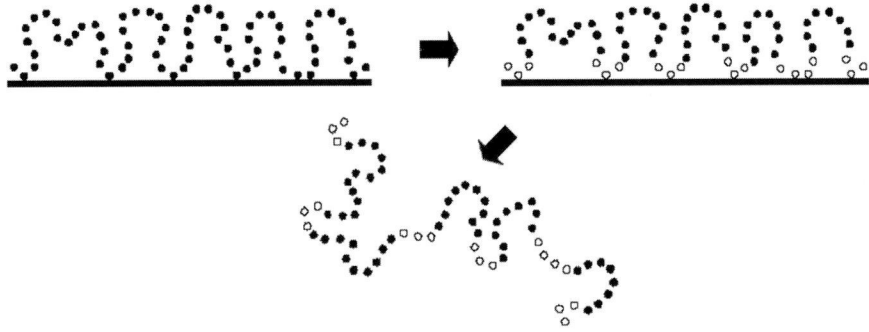

Fig. 11. – Preparation of an adsorption-tuned primary sequence.

In fig. 12 the plot of the average number of adsorbed B-units *vs.* the energy of their attraction to the surface, ε_B, is presented. It can be seen that the number of adsorbed segments (at a given value of ε_B) is always the highest for the designed AB-copolymers. In other words, due to the "memorizing" of some functional features of the "parent" conformation we have indeed obtained the AB-copolymer "tuned to adsorption on a plane surface".

One can imagine the analogous "coloring" procedure for the chain adsorbed on a small spherical colloidal particle (see fig. 13). In this case a copolymer chain with the primary sequence "tuned to the absorption" of a small droplet of organic solvent or a colloidal particle of given size could be obtained. Such a copolymer could be called "molecular dispenser". Indeed, imposed to the contact with the organic fluid such a copolymer will absorb a small droplet of this fluid with the volume approximately equal to the volume of the "parent" colloidal particle, because for such size of the droplet the maximal number of "hydrophobic" links will be in the contact with the fluid leading to maximal gain in the interaction energy per monomeric link. Being imposed to the contact with the solution of colloidal particles of different sizes such "molecular dispenser" will choose (select) the particles of size equal to that of the "parent" particle.

7. – Conclusions

In all cases described in this paper we generate AB-sequences with the help of some special *"parent"* conformation of a polymer chain. Such conformation-dependent sequence design can be called "engineering of AB-copolymers" using some remote analogy with protein engineering which deals with the design of the primary sequence of protein molecules.

It is important also to emphasize that in both cases some functional features of the "parent" conformation were *"memorized"* by the AB-copolymers generated according to our sequence design scheme. These features are then manifested in other conditions. Such an interrelation can be regarded as one of the possible mechanisms of molecular

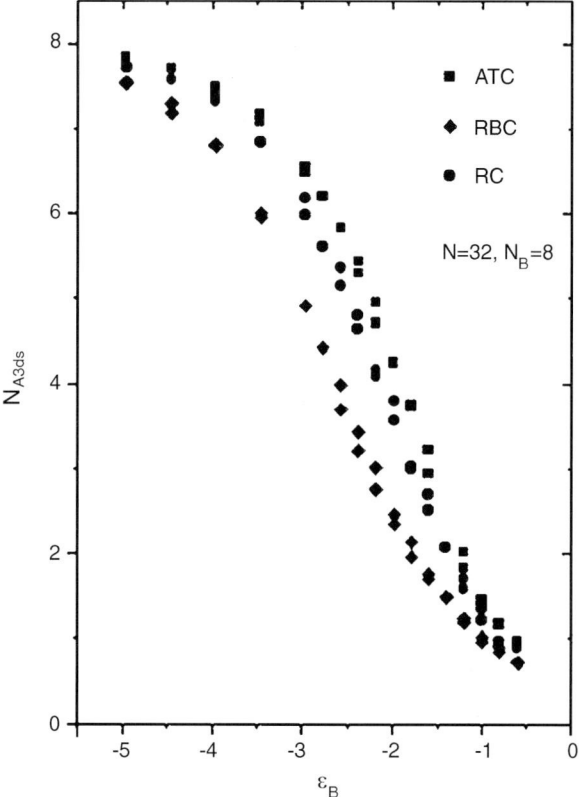

Fig. 12. – The average number of adsorbed type-B segments *vs.* the attraction energy to the surface, ε_B, for the adsorption-tuned (ATC), random-block (RBC), and random (RC) copolymer chains of length $N = 32$ with the number of type-B segments $N_B = 8$.

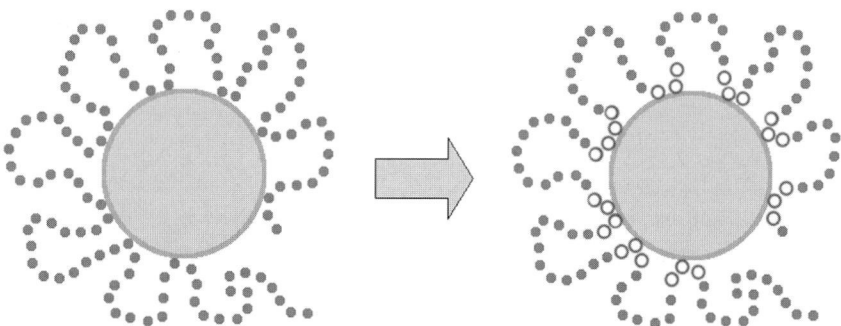

Fig. 13. – Preparation of the "molecular dispenser".

evolution: a polymer acquires some special primary sequence in the "parent" conditions and then (in other conditions) uses the fact that the primary structure is "tuned to perform certain functions".

REFERENCES

[1] LIFSHITZ I. M., GROSBERG A. YU. and KHOKHLOV A. R., *Rev. Mod. Phys.*, **50** (1978) 683.
[2] GROSBERG A. YU. and KHOKHLOV A. R., *Statistical Physics of Macromolecules* (American Insitute of Physics, New York) 1994.
[3] KHOKHLOV A. R. and KHALATUR P. G., *Physica A*, **249** (1998) 253.
[4] KHALATUR P. G., IVANOV V. I., SHUSHARINA N. P. and KHOKHLOV A. R., *Russ. Chem. Bull.*, **47** (1998) 855.
[5] KHOKHLOV A. R., IVANOV V. A., SHUSHARINA N. P. and KHALATUR P. G., *Engineering of Synthetic Copolymers: Protein-Like Copolymers*, in *The Physics of Complex Liquids*, edited by F. YONEZAWA, K. TSUJI, K. KAIJ, M. DOI and T. FUJIWARA (World Scientific, Singapore) 1998, p. 155.
[6] KHOKHLOV A. R. and KHALATUR P. G., *Phys. Rev. Lett.*, **82** (1999) 3456.
[7] PENG C.-K., BULDYREV S. V., GOLDBERGER A. L., HAVLIN S., SCIORTINO F., SIMONS M. and STANLEY H. E., *Nature*, **356** (1992) 168.
[8] BULDYREV S. V., GOLDBERGER S. V., HAVLIN S., MANTEGNA R. N., MATSA M. E., PENG C.-K., SIMONS M. and STANLEY H. E., *Phys. Rev. E*, **51** (1995) 5084.
[9] VISWANATHAN G. M., BULDYREV S. V., HAVLIN S. and STANLEY H. E., *Biophys. J.*, **72** (1997) 866.
[10] BULDYREV S. V., DOKHOLYAN N. V., GOLDBERGER A. L., HAVLIN S., PENG C.-K., STANLEY H. E. and VISWANATHAN G. M., *Physica A*, **249** (1998) 430.
[11] FELLER W., *An Introduction to Probability Theory and Its Applications*, 3rd edition (John Wiley & Sons, New York) 1970.
[12] SHLESINGER M. F., ZASLAVSKII G. M. and FRISCH U. (Editors), *Levy Flights and Related Topics in Physics, Lecture Notes Phys.* (Springer Verlag) 1996.
[13] GROSBERG A. YU., NECHAEV S. K. and SHAKHNOVICH E. I., *Biofizika*, **33** (1988) 247.
[14] CARMESIN and KREMER K., *Macromolecules*, **21** (1988) 2819.
[15] ZHELIGOVSKAYA E. A., KHALATUR P. G. and KHOKHLOV P. G., *Phys. Rev. E*, **59** (1999) 3071.

Tamara COUSSAERT
Unité de Physique des Polymères
CP 223 Campus Plaine
Université Libre de Bruxelles
Boulevard du Triomphe
1050 BRUSSELS
Belgium

Tel: +32-2-650-5741
Fax: +32-2-650-5675
tcouss@ulb.ac.be

Nikolay V. DOKHOLYAN
Department of Chemistry
and Chemical Biology
Harvard University
12 Oxford St.
CAMBRIDGE, MA 02138
USA

Tel: +1-617-496-5499
Fax: +1-617-495-3075
dokh@wild.harvard.edu

Pavel EMELIANOV
Gasheka 26-1-394
192288 ST. PETERSBURG
Russia

Tel: +7-812-1789504
egipet99@mail.ru

Giorgio FAVRIN
Complex Systems Division
Department of Theoretical Physics
Lund University
Sölvegatan 14A
S-223 62 LUND
Sweden

Tel: +46-46-2220667
Fax: +46-46-2229686

Carine GALLI
Solid State Physics
University of Fribourg
Ch. du Musée 3
1700 FRIBOURG
Switzerland

Tel: +41-26-3009076
Fax: +41-26-3009747
carine.galli@unifr.ch

Oxana V. GALZITSKAYA
Institute of Protein Research
Russian Academy of Sciences
Institutskaya st. 4
142292 PUSHCHINO
Moscow Region
Russia

Tel: +7-095-9240493
Fax: +7-095-9240493
ogalzit@vega.protres.ru

Patrizia GAMBADAURO
Dipartimento di Fisica
Università di Messina
Contrada Papardo
Villaggio S. Agata
98166 MESSINA
Italy

Tel: +39-090-6765120
Fax: +39-090-395004
mallamace@imeuniv.unime.it

Stefano GAROFOLI
Department of Chemistry
Brandeis University
P.O. Box 9110
WALTHAM, MA 02454-9110
USA

Tel: +1-781-7362567
Fax: +1-781-7362516
garofoli@brandeis.edu

Zeynep Nevin GEREK
Bogazici University
Polymer Research Centre
Bebek 80815
ISTANBUL
Turkey

Tel: +90-212-263-1540 Ext. 2002
Fax: +90-212-257-5032
nevin@prc.bme.boun.edu.tr

Gaddy GETZ
Department of Physics
of Complex Systems
The Weizmann Institute of Science
76100 REHOVOT
Israel

Tel: +972-8-9342433
Fax: +972-8-9344109
gaddyy@wicc.weizmann.ac.il

Stefano GIANNI
Dipartimento di Scienze Biochimiche
"A. Rossi Fanelli"
Università "La Sapienza"
Piazzale Aldo Moro, 5
00185 ROMA
Italy

Tel: +39-06-49910548
Fax: +39-06-4440062
stefanogianni@hotmail.com

Izabela GRGIC
Umeå University
90187 UMEÅ
Sweden

Tel: +46-90-7867918
Fax: +46-90-7867661
izabela_g@hotmail.com

Alessandro GROTTESI
Dipartimento di Biochimica
Università di Roma "La Sapienza"
Piazzale Aldo Moro, 5
00185 ROMA
Italy

Tel: +39-06-49913176
Fax: +39-06-490324
alex@cezanne.chem.uniroma.it

Maria Adriana IONESCU
National Institute of Research
& Development for Physics and
Nuclear Engineering "Horia Hulubei"
IFIN-HH
P.O. Box MG-6, R-76900
BUCURESTI-Magurele
Romania

Tel: +40-1-7807040/01-4231701
Fax: +40-1-4231701/+40-1-7896295
aionescu@ifin.nipne.ro

ELENCO DEI PARTECIPANTI

Alfonso JARAMILLO
Service de Conformation de
Macromolecules Biologiques
et de Bioinformatique
Université Libre de Bruxelles
Av F. D. Roosevelt 50
CP 160/16
1050 BRUXELLES
Belgium

Tel: +32-2-6505200/6502013
Fax: +32-2-6488954
alfonso@ucmb.ulb.ac.be

Alkan KABAKCIOGLU
Department of Physics
of Complex Systems
The Weizmann Institute of Science
76100 REHOVOT
Israel

Tel: +972-8-9343370
Fax: +972-8-9344109
alkan@wicc.weizmann.ac.il

Oksana KRAVCHENKO
University of "Kiev-Mohyla Academy"
Department of Physical and
Mathematical Sciences
2 Skovoroda St.
254070 KIEV
Ukraine

Tel: +38-044-2902926
Fax: +38-044-2902926
kravch—o@yahoo.com

Edo KUSSELL
Department of Biophysics
Harvard University
7 Springfield St.
CAMBRIDGE, MA 02139
USA

Tel: +1-617-441-9403
Fax: +1-617-495-3075
ekussell@fas.harvard.edu

Alexei LAZUTIN
Physics Department
Moscow State University
Vorob'evy Gory
117234 MOSCOW
Russia

Tel: +7-095-939-4756
Fax: +7-095-939-2988
lazutin@polly.phys.msu.su

Fabio LIBRIZZI
Dipartimento di Scienze
Fisiche ed Astronomiche
Via Archirafi 36
90123 PALERMO
Italy

Tel: +39-091-6234248
Fax: +39-091-6162461
dottorat@fisica.unipa.it

Roman E. LIMBERGER
Physics Department
Moscow State University
Vorob'evy Gory
117234 MOSCOW
Russia

Tel: +7-095-939-4756
Fax: +7-095-939-2988
roman@polly.phys.msu.su

Magnus LINDBERG
Umeå University
90187 UMEÅ
Sweden

Tel: +46 90 7867918
Fax: +46 90 7867661
magnus.lindberg@chem.umu.se

Emanuela LONARDI
Università di Padova
Dipartimento di Biologia
Via Trieste 75
35121 PADOVA
Italy

Tel: +39-049-8276340
Fax: +39-049-8276344
labsalv@civ.bio.unipd.it

E. Consuelo LOPEZ DIEZ
School of Informatics
University of Wales, Bangor
Dean Street
Gwynedd LL57 1UT
BANGOR
UK

Tel: +44-1248-382742
Fax: +44-1248-361429
consuelo@sees.bangor.ac.uk

Jon LOPEZ LLANO
Department of Biochemistry,
Molecular and Cellular Biology
University of Zaragoza
Pedro IV El Ceremonioso 10 11 A
50009 ZARAGOZA
Spain

Tel: +34-976-761277
Fax: +34-976-762123
jl_llano@yahoo.com/jon@posta.unizar.es

Fabio LUCIANI
Via Malaguti 1/5
40126 BOLOGNA
Italy

fabio.luciani@yahoo.com

Mauro MANNO
INFM
Dipartimento di Scienze
Fisiche ed Astronomiche
Università di Palermo
Via Archirafi, 36
90123 PALERMO
Italy

Tel: +39-091-6234203
Fax: +39-091-6173895

Ugo MAYOR
MRC Centre for Protein Engineering
Hills Road
CAMBRIDGE CB2 2QH
UK

Tel: +44-1-223-402141
Fax: +44-1-223-402140
umm@mrc-lmb.cam.ac.uk

Samat MOLDAKARIMOV
Physics Department
Moscow State University
MOSCOW 117234
Russia
and
Institute of Polymer Materials
and Technology
Satpaev Str. 18A
ALMATY 480013
Republic of Kazakhstan

Tel: +7-095-9394013
Fax: +7-095-9392988
samat@polly.phys.msu.su

Tel: +7-3272-627473

ELENCO DEI PARTECIPANTI

Matteo PALASSINI
Physics Department
University of California
Santa Cruz CA 95060
USA

Tel: +1-831-459-2684
Fax: +1-831-459-3043
matteo@physics.ucsc.edu

Claudio PASTORINO
Departamento de Física
Centro Atómico Constituyentes
Comisión Nacional de Energía Atómica
Av. Libertador 8250
1429 BUENOS AIRES
Argentina

Tel: +54-11-4754-7100
Fax: +54-11-4754-7121
clopasto@cnea.gov.ar

Alexey POLOTSKY
Macromolecular Compounds
Bol'shoy pr.31
ST. PETERSBURG 199004
Russia

polotsky@imc.macro.ru

Matej PRAPROTNIK
National Institute of Chemistry
Hajdrihova 19
1001 LJUBLJANA
Slovenia

Tel: +386-61-1760275
Fax: +386-61-1259244
praprot@hp11.cmm.ki.si

Marco PRETTI
Politecnico di Torino
Corso Duca degli Abruzzi, 24
10129 TORINO
Italy

Tel: +39-011-5647373
Fax: +39-011-5647399
pretti@athena.polito.it

Davide PROVASI
Dipartimento di Fisica
Università di Milano
Via Celoria 16
20133 MILANO
Italy

Tel: +39-02-2392225
davide.provasi@mi.infn.it

Nicoletta PUCELLO
ENEA
Casaccia Research Center
HPCN Project (SP59)
C.P. 2400
00100 ROMA
Italy

Tel: +39-06-30483084
Fax: +39-06-30484729
pucello@casaccia.enea.it

Emy PULSINELLI
Dipartimento di Biofisica
Università di Genova
C.so Europa 30
16132 GENOVA
Italy

Tel: +39-010-3538381
Fax: +39-010-3538346
emy@ibf.unige.it

Jan ROSSMEISL
Center for Atomic-scale Materials Physics
CAMP
Technical University of Denmark
Building 307
2800 LYNGBY
Denmark

Tel: +45-45-253177
Fax: +45-45-932399
jross@fysik.dtu.dk

Dmitry S. RYKUNOV
Institute of Protein Research
142290 PUSHCHINO
Moscow Region
Russia

Tel: +7-095-924-0493
Fax: +7-095-924-0493
rykunov@alpha.protres.ru

Paolo SCANNAPIECO
CRIBI Biotechnology Centre
Università di Padova
Viale G. Colombo 3
35121 PADOVA
Italy

Tel: +39-049-8276165
Fax: +39-049 8276159
paolo@eos.bio.unipd.it

Giovanni SETTANNI
SISSA
Via Beirut 2
34014 TRIESTE
Italy

Tel: +39-040-2240460
Fax: +39-040-3787528
settanni@sissa.it

Vahid SHAHREZAEI
Institute for Studies in Theoretical
Physics and Mathematics
P.O. Box 19395-5531
TEHRAN
Iran

Tel: +98-21-2280692
Fax: +98-21-2280415
shaahrez@theory.ipm.ac.ir

Yibing SHAN
Department of Physics
Drexel University
3141 Chestnut Street
PHILADELPHIA PA 19104
USA

Tel: +1-215-8952791
Fax: +1-215-8955934
shan@newton.physics.drexel.edu

Julia SHIFMAN
Institute of Protein Research
142290 PUSHCHINO
Moscow Region
Russia

Tel & Fax: +7-095-924-0493

Jun SHIMADA
Department of Chemistry
and Chemical Biology
12 Oxford St., 267
CAMBRIDGE, MA 02138
USA

Tel: +1-617-496-5499
Fax: +1-617-495-3075
shimada@whisky.harvard.edu

Nikos S. SKANTZOS
Department of Mathematics
King's College London
LONDON WC2R 2LS
UK

Tel: +44-0207-8481197
skantzos@mth.kcl.ac.uk

Erkan TUZEL
ISIK University
Department of Physics
80670 Faculty of Science and Letters
ISTANBUL
Turkey

etuzel@usa.net, tuzel@gursey.gov.tr

Sara M. VAIANA
INFM
Dipartimento di Scienze
Fisiche ed Astronomiche
Via Archirafi, 36
90123 PALERMO
Italy

Tel: +39-091-6234204
Fax: +39-091-6173895
sara@iaif.pa.cnr.it

Vadim VALUEV
Institute of Cytology and Genetic
Russian Academy of Sciences
Siberian Division
Lavrientieva ave. 10
630090 NOVOSIBIRSK
Russia

Tel: +7-3832-333119
Fax: +7-3832-331278
valuev@bionet.nsc.ru

Francesca VASILE
Dipartimento di Biofisica
Università di Genova
C.so Europa 30
16132 GENOVA
Italy

Tel: +39-010-3538381
Fax: +39-010-3538346
vasile@ibf.unige.it

Michele VENDRUSCOLO
OCMS
New Chemistry Laboratory
University of Oxford
South Parks Road 3 QT
OXFORD OX1
UK

Tel: +44-1865-275918
Fax: +44-1865-275921
michelev@bioch.ox.ac.uk

Stefan WALLIN
Department of Theoretical Physics
Lund University
Complex Systems
Sölveq 14A
223 62 LUND
Sweden

Tel: +46-46-222-9076
Fax: +46-46-222-9686
stefan@thep.lu.se

Peter YAKOLEV
Macromolecular Compounds,
Bolscoi pr.31
ST. PETERSBURG 199004
Russia

birshtei@imc.macro.ru

Semen YESILEVSKYY
The National University of Kiev
Mohyla Academy
The Faculty of Natural Sciences
Skavarty St. 2
KIEV
Ukraine

yesint@queen.ukma.kiev.ua

Observers

Renzo CAMPANELLA
Dipartimento di Fisica
Università degli Studi di Perugia
Via A. Pascoli 5
06100 PERUGIA
Italy

Tel: +39-075-5853034
Fax: +39-075-44666
renzo.campanella@fisica.unipg.it

Cristina CANTALE
ENEA
Via Anguillarese, 301
00060 ROMA
Italy

Tel: +39-06-30486602
Fax: +39-06-30484808
cantale@casaccia.enea.it

Harm Geert MULLER
Institute for Atomic & Molecular Physics
Krvislaan 407
1098 5J AMSTERDAM
The Netherlands

Tel: +31-20-6081234
Fax: +31-20-6684106
muller@amolf.nl

Aldona G. RAJEWSKA
Joint Institute for Nuclear Research
Laboratory of Neutron Physics
141980 DUBNA
Moskow region
Russia

Tel: +7-09-621-66977
Fax: +7-09-621-65882
aldonar@nf.jinr.ru

Enrico VIGEZZI
Istituto Nazionale di Fisica Nucleare
Via Celoria 16
20133 MILANO
Italy

Tel: +39-02-2392251
Fax: +39-02-2392487

PROCEEDINGS OF THE INTERNATIONAL SCHOOL OF PHYSICS «ENRICO FERMI»

Course I (1953)
Questioni relative alla rivelazione delle particelle elementari, con particolare riguardo alla radiazione cosmica
edited by G. PUPPI

Course II (1954)
Questioni relative alla rivelazione delle particelle elementari, e alle loro interazioni con particolare riguardo alle particelle artificialmente prodotte ed accelerate
edited by G. PUPPI

Course III (1955)
Questioni di struttura nucleare e dei processi nucleari alle basse energie
edited by C. SALVETTI

Course IV (1956)
Proprietà magnetiche della materia
edited by L. GIULOTTO

Course V (1957)
Fisica dello stato solido
edited by F. FUMI

Course VI (1958)
Fisica del plasma e relative applicazioni astrofisiche
edited by G. RIGHINI

Course VII (1958)
Teoria della informazione
edited by E. R. CAIANIELLO

Course VIII (1958)
Problemi matematici della teoria quantistica delle particelle e dei campi
edited by A. BORSELLINO

Course IX (1958)
Fisica dei pioni
edited by B. TOUSCHEK

Course X (1959)
Thermodynamics of Irreversible Processes
edited by S. R. DE GROOT

Course XI (1959)
Weak Interactions
edited by L. A. RADICATI

Course XII (1959)
Solar Radioastronomy
edited by G. RIGHINI

Course XIII (1959)
Physics of Plasma: Experiments and Techniques
edited by H. ALFVÉN

Course XIV (1960)
Ergodic Theories
edited by P. CALDIROLA

Course XV (1960)
Nuclear Spectroscopy
edited by G. RACAH

Course XVI (1960)
Physicomathematical Aspects of Biology
edited by N. RASHEVSKY

Course XVII (1960)
Topics of Radiofrequency Spectroscopy
edited by A. GOZZINI

Course XVIII (1960)
Physics of Solids (Radiation Damage in Solids)
edited by D. S. BILLINGTON

Course XIX (1961)
Cosmic Rays, Solar Particles and Space Research
edited by B. PETERS

Course XX (1961)
Evidence for Gravitational Theories
edited by C. MØLLER

Course XXI (1961)
Liquid Helium
edited by G. CARERI

Course XXII (1961)
Semiconductors
edited by R. A. SMITH

Course XXIII (1961)
Nuclear Physics
edited by V. F. WEISSKOPF

Course XXIV (1962)
Space Exploration and the Solar System
edited by B. ROSSI

Course XXV (1962)
Advanced Plasma Theory
edited by M. N. ROSENBLUTH

Course XXVI (1962)
Selected Topics on Elementary Particle Physics
edited by M. CONVERSI

Course XXVII (1962)
Dispersion and Absorption of Sound by Molecular Processes
edited by D. SETTE

Course XXVIII (1962)
Star Evolution
edited by L. GRATTON

Course XXIX (1963)
Dispersion Relations and their Connection with Causality
edited by E. P. WIGNER

Course XXX (1963)
Radiation Dosimetry
edited by F. W. SPIERS and G. W. REED

Course XXXI (1963)
Quantum Electronics and Coherent Light
edited by C. H. TOWNES and P. A. MILES

Course XXXII (1964)
Weak Interactions and High-Energy Neutrino Physics
edited by T. D. LEE

Course XXXIII (1964)
Strong Interactions
edited by L. W. ALVAREZ

Course XXXIV (1965)
The Optical Properties of Solids
edited by J. TAUC

Course XXXV (1965)
High-Energy Astrophysics
edited by L. GRATTON

Course XXXVI (1965)
Many-Body Description of Nuclear Structure and Reactions
edited by C. BLOCH

Course XXXVII (1966)
Theory of Magnetism in Transition Metals
edited by W. MARSHALL

Course XXXVIII (1966)
Interaction of High-Energy Particles with Nuclei
edited by T. E. O. ERICSON

Course XXXIX (1966)
Plasma Astrophysics
edited by P. A. STURROCK

Course XL (1967)
Nuclear Structure and Nuclear Reactions
edited by M. JEAN and R. A. RICCI

Course XLI (1967)
Selected Topics in Particle Physics
edited by J. STEINBERGER

Course XLII (1967)
Quantum Optics
edited by R. J. GLAUBER

Course XLIII (1968)
Processing of Optical Data by Organisms and by Machines
edited by W. REICHARDT

Course XLIV (1968)
Molecular Beams and Reaction Kinetics
edited by CH. SCHLIER

Course XLV (1968)
Local Quantum Theory
edited by R. JOST

Course XLVI (1969)
Physics with Intersecting Storage Rings
edited by B. TOUSCHEK

Course XLVII (1969)
General Relativity and Cosmology
edited by R. K. SACHS

Course XLVIII (1969)
Physics of High Energy Density
edited by P. CALDIROLA and H. KNOEPFEL

Course IL (1970)
Foundations of Quantum Mechanics
edited by B. D'ESPAGNAT

Course L (1970)
Mantle and Core in Planetary Physics
edited by J. COULOMB and M. CAPUTO

Course LI (1970)
Critical Phenomena
edited by M. S. GREEN

Course LII (1971)
Atomic Structure and Properties of Solids
edited by E. BURSTEIN

Course LIII (1971)
Developments and Borderlines of Nuclear Physics
edited by H. MORINAGA

Course LIV (1971)
Developments in High-Energy Physics
edited by R. R. GATTO

Course LV (1972)
Lattice Dynamics and Intermolecular Forces
edited by S. CALIFANO

Course LVI (1972)
Experimental Gravitation
edited by B. BERTOTTI

Course LVII (1972)
History of 20th Century Physics
edited by C. WEINER

Course LVIII (1973)
Dynamics Aspects of Surface Physics
edited by F. O. GOODMAN

Course LIX (1973)
Local Properties at Phase Transitions
edited by K. A. MÜLLER and A. RIGAMONTI

Course LX (1973)
C*-Algebras and their Applications to Statistical Mechanics and Quantum Field Theory
edited by D. KASTLER

Course LXI (1974)
Atomic Structure and Mechanical Properties of Metals
edited by G. CAGLIOTI

Course LXII (1974)
Nuclear Spectroscopy and Nuclear Reactions with Heavy Ions
edited by H. FARAGGI and R. A. RICCI

Course LXIII (1974)
New Directions in Physical Acoustics
edited by D. SETTE

Course LXIV (1975)
Nonlinear Spectroscopy
edited by N. BLOEMBERGEN

Course LXV (1975)
Physics and Astrophysics of Neutron Stars and Black Holes
edited by R. GIACCONI and R. RUFFINI

Course LXVI (1975)
Health and Medical Physics
edited by J. BAARLI

Course LXVII (1976)
Isolated Gravitating Systems in General Relativity
edited by J. EHLERS

Course LXVIII (1976)
Metrology and Fundamental Constants
edited by A. FERRO MILONE, P. GIACOMO and S. LESCHIUTTA

Course LXIX (1976)
Elementary Modes of Excitation in Nuclei
edited by A. BOHR and R. A. BROGLIA

Course LXX (1977)
Physics of Magnetic Garnets
edited by A. PAOLETTI

Course LXXI (1977)
Weak Interactions
edited by M. BALDO CEOLIN

Course LXXII (1977)
Problems in the Foundations of Physics
edited by G. TORALDO DI FRANCIA

Course LXXIII (1978)
Early Solar System Processes and the Present Solar System
edited by D. LAL

Course LXXIV (1978)
Development of High-Power Lasers and their Applications
edited by C. PELLEGRINI

Course LXXV (1978)
Intermolecular Spectroscopy and Dynamical Properties of Dense Systems
edited by J. VAN KRANENDONK

Course LXXVI (1979)
Medical Physics
edited by J. R. GREENING

Course LXXVII (1979)
Nuclear Structure and Heavy-Ion Collisions
edited by R. A. BROGLIA, R. A. RICCI and C. H. DASSO

Course LXXVIII (1979)
Physics of the Earth's Interior
edited by A. M. DZIEWONSKI and E. BOSCHI

Course LXXIX (1980)
From Nuclei to Particles
edited by A. MOLINARI

Course LXXX (1980)
Topics in Ocean Physics
edited by A. R. OSBORNE and P. MALANOTTE RIZZOLI

Course LXXXI (1980)
Theory of Fundamental Interactions
edited by G. COSTA and R. R. GATTO

Course LXXXII (1981)
Mechanical and Thermal Behaviour of Metallic Materials
edited by G. CAGLIOTI and A. FERRO MILONE

Course LXXXIII (1981)
Positrons in Solids
edited by W. BRANDT and A. DUPASQUIER

Course LXXXIV (1981)
Data Acquisition in High-Energy Physics
edited by G. BOLOGNA and M. VINCELLI

Course LXXXV (1982)
Earhquakes: Observation, Theory and Interpretation
edited by H. KANAMORI and E. BOSCHI

Course LXXXVI (1982)
Gamow Cosmology
edited by F. MELCHIORRI and R. RUFFINI

Course LXXXVII (1982)
Nuclear Structure and Heavy-Ion Dynamics
edited by L. MORETTO and R. A. RICCI

Course LXXXVIII (1983)
Turbulence and Predictability in Geophysical Fluid Dynamics and Climate Dynamics
edited by M. GHIL, R. BENZI and G. PARISI

Course LXXXIX (1983)
Highlights of Condensed-Matter Theory
edited by F. BASSANI, F. FUMI and M. P. TOSI

Course XC (1983)
Physics of Amphiphiles: Micelles, Vesicles and Microemulsions
edited by V. DEGIORGIO and M. CORTI

Course XCI (1984)
From Nuclei to Stars
edited by A. MOLINARI and R. A. RICCI

Course XCII (1984)
Elementary Particles
edited by N. CABIBBO

Course XCIII (1984)
Frontiers in Physical Acoustics
edited by D. SETTE

Course XCIV (1984)
Theory of Reliability
edited by A. SERRA and R. E. BARLOW

Course XCV (1985)
Solar-Terrestrial Relationships and the Earth Environment in the Last Millennia
edited by G. CINI CASTAGNOLI

Course XCVI (1985)
Excited-State Spectroscopy in Solids
edited by U. M. GRASSANO and N. TERZI

Course XCVII (1985)
Molecular-Dynamics Simulations of Statistical-Mechanical Systems
edited by G. CICCOTTI and W. G. HOOVER

Course XCVIII (1985)
The Evolution of Small Bodies in the Solar System
edited by M. FULCHIGNONI and L̆. KRESÁK

Course XCIX (1986)
Synergetics and Dynamic Instabilities
edited by G. CAGLIOTI and H. HAKEN

Course C (1986)
The Physics of NMR Spectroscopy in Biology and Medicine
edited by B. MARAVIGLIA

Course CI (1986)
Evolution of Interstellar Dust and Related Topics
edited by A. BONETTI and J. M. GREENBERG

Course CII (1986)
Accelerated Life Testing and Experts Opinions in Reliability
edited by C. A. CLAROTTI

Course CIII (1987)
Trends in Nuclear Physics
edited by P. KIENLE, R. A. RICCI and A. RUBBINO

Course CIV (1987)
Frontiers and Borderlines in Many-Particle Physics
edited by R. A. BROGLIA and J. R. SCHRIEFFER

Course CV (1987)
Confrontation between Theories and Observations in Cosmology: Present Status and Future Programmes
edited by J. AUDOUZE and F. MELCHIORRI

Course CVI (1988)
Current Trends in the Physics of Materials
edited by G. F. CHIAROTTI, F. FUMI and M. TOSI

Course CVII (1988)
The Chemical Physics of Atomic and Molecular Clusters
edited by G. SCOLES

Course CVIII (1988)
Photoemission and Absorption Spectroscopy of Solids and Interfaces with Synchrotron Radiation
edited by M. CAMPAGNA and R. ROSEI

Course CIX (1988)
Nonlinear Topics in Ocean Physics
edited by A. R. OSBORNE

Course CX (1989)
Metrology at the Frontiers of Physics and Technology
edited by L. CROVINI and T. J. QUINN

Course CXI (1989)
Solid-State Astrophysics
edited by E. BUSSOLETTI and G. STRAZZULLA

Course CXII (1989)
Nuclear Collisions from the Mean-Field into the Fragmentation Regime
edited by C. DETRAZ and P. KIENLE

Course CXIII (1989)
High-Pressure Equation of State: Theory and Applications
edited by S. ELIEZER and R. A. RICCI

Course CXIV (1990)
Industrial and Technological Applications of Neutrons
edited by M. FONTANA and F. RUSTICHELLI

Course CXV (1990)
The Use of EOS for Studies of Atmospheric Physics
edited by J. C. GILLE and G. VISCONTI

Course CXVI (1990)
Status and Perspectives of Nuclear Energy: Fission and Fusion
edited by R. A. RICCI, C. SALVETTI and E. SINDONI

Course CXVII (1991)
Semiconductor Superlattices and Interfaces
edited by A. STELLA

Course CXVIII (1991)
Laser Manipolation of Atoms and Ions
edited by E. ARIMONDO, W. D. PHILLIPS and F. STRUMIA

Course CXIX (1991)
Quantum Chaos
edited by G. CASATI, I. GUARNERI and U. SMILANSKY

Course CXX (1992)
Frontiers in Laser Spectroscopy
edited by T. W. HÄNSCH and M. INGUSCIO

Course CXXI (1992)
Perspectives in Many-Particle Physics
edited by R. A. BROGLIA, J. R. SCHRIEFFER and P. F. BORTIGNON

Course CXXII (1992)
Galaxy Formation
edited by J. SILK and N. VITTORIO

Course CXXIII (1992)
Nuclear Magnetic Double Resonance
edited by B. MARAVIGLIA

Course CXXIV (1993)
Diagnostic Tools in Atmospheric Physics
edited by G. FIOCCO and G. VISCONTI

Course CXXV (1993)
Positron Spectroscopy of Solids
edited by A. DUPASQUIER and A. P. MILLS jr.

Course CXXVI (1993)
Nonlinear Optical Materials: Principles and Applications
edited by V. DEGIORGIO and C. FLYTZANIS

Course CXXVII (1994)
Quantum Groups and their Applications in Physics
edited by L. CASTELLANI and J. WESS

Course CXXVIII (1994)
Biomedical Applications of Synchrotron Radiation
edited by E. BURATTINI and A. BALERNA

Course CXXIX (*) (1994)
Observation, Prediction and Simulation of Phase Transitions in Complex Fluids
edited by M. BAUS, L. F. RULL and J.-P. RYCKAERT

Course CXXX (1995)
Selected Topics in Nonperturbative QCD
edited by A. DI GIACOMO and D. DIAKONOV

Course CXXXI (1995)
Coherent and Collective Interactions of Particles and Radiation Beams
edited by A. ASPECT, W. BARLETTA and R. BONIFACIO

Course CXXXII (1995)
Dark matter in the Universe
edited by S. BONOMETTO and J. PRIMACK

Course CXXXIII (1996)
Past and Present Variability of the Solar-Terrestrial System: Measurement, Data Analysis and Theoretical Models
edited by G. CINI CASTAGNOLI and A. PROVENZALE

Course CXXXIV (1996)
The Physics of Complex Systems
edited by F. MALLAMACE and H. E. STANLEY

Course CXXXV (1996)
The Physics of Diamond
edited by A. PAOLETTI and A. TUCCIARONE

Course CXXXVI (1997)
Models and Phenomenology for Conventional and High-Temperature Superconductivity
edited by G. IADONISI, J. R. SCHRIEFFER and M. L. CHIOFALO

(*) This course belongs to the *NATO ASI Series C*, Vol. 460 (Kluwer Academic Publishers).

Course CXXXVII (1997)
Heavy Flavour Physics: a Probe of Nature's Grand Design
edited by I. BIGI and L. MORONI

Course CXXXVIII (1997)
Unfolding the Matter of Nuclei
edited by A. MOLINARI and R. A. RICCI

Course CXXXIX (1998)
Magnetic Resonance and Brain Function: Approaches from Physics
edited by B. MARAVIGLIA

Course CXL (1998)
Bose-Einstein Condensation in Atomic Gases
edited by M. INGUSCIO, S. STRINGARI and C. E. WIEMAN

Course CXLI (1998)
Silicon-Based Microphotonics: From Basics to Applications
edited by O. BISI, S. U. CAMPISANO, L. PAVESI and F. PRIOLO

Course CXLII (1999)
Plasma Astrophysics
edited by B. COPPI, A. FERRARI and E. SINDONI

Course CXLIII (1999)
New Directions in Quantum Chaos
edited by G. CASATI, I. GUARNERI and U. SMILANSKY

Course CXLIV (2000)
Nanometer Scale Science and Technology
edited by M. ALLEGRINI, N. GARCIA and O. MARTI